Exclusive and Semi-Exclusive Processes at High Momentum Transfer

PROCEEDINGS FROM THE INSTITUTE FOR NUCLEAR THEORY

Series Editors: Wick C. Haxton *(Univ. of Washington)*
Ernest M. Henley *(Univ. of Washington)*

Published

Vol. 1: Nucleon Resonances and Nucleon Structure
ed. G. A. Miller

Vol. 2: Solar Modeling
eds. A. B. Balantekin and J. N. Bahcall

Vol. 3: Phenomenology and Lattice QCD
eds. G. Kilcup and S. Sharpe

Vol. 4: N* Physics
eds. T.-S. H. Lee and W. Roberts

Vol. 5: Tunneling in Complex Systems
ed. S. Tomsovic

Vol. 6: Nuclear Physics with Effective Field Theory
eds. M. J. Savage, R. Seki and U. van Kolck

Vol. 7: Quarkonium Production in High-Energy Nuclear Collisions
eds. B. Jacak and X.-N. Wang

Vol. 9: Nuclear Physics with Effective Field Theory II
eds. P. F. Bedaque, M. J. Savage, R. Seki and U. van Kolck

Vol. 10: Exclusive and Semi-Exclusive Processes at High Momentum Transfer
eds. C. Carlson and A. Radyushkin

Forthcoming

Vol. 8: Quark Confinement and the Hadron Spectrum
eds. A. Radyushkin and C. Carlson

Proceedings from the Institute for Nuclear Theory — Vol. 10

Exclusive and Semi-Exclusive Processes at High Momentum Transfer

Newport News, VA, USA May 19–22, 1999

Editors

Carl Carlson
College of William & Mary

Anatoly Radyushkin
Old Dominion University / Jefferson Laboratory

World Scientific
Singapore • New Jersey • London • Hong Kong

Published by

World Scientific Publishing Co. Pte. Ltd.

P O Box 128, Farrer Road, Singapore 912805

USA office: Suite 1B, 1060 Main Street, River Edge, NJ 07661

UK office: 57 Shelton Street, Covent Garden, London WC2H 9HE

British Library Cataloguing-in-Publication Data
A catalogue record for this book is available from the British Library.

EXCLUSIVE AND SEMI-EXCLUSIVE PROCESSES AT HIGH MOMENTUM TRANSFER

ISBN 981-02-4355-3

Printed in Singapore by World Scientific Printers

INTERNATIONAL ADVISORY COMMITTEE

PREFACE

These are the proceedings of the Workshop on Exclusive and Semiexclusive Processes at High Momentum Transfer, which was held 20-22 May, 1999, at the Thomas Jefferson National Accelerator Facility (Jefferson Lab) in Newport News, Virginia, USA. The Workshop was jointly sponsored by the Institute for Nuclear Theory, Seattle, Washington, and by Jefferson Lab.

The workshop was open to all interested scientists, and embraced both experimenters and theorists. Many thanks are due the speakers and participants at the conference for the stimulating atmosphere that prevailed. We trust the written versions of the talks will reproduce some of that feeling.

Thanks are also due to the people who worked hard to organize the conference and make it a success. The Conference Services group at Jefferson Lab was indispensable. Allow us to give special thanks to Martha Hightower and Ruth Bizot, and even more special thanks to Susan Ewing for all her efforts before, during, and since the conference.

Carl E. Carlson

Anatoly V. Radyushkin

December 1999

SERIES PREFACE

The National Institute for Nuclear Theory Series

The National Institute for Nuclear Theory (INT) was established by the U.S. Department of Energy in March 1990. The goals of the INT include:

1. Creating a productive research environment where visiting scientists can focus their energies and exchange ideas on key issues facing the field of nuclear physics including those crucial to the success of existing and future experimental facilities;
2. Encouraging interdisciplinary research at the intersections of nuclear physics with related subfields, including particle physics, astrophysics, atomic physics, and condensed matter;
3. Furthering the development and advancement of physicists with recent Ph.D.s;
4. Contributing to scientific education through graduate student research, INT summer schools, undergraduate summer research programs, and graduate student participation in INT workshops and programs;
5. Strengthening international cooperation in physics research through exchanges and other interactions.

While the INT strives to achieve these goals in a variety of ways, its most important efforts are the three-month programs, workshops, and schools it sponsors. These typically attract 300 visitors to the INT each year. In order to make selected INT workshops and summer schools available to a wider audience, the INT and World Scientific established the series of books to which this volume belongs.

The Thomas Jefferson National Accelerator Facility and the INT inaugurated a workshop series in 1992. The motivation was to provide an informal forum for discussions between theorists and experimentalists, with the goal of optimizing the physics coming from this new accelerator. These joint workshops have traditionally encouraged the participation of younger physicists, both as speakers and as organizers. The sixth workshop in this series took place in May, 1999, and focused on Exclusive and Semi-exclusive Processes at High Momentum Transfer. Organized by Carl Carlson and Anatoly Radyushkin and hosted by Jefferson Lab, the workshop attracted approximately 100 participants to debate topics such as off-forward parton distributions and virtual Compton scattering. This tenth volume of the INT series is dedicated to the proceedings of this workshop.

Earlier series volumes include the proceedings of the 1991 and 1993 Uehling summer schools on Nucleon Resonances and Nucleon Structure and on Phenomenology and Lattice QCD; the 1994 INT workshop on Solar Modeling; the 1997 Jefferson Lab/INT workshop on Nucleon Resonance Physics; the tutorials of the spring 1997 INT program on Tunneling in Complex Systems; the 1998 Caltech/INT workshop on Nuclear Physics with Effective Field Theory and the follow-up workshop held at the INT in 1999; the proceedings of the 1998 RHIC Winter Workshop on Quarkonium Production in Relativistic Nuclear Collisions; and the proceedings of the Quark Confinement and Hadron Spectrum III workshop. We will continue to publish those proceedings of INT workshops and schools which we judge to be of broad interest to the physics community.

Wick C. Haxton and Ernest Henley
Seattle, Washington, February , 2000

CONTENTS

SESSION PRESENTATIONS

SOFT PERTURBATIVE QCD[a]

PAUL HOYER
Nordita, Blegdamsvej 17
DK-2100 Copenhagen, Denmark
E-mail: hoyer@nordita.dk

There are indications of a faith transition in QCD: The strong coupling may freeze at a sufficiently low value to make PQCD relevant even in the confining regime. The properties of PQCD at low Q^2 depend on the asymptotic field configuration assumed at $t = \pm\infty$. A change in the $i\epsilon$ prescription of the free propagators is equivalent to including quarks and gluons in the initial and final states. Perturbative expansions thus generated are formally as justified as standard PQCD and may turn out to be phenomenologically relevant. I discuss examples of the effects of such propagator modifications in QED and QCD.

1 The Faith Transition

Comparisons of physical gauge theories with data rely on perturbation theory. In particular, the successful comparisons of perturbative QCD (PQCD) with data on hard processes has established QCD as the correct theory of the strong interaction.

PQCD does not correctly describe soft processes, where quarks get dressed to massive constituent quarks that are confined to hadrons. It is commonly assumed that this physics is related to a strongly coupled non-perturbative sector of QCD. Evidence is, however, building up that this assumption may be wrong. Phenomenological analyses point to a situation where the strong coupling $\alpha_s(Q^2)$ 'freezes' for low Q^2 at a moderate value $\alpha_s(0)/\pi = 0.14\ldots0.17$. The insight that the physics of confinement may be described perturbatively has been dubbed the 'QCD faith transition' by Dokshitzer [2].

The following examples indicate a close empirical connection between hard and soft processes. PQCD continues to work even at low Q^2 provided a regularization of the form $Q^2 \to Q^2 + Q_0^2$ is made, where Q_0 is a hadronic scale.

- *DIS scaling.* D. Haidt [3] has shown that HERA data on the structure function $F_2(x, Q^2)$ in the range $x \leq .001$ can be fit by the simple function $F_2 = 0.41\log(0.04/x)\log(1 + Q^2/Q_0^2)$, where $Q_0^2 = 0.5$ GeV^2. This parametrization works for all available Q^2, $0.1 \lesssim Q^2/\text{GeV}^2 \lesssim 35$, with no sign of a PQCD to NPQCD 'phase transition'.

[a]An unabridged version of this talk may be found in Ref. 1.

- *Exclusive scaling.* Cross sections of exclusive processes obey dimensional scaling down to remarkably low momentum transfers. A striking example is [4] $\sigma(\gamma + d \to p + n) \propto E_{CM}^{22}$ at $\theta_{CM} = 89°$ for $E_\gamma = 1 \ldots 4$ GeV. The PQCD prediction is proportional to α_s^{10} and agrees with dimensional scaling only provided the coupling is frozen at low momentum transfers.

- *Local Parton Hadron Duality.* The inclusive *hadron* distribution in $e^+e^- \to h+X$ and in $eN \to h+X$ appears [5] to faithfully track the inclusive *gluon* distribution calculated using PQCD, down to momenta of $\mathcal{O}(200$ MeV).

Fazed by the intriguing possibility that QCD is a weakly coupled theory also in the confinement regime it seems imperative to search for perturbative expansions which can describe data even at low momentum transfers. In the examples above, a hadronic scale was added to standard PQCD 'by hand', without theoretical justification.

PQCD is uniquely defined by the QCD lagrangian and by the boundary condition at asymptotic times, $t = \pm\infty$. Formally, the choice of boundary state $|\Omega_0\rangle$ does not affect the exact all-orders sum, provided that it has an overlap with the true ground state $|\Omega\rangle$:

$$\langle\Omega|\Omega_0\rangle \neq 0 \quad \Longrightarrow \quad \lim_{t\to\infty(1-i\epsilon)} |\Omega_0, t\rangle \propto |\Omega\rangle \tag{1}$$

In standard PQCD we choose $|\Omega_0\rangle = |0\rangle$, the empty 'perturbative vacuum'. This choice works well in QED. Apparently the true QED vacuum, even though extremely complicated, is sufficiently close to $|0\rangle$ for the deviations to be treated perturbatively. The long-distance regime of QCD is, on the other hand, believed to be influenced by a non-trivial 'gluon condensate' vacuum [6]. We should then consider other choices for $|\Omega_0\rangle$, which are sufficiently simple to allow analytic calculations yet are better models of the true QCD ground state than the perturbative vacuum.

Standard PQCD has the elegant feature of being exactly Lorentz covariant at each order of α_s. S-matrix elements that have free (quark and gluon) asymptotic states manifest their full symmetries at each order of perturbation theory. Conversely, S-matrices whose asymptotic states are bound (hadrons) require infinite orders of α_s just to generate the bound state poles. It is unlikely that such resummations can preserve explicit Lorentz covariance at each order of the approximation.

Bound states wave functions are defined on equal time or light-like surfaces. As emphasized by Dirac [7], Lorentz boost generators that do not leave those surfaces invariant have non-trivial representations which involve interactions.

Consider the Fock expansion of the true QCD vacuum (disregarding quarks),

$$|\Omega\rangle = \phi_0|0\rangle + \sum_{n=2}^{\infty} \int \left[\prod_i^n \frac{d^3\vec{p}_i}{(2\pi)^3}\right] \phi_n(\vec{p}_i)|g(\vec{p}_1)\cdots g(\vec{p}_n)\rangle \tag{2}$$

The gluon condensate contributes primarily to the long distance Fock amplitudes $\phi_n(|\vec{p}_i| \lesssim \Lambda_{QCD})$. The boost invariance of the ground state,

$$\hat{M}^{0k}|\Omega\rangle \propto |\Omega\rangle \tag{3}$$

implies that the condensate appears in the low momentum Fock components $|\vec{p}_i| \lesssim \Lambda_{QCD}$ *in all frames.* This is possible since the exact boost $\hat{M}^{0k}$ contains interactions. A free boost $\hat{M}_0^{0k}$ would simply translate the momenta in each Fock component of Eq. (2), resulting in a very different state.

Describing the boost invariance (3) of the true QCD vacuum is a task of the same magnitude as showing that it is an eigenstate of the exact hamiltonian. Approximate, perturbative calculations satisfy neither exact time translation nor boost invariance. PQCD asymptotic states $|\Omega_0\rangle$ which contain particles with momenta of $\mathcal{O}(\Lambda_{QCD})$ will not be invariant even under free boosts.

The general property (1) guarantees exact Lorentz covariance of the full answer, even for non-covariant asymptotic states $|\Omega_0\rangle$. Only the *rate of convergence* of the perturbative series may be frame dependent. This can be illustrated for $\alpha_s = 0$ by choosing $|\Omega_0\rangle$ to be a superposition of the perturbative vacuum and a state with a gluon of momentum $\vec{k}$,

$$|\Omega_0\rangle = \phi_0|0\rangle + \phi_1|g(\vec{k})\rangle \tag{4}$$

This state evolves in time t into

$$e^{-iH_0t}|\Omega_0\rangle = \phi_0|0\rangle + \phi_1 e^{-i|\vec{k}|t}|g(\vec{k})\rangle \tag{5}$$

The gluon state $|g(\vec{k})\rangle$ is exponentially suppressed wrt. the perturbative vacuum $|0\rangle$ in the asymptotic time limit $t \to \infty(1 - i\epsilon)$.

The above arguments are formal, but so is the justification of standard PQCD. The central point is that it is theoretically justified to consider perturbative expansions with alternative boundary states $|\Omega_0\rangle$. A choice different from $|0\rangle$ is motivated by the known properties of the QCD ground state. The study and interpretation of PQCD for non-trivial $|\Omega_0\rangle$ is theoretically fascinating and could turn out to be phenomenologically relevant.

2 The Perturbative Gluon Condensate

In considering boundary states $|\Omega_0\rangle$ that might model the gluon condensate, it is instructive to start with perturbation theory in a fermion condensate. Denote the free fermion propagator evaluated between states $|\bar{f}(\vec{k},\frac{1}{2})\bar{f}(\vec{k},-\frac{1}{2})\rangle$ of two antifermions with 3-momenta $\vec{k}$ and spin $\pm\frac{1}{2}$ by

$$iS_k(x-y) \equiv \langle \bar{f}\bar{f}|T\{\psi(x)\bar{\psi}(y)\}|\bar{f}\bar{f}\rangle/\langle k|k\rangle^2 \qquad (6)$$

where $\langle k|k\rangle = (2\pi)^3 2E_k \delta^3(\vec{0})$. In momentum space

$$S_k(p) = \begin{cases} S_F(p) & (\vec{p} \neq -\vec{k}) \\ S_E(p) & (\vec{p} = -\vec{k}) \end{cases} \qquad (7)$$

where

$$S_E(p) = \frac{\not{p}+m}{(p^0 - E_k + i\epsilon)(p^0 + E_k + i\epsilon)} \qquad (8)$$

differs from the standard Feynman propagator $S_F(p)$ only in the $i\epsilon$ prescription at[b] $p^0 = -E_k$. Eq. (7) may also be expressed as

$$S_k(p) = S_F(p) + i\frac{\not{p}+m}{\langle k|k\rangle}(2\pi)^4\delta^4(p+k) \qquad (9)$$

where $k^0 = E_k$. Filling all antifermion states with $|\vec{k}| < \Lambda$ gives a propagator that equals $S_E(p)$ for all $|\vec{p}| < \Lambda$.

The equivalence between adding fermions to the boundary states and a change of $i\epsilon$ prescription generalizes to Green functions at arbitrary order in perturbation theory [8]:

Any Green function (to arbitrary order in the coupling α_s) evaluated with the S_k fermion propagator gives the same result as a standard calculation using S_F with antifermions of momentum $\vec{k}$ added to the initial and final states.

Modifying the pole prescription of the free propagator for a range of 3-momenta $\vec{k}$ thus allows to include the effect of large number of 'background' fermions. The apparent non-causality of the propagation stems from an interference between the background fermions and those propagating (on-shell) in the Green function. The gauge invariance of the standard (S_F) calculation with external fermions guarantees the gauge invariance of the calculation using S_k.

---

[b] Adding fermions rather than antifermions to the asymptotic states would change the $i\epsilon$ prescription at the $p^0 = +E_k$ pole.

Due to the Pauli principle a fermion condensate prevents the creation of (anti-)fermions with the same momenta as the external ones. In a calculation using S_k this is ensured by the $i\epsilon$ sign change, which puts both poles of the propagator on the same side of the real axis, thus preventing a pinch.

It seems natural to ask whether a corresponding change of $i\epsilon$ sign in *boson* propagators is also equivalent to standard (Feynman) perturbation theory in the presence of external bosons. This can then serve as a prescription for a 'perturbative gluon condensate' $|\Omega_0\rangle$, which incorporates effects of multiple gluons already at lowest order of PQCD. In particular, such a procedure implies an exclusion principle for bosons, preventing the emission of soft gluons.

As is often the case, the situation for bosons is analogous to that for fermions, with a twist. For a scalar boson that is identical to its antiparticle the Green functions do not depend on whether the sign of $i\epsilon$ is changed at the positive or negative energy pole of the propagator. Modifying the prescription at the negative energy pole for a single momentum $\vec{k}$ gives a scalar propagator analogous to Eq. (9)

$$D_k(p) = D_F(p) + \frac{i}{\langle k|k\rangle}(2\pi)^4\delta^4(p+k) \tag{10}$$

In a realistic application one would change the prescription for a range of momenta, say for $|\vec{k}| < \Lambda$.

Perturbation theory based on D_k is again equivalent to the Feynman (D_F) one, in the following sense:

Any Green function evaluated (at arbitrary order in the perturbative expansion) using the modified scalar propagator D_k is equivalent to a specific superposition of Feynman Green functions with $0, 1, 2, \ldots$ *scalars of momentum $\vec{k}$ in the initial and final states.*

A derivation of this, together with the weights for each term in the superposition, may be found in Ref. [8]. There are only diagonal terms, *i.e.*, as many scalars of momentum $\vec{k}$ in the incoming as in the outgoing state. The absence of cross terms means that there is no unique boundary state $|\Omega_0\rangle$. However, the arguments based on Eq. (1) may be applied to each term in the superposition separately, to show that it generates a formally correct perturbative series. Many of the terms will have non-vanishing total 3-momentum and hence no overlap with the true vacuum – their contributions will be exponentially suppressed as in Eq. (5).

Similar reasoning can be applied to QCD. The *perturbative gluon condensate* (PGC) proposal [8] is to modify the Feynman gluon propagator at low

3-momenta, $|\vec{p}| < \Lambda_{QCD}$:

$$D_\Lambda^{\mu\nu}(p) = D_F^{\mu\nu}(p) - \frac{i\pi g^{\mu\nu}}{2|\vec{p}|}\left[\delta(p^0 - |\vec{p}|) + \delta(p^0| + \vec{p}|)\right]\theta(\Lambda_{QCD} - |\vec{p}|) \qquad (11)$$

This is written in Feynman gauge with the tacit assumption that the contributions from unphysical gluon polarizations and ghosts, which appear as external particles in the equivalent Feynman calculation, cancel among themselves (as they do in internal lines of standard PQCD). A corresponding modification is then required for the ghost propagator,

$$D_\Lambda^{ghost}(p) = D_F^{ghost}(p) + \frac{i\pi}{2|\vec{p}|}\left[\delta(p^0 - |\vec{p}|) + \delta(p^0| + \vec{p}|)\right]\theta(\Lambda_{QCD} - |\vec{p}|) \quad (12)$$

Averaging over the $i\epsilon$ modification at the positive and negative energy poles ensures ghost-antighost symmetry. The same average should be taken in the gluon propagator to satisfy Ward identities.

In the examples of the next section the propagator modifications do, as expected, preserve gauge invariance. It would be important to have an all orders proof of this[c].

3 Examples

3.1 Massless QED in 1+1 Dimensions

As shown in [1], the exact Schwinger model solution is unchanged (at least to $\mathcal{O}(e^4)$) when the electron propagator is modified according to Eq. (9). This illustrates that the non-covariant modification (9) is compatible with fully Lorentz covariant Green functions.

3.2 Bound States of Weakly Coupled QED$_2$

Bound state equations for wave functions taken at equal time in all frames are not explicitly Lorentz covariant. QED$_2$ in the limit $e/m \ll 1$, where positronium has only an $|e^+e^-\rangle$ component, provides an explicit example of how the solution can nevertheless have the proper relativistic dependence on the CM momentum. For details see [1,11].

---

[c] An alternative procedure would be to change the $i\epsilon$ prescription only for physical, transverse gluons (and not for ghosts). Gauge invariance should then be guaranteed since only physical gluons appear as external particles in the equivalent formulation in terms of Feynman propagators.

3.3 Scalar QED in 3+1 Dimensions

Scalar QED offers an opportunity for studying the effect of the boson propagator modification (10) on perturbation theory in a setting that has many similarities with PQCD. Defining the photon self-energy correction $\Pi(q)$ through

$$\Pi^{\mu\nu}(q) = \left(q^2 g^{\mu\nu} - q^\mu q^\nu\right) \Pi(q) \tag{13}$$

the one-loop result using Feynman propagators is

$$\Pi_F(q) = \frac{\alpha}{8\pi} \int_0^1 dx \sqrt{1-x} \log\left(1 - \frac{q^2 x}{4m^2}\right) \tag{14}$$

The modified propagator (10) gives a two-point function $\Pi_k^{\mu\nu}(q)$ which differs from the one obtained using Feynman propagators by

$$\Delta_k \Pi^{\mu\nu}(q) = \frac{e^2}{\langle k|k\rangle} \sum_{\sigma=\pm 1} \left[\frac{(2k+\sigma q)^\mu (2k+\sigma q)^\nu}{q^2 + 2\sigma k \cdot q} - g^{\mu\nu}\right] \tag{15}$$

The contribution from the product of δ-functions has been left out in Eq. (15). At the exceptional values of q where both propagators are modified at the same loop momentum the loop integral vanishes, since all $i\epsilon$'s have the same sign.

The modification (15) is readily seen to satisfy the gauge invariance condition $q_\mu \Delta_k \Pi^{\mu\nu}(q) = 0$. Without loss of generality we may take $\vec{q}$ along the 3-axis, $q = (q^+, q^-, \vec{0}_\perp)$, with $q^\pm = q^0 \pm q^3$. Gauge invariance then constrains $\Pi^{\mu\nu}$ to the form (13) for $\mu, \nu = \pm$, even without assuming Lorentz invariance. One finds,

$$\Delta_k \Pi_L = \frac{2e^2}{\langle k|k\rangle} \frac{4(m^2 + k_\perp^2) - q^2}{q^4 - 4(q\cdot k)^2} \tag{16}$$

where the subscript L indicates that $\Delta_k \Pi_L$ refers to Eqs. (13), (15) for $\mu, \nu = \pm$ only. The transverse $(i, j = 1, 2)$ components of Eq. (15) give

$$\Delta_k \Pi^{ij} = \frac{2e^2}{\langle k|k\rangle} \left(\delta^{ij} + \frac{4k^i k^j q^2}{q^4 - 4(q\cdot k)^2}\right) \tag{17}$$

There are also cross terms $\Delta_k \Pi^{\pm,i} \propto k^i$. When summed over all directions of $\vec{k}$ as in the PGC case (11) such terms will cancel. For the present discussion it is sufficient to assume that an integral over the transverse direction (azimuthal angle) of $\vec{k}$ is done. This implies $\Delta_k \Pi^{\pm,i} = 0$ and $\Delta_k \Pi^{ij} = -q^2 \delta^{ij} \Delta_k \Pi_T$, where

$$\Delta_k \Pi_T = -\frac{\alpha}{\langle k|k\rangle} \left[\frac{1}{q^2} + \frac{2k_\perp^2}{q^4 - 4(q\cdot k)^2}\right] \tag{18}$$

The geometric sum of the 1PI contribution (13) gives the photon propagator

$$\Sigma^{\mu\nu}(q) = -\frac{g^{\mu\nu} - q^\mu q^\nu/q^2}{q^2}\frac{1}{1-\Pi(q)} \tag{19}$$

The usual argument that gauge and Lorentz invariance implies a vanishing of the photon mass in perturbation theory follows from Eq. (19) given that $\Pi(q)$ is finite at $q^2 = 0$. This is verified for Feynman propagators from the expression (14).

With the modification (10) the geometric sum must be done separately for the longitudinal and transverse components (which do not mix as explained above). The function $\Pi(q)$ in Eq. (19) then depends on the photon polarization,

$$\begin{aligned} \Pi_L(q) &= \Pi_F(q) + \Delta_k\Pi_L(q) \\ \Pi_T(q) &= \Pi_F(q) + \Delta_k\Pi_T(q) \end{aligned} \tag{20}$$

From Eq. (16) it may be seen that Π_L is regular at $q^2 = 0$, hence longitudinally polarized photons are massless. However, $\Delta_k\Pi_T$ of Eq. (18) has a pole at $q^2 = 0$, implying a finite mass for transverse photons. This is an 'effective' mass, which depends on the photon momentum q. Moreover, it has an imaginary part, as we shall next discuss.

The singularities of loop integrals arise from pinches between propagator poles that have opposite $i\epsilon$ prescriptions. For Feynman propagators pinches can occur between positive and negative energy poles, and correspond to particle-antiparticle pair production. A change of $i\epsilon$ prescription at the negative energy pole as in Eq. (10) removes some particle-antiparticle production – for fermions this ensures the Pauli exclusion principle. Instead, 'pseudothreshold' singularities arise from pinches between two negative energy poles. These correspond to scattering of the propagating photon on the background scalars. The pseudothreshold singularities occur at $q^2 = \pm 2q \cdot k$, *i.e.*, at

$$q^+q^- = \pm(q^+k^- + q^-k^+) \tag{21}$$

This condition has solutions for both $q^2 > 0$ and $q^2 < 0$, with q^+ and/or q^- of order $k^\pm$. *E.g.*, for $k^- = 0$ we have $q^+ = \pm k^+$ and no condition on q^-.

In the PGC prescription (11) the background gluon momenta are summed over $|\vec{k}| < \Lambda_{QCD}$ and the pinch singularities (21) turn into cuts. The imaginary part that thus arises in the gluon propagator reflects a finite propagation length of the gluon due to its scattering on the background particles. The physical QCD condensate will similarly disperse gluons. It is reasonable to assume that the true eigenmodes of propagation in the QCD vacuum will be color singlets, which do not suffer arbitrarily soft interactions. This is in fact what is meant by the term 'color confinement'. The perturbative gluon condensate offers an opportunity for studying these issues.

3.4 A QCD Ward Identity

QCD Ward identitities follow from the requirement that the BRS transformation of Green functions vanish[12]. The two-point function $\langle 0|T\{\bar{c}_a(x_1)A_b^\mu(x_2)\}|0\rangle$, where c and A are ghost and gluon fields, respectively, yields the identity

$$\langle 0|T\left\{\left[\frac{\partial}{\partial x_1^\nu}A_a^\nu(x_1)\right]A_b^\mu(x_2)-\bar{c}_a(x_1)\left[\frac{\partial}{\partial x_{2\mu}}\delta_{bc}+gf_{bcd}A_d^\mu(x_2)\right]c_c(x_2)\right\}|0\rangle=0 \tag{22}$$

As discussed in[1], this identity remains valid at $\mathcal{O}(g^2)$ after an $i\epsilon$ modification of the gluon and ghost propaagators. While this check of QCD gauge invariance is non-trivial, it does not obviate the need for a general proof.

4 Summary

My proposal to consider modifications of the $i\epsilon$ prescription in PQCD is unusual. There seems, however, to be good reasons for pursuing such an approach.

- Perturbation theory is a central tool for comparing physical gauge theories with data. The properties of formally equivalent perturbative series, obtained using different asymptotic states at $t = \pm\infty$, thus merit investigation.
- There are phenomenological indications that $\alpha_s(Q^2)$ freezes at low Q^2. PQCD appears to be relevant even in the confining, long-distance regime.
- Relativistic corrections are indistinguishable from other higher order effects for bound states, since $\alpha_s = v/c$. Exact Lorentz covariance is not required at each order of the approximation[13].
- The PGC prescription (11) provides a gauge-invariant framework for studying the influence of background color fields on quark and gluon interactions. All short distance properties of PQCD are preserved.

The QCD ground state is believed to have also a quark condensate, related to the breaking of chiral symmetry[6]. It may thus be interesting to consider asymptotic states $|\Omega_0, t = \pm\infty\rangle$ which contain also $q\bar{q}$ pairs, as a 'Perturbative Quark Condensate' model of the true vacuum. The free fermion propagator between states which have a fermion-antifermion pair of momentum $\vec{k}$ differs from the Feynman propagator in the $i\epsilon$ prescription at *both* the positive and negative energy poles (*cf.* the discussion at the beginning of Section 2). An analysis of the type outlined above for gluons should thus be carried out also for quarks.

Acknowledgments

I am grateful to the organizers for their invitation to this interesting workshop, and for helpful discussions with S. J. Brodsky, G. W. Carter, C. S. Lam and H. Weigert. This work was supported in part by the EU/TMR contract EBR FMRX-CT96-0008.

References

1. P. Hoyer, Talk at the workshop on *Exclusive and Semi-exclusive Processes at High Momentum Transfer,* Jefferson Lab, USA, May 1999, hep-ph/9908501.
2. Yu. L. Dokshitzer, Talk at ICHEP 98 (Vancouver), hep-ph/9812252.
3. D. Haidt, Talk at DIS 99 (Zeuthen), www.ifh.de/ dis99p/WGROUP1 /wg1-32.ps.gz .
4. C. Bochna *et al.*, *Phys. Rev. Lett.* **81** (1998) 4576.
5. V. A. Khoze and W. Ochs, *Int. J. Mod. Phys.* **A12** (1997) 2949, hep-ph/9701421;
   V. A. Khoze, S. Lupia and W. Ochs, *Eur. Phys. J.* **C5** (1998) 77-90, hep-ph/9711392.
6. M. A. Shifman, A. I. Vainshtein and V. I. Zakharov, *Nucl. Phys.* **B 147** (1979) 385, 448, 519.
7. P. A. M. Dirac, *Rev. Mod. Phys.* **21** (1949) 392.
8. P. Hoyer, hep-ph/9610270 and Proc. APCTP-ICTP Int. Conf. in Seoul, Korea (1997), Y. M. Cho and M. Virasoro, Eds., World Scientific (1998), p. 148, hep-ph/9709444.
9. J. Schwinger, *Phys. Rev.* **128** (1962) 2425.
10. S. Coleman, *Ann. Phys. (N.Y.)* **101** (1976) 239.
11. P. Hoyer, *Phys. Lett.* **172B** (1986) 101.
12. G. Sterman, *An Introduction to Quantum Field Theory,* Cambridge University Press, 1993.
13. W. E. Caswell and G.P. Lepage, *Phys. Rev.* **A18** (1978) 810.

EXCLUSIVE REACTIONS IN FEW BODY SYSTEMS

J.-M. LAGET
CEA/Saclay, DAPNIA/SPhN,
F91191 GIf-Sur-Yvette Cedex, France
E-mail: jlaget@cea.fr

Exclusive reactions at high momentum and energy transfers appear to be one of the best ways to "see" the mechanisms of confinement at work. On free nucleon targets, they allow to resolve the exchanged Pomeron or Reggeons in terms of their partonic constituents, and access Off Forward Parton Distribution at high Q^2 and Valence Quark Distribution Amplitudes (wave functions) at high t. On few body targets, they allow to access their short range structure, and to study the propagation in the nuclear medium of configurations of few partons with small transverse size.

1 Introduction

One of the fascinating issues in Hadronic Physics is the understanding of the structure of Hadronic Matter at the Confinement Scale. The way which appears to be very promising to address it consists to select the simplest quarks and gluons configurations (three valence quarks in baryons, two valence quarks in mesons, three quarks and one or two gluons, etc ...) which are thought to be more amenable to calculations than the full structure of hadrons. This is achieved in *exclusive* reactions involving a large momentum transfer which forces a minimal configuration of partons to be present in the small interaction volume. The study of their subsequent evolution toward a full fledge normal hadron in the nuclear medium provides us with a way to constrain their structure.

In this talk, I will deal with reactions induced by Real Photons (in which case the hard scale is set by the momentum transfer t, between the incoming photon and the detected meson) as well as with reactions induced by Virtual Photons (in which case the hard scale is set by the mass squared Q^2 of the virtual photon). I will start by a survey of our understanding of exclusive reactions induced on nucleon targets, then investigate how the study of exclusive reactions induced in the Nuclear Medium may help us to disentangle Hard and Soft mechanisms, and devote the last part of my talk to the study of short range behavior of Few Body Systems.

Needless to say, most of the material in this talk results from an intense activity, over the past few years, to sharpen the physics case and the feasibility

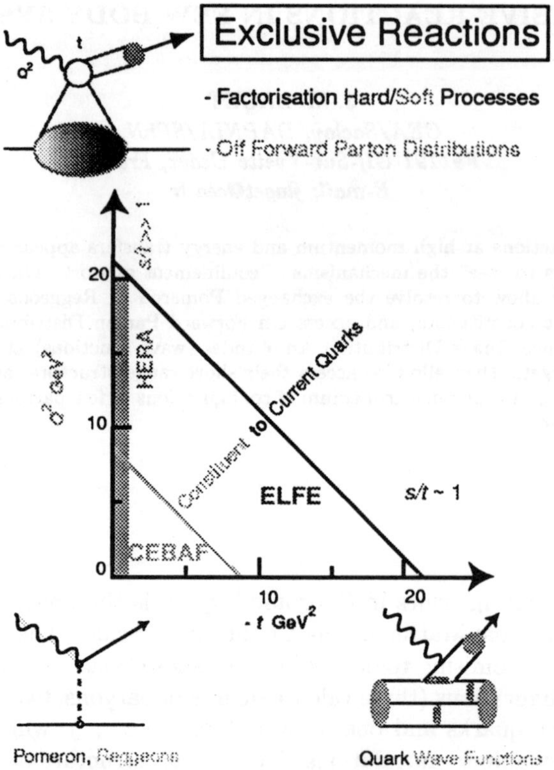

Figure 1: The Physics accessible for various combinations of the squared four momentum transfer t and the squared mass of the virtual photons Q^2.

of ELFE. I refer the reader to the proceedings of the last Saint-Malo meeting[1] for a more detailed discussion.

2 Exclusive Reactions on a Nucleon Target

Four kinematical regimes are relevant for exclusive reactions induced on nucleon targets, and schematically depicted in Fig. 1:

- Low Q^2 and low t. Below 1 or 2 GeV^2, the differential cross sections and the various spin observables are well accounted for by the exchange of a few Regge trajectories.
- High Q^2 but low t. Keeping t below 1 GeV^2, but increasing Q^2 up to

20–30 GeV^2 allows to reach a domain where one can factorize the hard scattering of the photon on the valence constituents of the emitted meson and the non perturbative Off Forward Parton Distributions in the target nucleon [2].

- High t but low Q^2. The other extreme consists to keep Q^2 small (near or at the real photon point) and increase t up to 20–30 GeV^2. Here, one can factorize the hard scattering of the incoming photon with the valence quarks, of both the target nucleon and the emitted meson, and their non perturbative wave functions.
- High Q^2 and high t. This really perturbative regime may turns out to be almost impossible to reach, due to the smallness of the corresponding exclusive cross-sections.

When going from high to low momentum transfers one really probe the evolution from a description of reaction mechanisms and of the structure of hadrons in terms of valence quarks toward a description in terms of constituent quarks, and hopefully see the confinement mechanisms at work.

Let me illustrate this strategy with a few typical examples.

2.1 Production of Mesons at Large Q^2

Fig. 2 shows the evolution of the cross-section of the exclusive electroproduction of ρ mesons with the energy available in the c.m. system, and with the mass squared of the virtual photon. The data set spans the range of energies from SLAC and early experiments at DESY (W below 7 GeV) to the present HERA experiments (W up to 200 GeV), covering experiments at CERN and Fermi Lab. (W around 10–20 GeV). While Pomeron exchange (dashed lines) reproduces fairly well the high energy behavior of the cross-section, f_2 exchange (dotted lines) and to a lesser extent σ exchange (full lines) account for the low energy part.

However this very simple and elegant picture is successful near the real photon point (Q^2 up to 2–3 GeV^2) but fails to reproduce the increase of the rise of the cross-section with the energy W, observed at high Q^2. Here, one enters a new domain where the cross sections are rather described by the exchange of two perturbative partons: two gluons at high energy [3] or two quarks at low energy [4]. Fig. 3 illustrates this point when reasonable assumptions are made on the Off Forward Parton Distributions which modelize the probability of finding two partons in the target nucleon.

To day these OFPD's are under lively discussions, as they generalize the parton distributions which are usually extracted from the analysis of inclusive

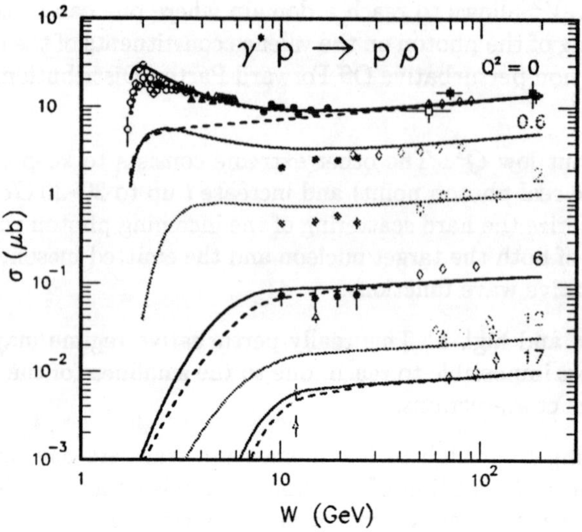

Figure 2: The cross-section of exclusive electroproduction of ρ meson on nucleon is plotted against the energy in the c.m. system W, for selected values of the four momentum transfer Q^2.

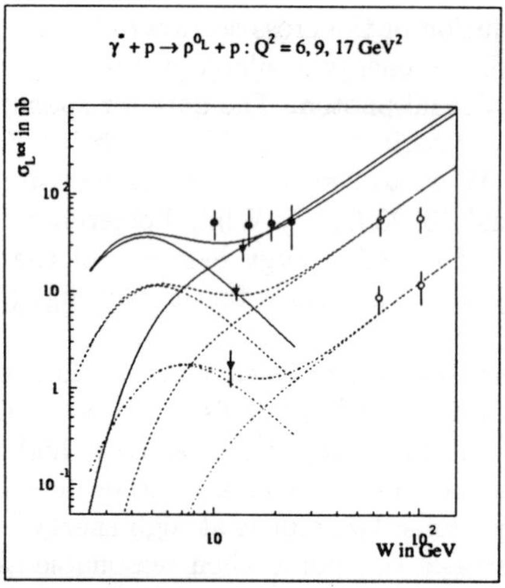

Figure 3: A zoom of Fig. 2, at large four momentum transfer Q^2: $6GeV^2$ (full line), $9GeV^2$ (dashed line) and $17GeV^2$ (dot-dashed line). Quark exchange dominates the low energy sector while gluon exchange dominates the high energy sector.

cross-sections in the Deep Inelastic Scattering regime. Their study can be already started at Cebaf, around 6 GeV, will greatly benefit of an energy upgrade to 12 GeV, but will be completed with the advent of a 30 GeV beam.

2.2 Production of Mesons at Large t

The other path toward the hard scattering regime lies close or at the real photon point. Here the scale and the hardness of the interaction is set by the momentum transfer t between the incoming photon and the emitted hadrons. This is illustrated in Fig. 4, which shows the angular distribution of the ϕ meson in the exclusive reaction $p(\gamma,\phi)p$.

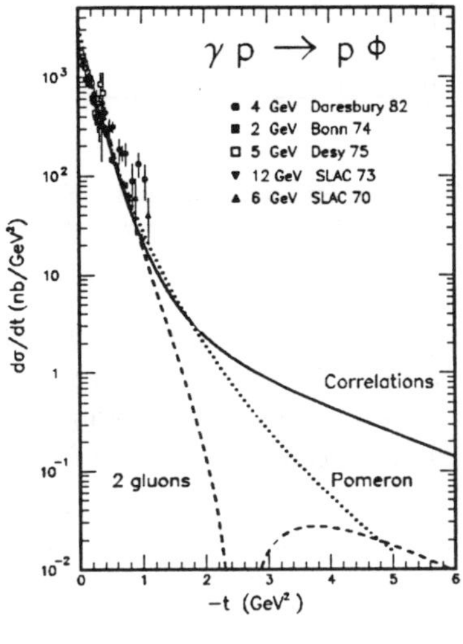

Figure 4: The cross-section of exclusive photoproduction of ϕ meson on nucleon is plotted against the momentum transfer t, in the energy range 4–10 GeV. Two gluons exchange with and without quark correlations, in the proton ground state, correspond respectively to the full and dashed lines. The dotted line corresponds to the Pomeron exchange mechanism.

At low momentum transfer t, the exponential slope is fairly well reproduced by the exchange of the Pomeron. At high momentum transfer, one may expect to resolve the structure of the Pomeron into its simplest constituents. Due to the $s\bar{s}$ structure of the ϕ meson, two gluons is the simplest system which can be exchanged in the t channel, provided that no strangeness exists in the nucleon ground state. Two gluon exchange[5] nicely matches Pomeron exchange around $|t| = 1\,GeV^2$ and exhibits a characteristic node around $|t| = 3\,GeV^2$. At such a high momentum, the s and $\bar{s}$ quarks are close enough to allow each

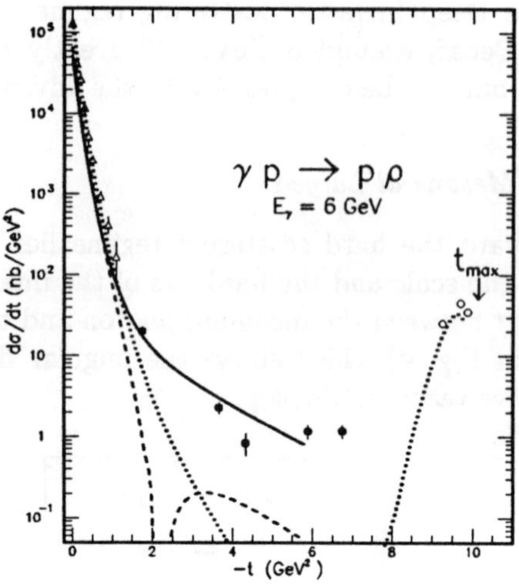

Figure 5: The same as in Fig. 4, but for ρ production at 6 GeV. At the highest t, the dotted curve corresponds to the exchange of baryons in the u-channel.

exchanged gluon to couple either to the same quark or to different quarks inside the ϕ meson. The color singlet nature of the $s\bar{s}$ pair imposes a destructive interference between the two corresponding amplitudes. Note that such a cancellation is at the origin of the mechanism of color transparency (see for instance Ref. [6]).

Each gluon can also couple to a different quark inside the nucleon, giving access to correlations between quarks in its ground state. A straightforward estimate[7], based on a simple model of the quark wave function in the proton, shows that this mechanism dominates the photoproduction of ϕ at large t. Such a picture seems to be supported by the only data[8] available in this range of momentum transfer: Fig. 5 compares the experimental and the theoretical angular distributions of the reaction $p(\gamma,\rho)p$ at 6 GeV.

However, besides two gluons exchange, quark interchange mechanisms may contribute to this channel. Also, baryon exchange mechanisms are more important at backward angles in the ρ than in the ϕ production channel (since the ρ-Nucleon coupling constant is much larger than the ϕ-Nucleon one). Finally, the available data set have been obtained with continuous bremstralhung beams. For all these reasons a comprehensive experimental study of vector meson production has been completed at 4 GeV in June 98 and will be purshued at 5.5 GeV in July 99 at Cebaf[9]. Its extension in the range 10–12 GeV will eventually lead to a better understanding of the Pomeron structure (complementary to diffractive physics at Hera) and provides us with a way to access quark correlations in the nucleon ground state.

In the pseudo-scalar meson sector, quark interchange mechanisms are thought to be responsible for the reaction cross-sections at large transfer t and u. Here (around 90°) the cross-section of the various photoproduction exclusive channels [8] exhibits a scaling behavior (s^{-N}): this is a necessary condition for being in the hard scattering regime. However this not sufficient and no models have been able yet to put such a hard picture on solid grounds. Quark–diquark modelizations of the hadron wave function lead to a fair agreement with the data at the expense of a few reasonable parameters [10]. The only direct calculation [11] of hard scattering mechanisms suffers from inaccuracy in the treatment of the singularities arising from the on-shell propagation of the partons in the hard scattering amplitude.

Compton Scattering is the only channel where these issues have been addressed in a comprehensive way [12,13,14], as discussed elsewhere in these proceedings.

3 Exclusive Reactions in Nuclei

The key question is therefore: At which momentum transfer hard scatterings dominate the reaction amplitude? This is still an open issue which is under fierce debates. Color Transparency may prove to be the most direct way to answer it.

At other extreme, one may ask how to address issues related to the short range structure of nuclear matter by mean of hard scattering mechanisms.

3.1 Color Transparency

The concept of "Color Transparency" follows from the underlying structure of QCD: interactions between "white" objects depend on their transverse size. A hard scattering of the probe produces recoiling particles with small transverse size, whose the subsequent interactions in Nuclear Matter are reduced. There is no doubt that Color Transparency should occur. The question is where and when.

The idea is to select one of the simplest configurations of a nucleon (or a hadron) in a nucleus and to see how it evolves toward its asymptotic wave function. The study of the interaction of the outgoing hadron with the nuclear medium, as a function of the size of the nucleus, will give us informations on the corresponding evolution. However, the characteristic scale of the evolution should be larger, or comparable, to the size of the largest nuclei.

To date there are no convincing evidences for Color Transparency. The reason is that most of the attempts were performed in quasi elastic kinematics. In the A(e,e'p) reactions, for instance, it is very likely that the values of Q^2 are too low to observe color transparency in the quasi-free kinematics channels, where the energy of the ejected nucleon T_p and the photon four-momentum are not independent ($T_p = Q^2/2M$). In the range of reasonable values of Q^2, for which the cross section does not vanish, the life time of the small object is of the order of the distance between nucleons rather than the nuclear radius. For instance, at the highest $Q^2 = 6$ GeV^2 where data exist, the energy of the outgoing proton is only 3 GeV and its characteristic evolution distance is no more than 1.5 fm.

3.2 Color Transparency in Few Body Systems

The way to overcome this difficulty is to study reactions induced by photons or electrons in few body systems: Exclusive reactions allow to adjust the formation length of the hadron to the distance between nucleons. The kinematics should be chosen such that the interactions of the emerging hadron with a second nucleon are maximal. This occurs when the produced hadron propagates on-shell (triangular logarithmic singularity). A clear signal for color transparency would be the suppression of final state interactions when the momentum transfer increases.

As an example, Fig. 6 shows such a signal in the case of the $D(e,e'p)n$ reaction. When the recoiling neutron momentum is high enough (about 400 MeV/c or higher), the quasi free process is suppressed and Nucleon-Nucleon rescattering dominates the cross-section: it is maximum when $X = Q^2/2m\nu$ $= 1$. This is a clear signal which depends only upon on-shell matrix elements and upon the low momentum components of the nuclear wave function: The starting point is founded on solid grounds. When Q^2 increases, one select configurations in the nucleon with smaller and smaller transverse extension, which interact less likely with the second nucleon: it is expected that the on-shell rescattering peak disappears above a certain momentum transfer.

This situation is more comfortable than in the more classical study of quasi elastic scattering of electrons from heavy nuclei, where one look for a change of a flat level of attenuation of the outgoing nucleon, instead of the evolution of a well defined peak.

Counting rate estimates[15] have been performed in the context of the ELFE project. They show that, for an energy around 30 GeV, the signal can be extracted up to $Q^2 \simeq 20$ GeV^2, provided a luminosity of $\mathcal{L} = 10^{38} cm^{-2} s^{-1}$ is achieved when two well shielded magnetic spectrometers are used, or $\mathcal{L} =$

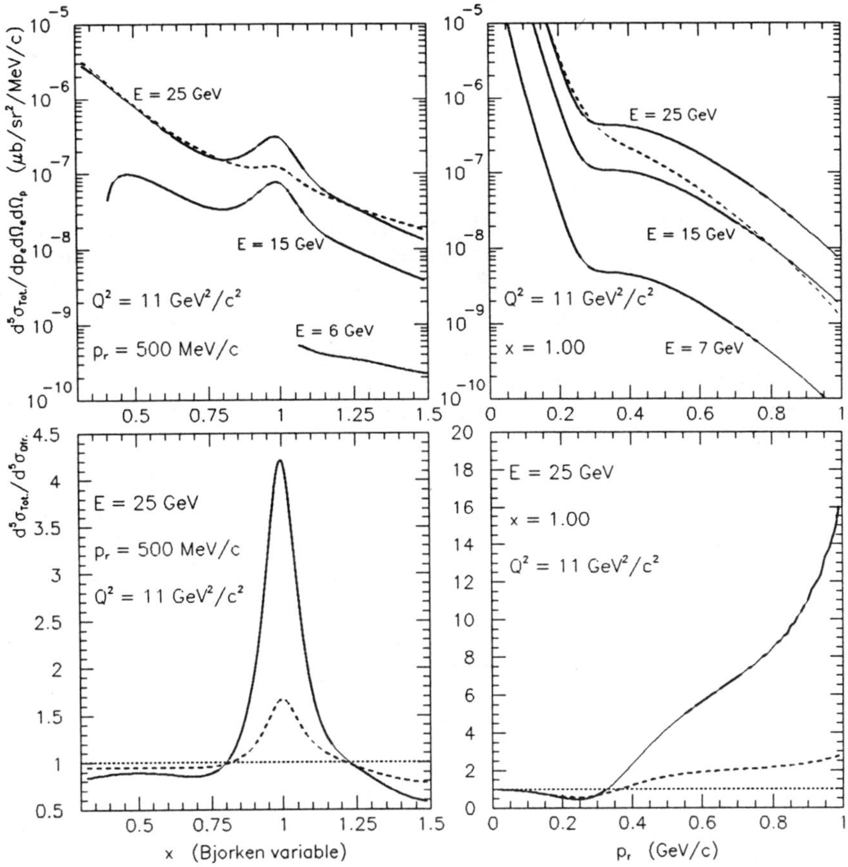

Figure 6: Fully differential cross-section (top) and ratio of the total to the quasi-free cross section (bottom) of the $D(e,e'p)n$ reaction. The left panels correspond to constant neutron recoil momentum P_r. The right panels correspond to constant $X = q^2/2M\nu$. The full lines and dashed lines include nucleon rescattering without and with color transparency effect, respectively. The dotted curves correspond to the quasi-free process.

$10^{35} cm^{-2} s^{-1}$ in the case of a 4π detector. An experiment has been approved at CEBAF [16], but only $Q^2 = 6$ GeV^2 will be reached there.

The onset of Color Transparency will indicate the range of momentum transfer where hard scatterings dominate the reaction amplitude: it may well depend on each channel.

3.3 Production of Pseudoscalar Mesons

It may occur earlier in Photo and Electroproduction of Mesons. The reason is that these channels lie in between Deep Inelastic Scattering (DIS) and Elastic Scattering (Form Factors). In DIS the energy transfer is large enough to break-up the target, making a hard scattering description good enough at momentum transfers as low as $Q^2 \simeq 1$ GeV^2. On the contrary, the energy transfer is vanishing in form factor measurements, rending questionable a hard scattering description at low momentum transfers, in the range of few GeV^2. In exclusive mesons production reactions, both the energy and the momentum transfers are large enough to allow for a perturbative treatment in the momentum transfer range of few GeV^2.

A good example is the reaction $D(\gamma, p\pi^-)p$ in the energy range $4 < E_\gamma <$ 10 GeV. For real photon the momentum transfer t, between the incoming photon and the outgoing pion, sets the size of the interaction volume. As can be seen in Fig. 7, the on shell rescattering peaks corresponding to πp or pp interactions are clearly separated. At the top of each peak, the rescattering amplitude is dominated by low momentum components of the deuteron wave function and on mass shell elementary reaction amplitudes (see Ref. [17]): Such a Logarithmic singularity has already been observed at lower energies [18]. Furthermore, the elementary reaction $n(\gamma, \pi^-)p$ is almost flat, in the range of t around 3 to 10 GeV^2, and exhibits, around $\theta_\pi = 90°$, a scaling behavior which is a necessary condition for hard scatterings. Here, the nucleon cross section is well reproduced by a model based on the exchange of saturating Regge trajectories [19].

Figs. 8 and 9 show the evolution of the cross-sections at the top of the πp rescattering peak against the recoil momentum P_R and the momentum transfer t, respectively. The expected effect of Color Transparency is clearly apparent. A toy model, based on a geometrical expansion of the mini configuration of the ejected hadron, has been used and is meant as a guide: Only experiment will tell us what is the relevant nature of the process which governs its evolution and formation.

Counting rate estimates show that this channel is already accessible in the CLAS detector at CEBAF, in the range of t around 3 to 5 GeV^2. A beam of

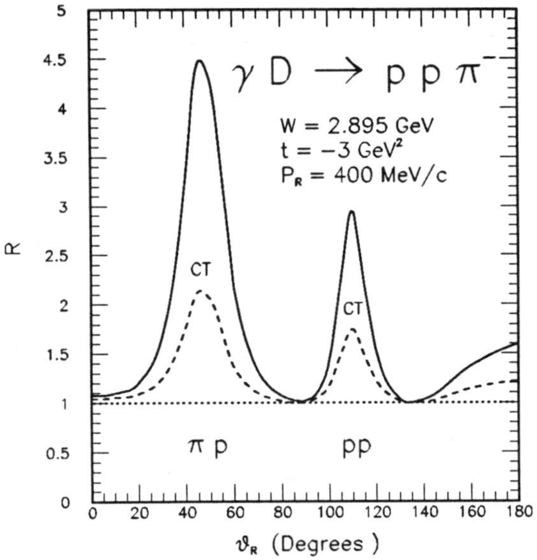

Figure 7: Ratio of the total to the quasi-free cross section of the $D(\gamma, \pi^- p)p$ reaction against the angle of the recoiling proton whose the momentum is kept constant at 400 MeV/c. The peak at the left corresponds to πp rescattering, while the peak at the right corresponds to pp rescattering. The full lines and dashed lines include hadron rescattering without and with color transparency effect, respectively. The dotted curves correspond to the quasi-free process.

real photons of 10 GeV would be ideal to map out the t dependence up to 10 GeV^2. Above, the cross sections are decreasing too fast (as s^{-7}), making the measurement almost impossible.

3.4 Production of Vector Mesons

Exclusive Vector Mesons production on few body systems is certainly very promising. It allows to prepare a pair of quarks, with an adjustable transverse size, and to study its interaction with a nucleon in well defined kinematics. Furthermore, the coherence time (during which the incoming photon oscillates into a $q\overline{q}$ pair) and the formation time (after which this pair recombines into the final meson) can be adjusted independently to the internucleon distance. Such a study may end up with a better understanding of the formation of vector mesons in cold hadronic matter, and is complementary to the study of

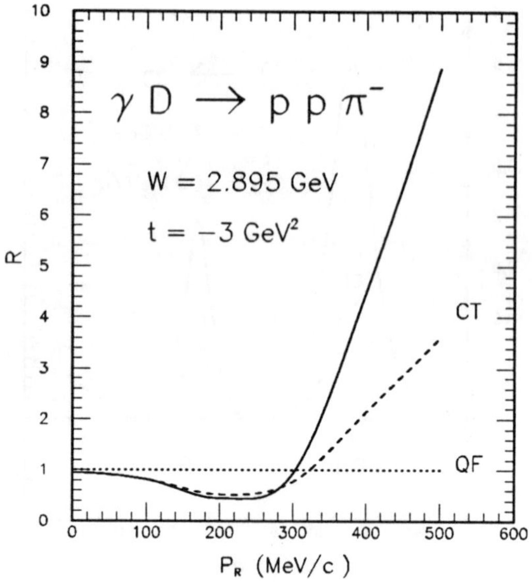

Figure 8: Ratio of the total to the quasi-free cross section of the $D(\gamma, \pi^- p)p$ reaction against the momentum of the recoiling proton, at the top of the πp rescattering peak. The full lines and dashed lines include πp scattering without and with color transparency effect, respectively. The dotted curves correspond to the quasi-free process.

vector meson production in heavy ion collisions. Although more work has to be done to put it on the same footing as the previously discussed channels, the strategy will follow the same lines: the elementary cross sections are of same order of magnitude.

A special emphasis should be put on ϕ and J/Ψ mesons production: not only these narrow states are more easy to identify, but their flavor content, different from that of the ground state of cold hadronic matter, makes them a promising probe.

3.5 Short Range Properties of Nuclei

Instead of using the well established long range properties of a nucleus as a filter to pin down hard scattering mechanisms, one may alternatively take advantage of the short range nature of hard processes to unravel some aspects of the short distance behavior of nuclei. Among many possibilities, let me emphasize three axis of research.

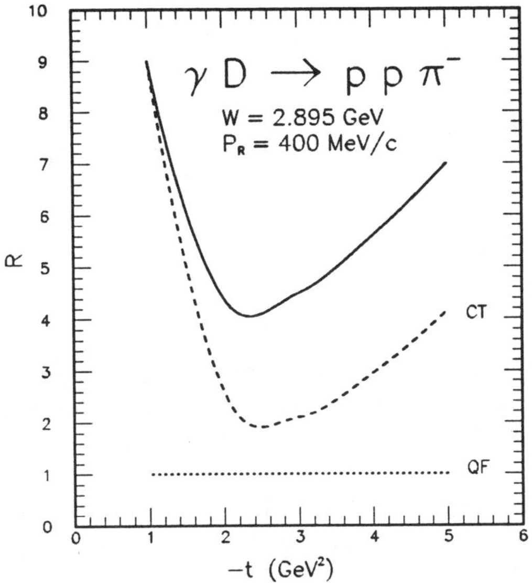

Figure 9: Same as in Fig. 8, but against the momentum transfer.

One of the fascinating issues in Nuclear Physics is that the conventional constituents of nuclei appear to keep their identity, even if the nucleus is shaken by a photon of very high energy. The best example is the two body photodisintegration of the deuterium, of which the scaling of the cross-section with the energy, previously observed at SLAC [20] up to 2 GeV, has recently been confirmed at energies above 3 GeV at Cebaf[21]. Of course, this occurs in a tiny part of the available phase space, but corresponds to the excitation of a highly excited intermediate state which couples to the short range part of the initial deuteron and the final two nucleon system. Understanding such a mechanism requires a comprehensive study not only of the unpolarized cross-section of the photodisintegration of the few body systems, but also of the various spin observables.

The second issue concerns exotic configurations which may occur at short range in the few body systems. In deuterium, for instance, instead of projecting onto two white clusters (the nucleons) the six quarks may also project onto two colored clusters which eventually couple to a white deuterium. Such a hidden color configuration may be revealed in ϕ photoproduction off deuterium: each of the two exchanged gluons can couple to a different colored cluster, transforming it in a white nucleon which is detected in the final state. The physics case is developed in Ref. [5] which I refer the reader to.

Last, the response of nuclei at $X > 1$ is still an open issue. Here, for kinematical reasons, only nucleons can be ejected, but free nucleon kinematics are forbidden: this opens a window on short range mechanisms. A comprehensive study requires not only electron beams of high intensity, since the corresponding cross-sections are very small, in the 12 GeV range [22] but will benefit greatly of the study of selected spin observables.

4 Conclusion

To conclude, exclusive reactions induced on nucleon targets will offer us with an unique window to track (to "see") confinement mechanisms at work: either in the exchange of, free or bound, partonic systems or in selected few parton systems.

Exclusive reactions induced in the few body systems may well be the best place to study the onset of color transparency and its relationship to hard scattering mechanisms. The key issue is to select kinematics where rescatterings between hadrons are maximized and are described without any uncertainty: this provides a safe starting point for determining the evolution of scattering cross sections with the momentum transfer.

An intense continuous beam of *real photons* of about 10 GeV will be ideally suited: above the exclusive reaction cross-sections decrease too quickly to be measured. However, a continuous beam of *electrons* of 30 GeV is necessary to map out the Q^2 dependency of the various channels.

The corresponding experimental program can be started at CEBAF, in its present 6 GeV configuration, and fully justifies an upgrade up to 12 GeV.

The step toward 30 GeV requires a coherent effort of the community. To day, two paths are open: either continue the energy increase on the CEBAF site, or take advantage of the opportunity to built ELFE in Europe, either at DESY [1] or at CERN [23]. It is very likely that the final choice will result from a world-wide decision making process.

Acknowledgements

This work was supported by the French Atomic Energy Commission (CEA) and in part by the European Commission (EC) under contract ERB FMRX-CT96-0008.

References

1. N. d'Hose et al., Eds. "Prospects of Hadrons and Quark Physics with Electromagnetic Probes". Saint-Malo, France, Sept. 23–27, 1996. Nucl. Phys. A622 (1997) Nos 1,2.
2. J. C. Collins, L. Frankfurt and M. Strikman, Phys. Rev. D56 (1997) 2982.
3. L. Frankfurt, W. Koepf and M. Strikman, Phys. Rev. D54 (1996) 3194; S. J. Brodsky et al., Phys. Rev. D50 (1994) 3134.
4. M. Vanderhaeghen, P.A.M. Guichon and M. Guidal, Phys. Rev. Lett. 80 (1998) 5064.
5. J.-M. Laget and R. Mendez-Galain, Nucl. Phys. A581 (1995) 397.
6. J.F. Gunion and D. Soper, Phys. Rev. D15 (1977) 2617.
7. J.-M. Laget, to be published.
8. R.L. Anderson et al., Phys. Rev. D4 (1971) 1937.
9. G. Audit et al., Cebaf Experiment 93–031.
10. P. Kroll et al., Phys. Rev. D55 (1997) 4315.
11. G. Farrar et al., Nucl. Phys. B349 (1991) 655.
12. M. Vanderhaeghen, P. Guichon and J. Van de Wiele, Nucl. Phys. A622 (1997) 144c.
13. A. Radyushkin, JLAB-THY-98-10, hep-ph/9803316.
14. P. Kroll, these proceedings.
15. E. Voutier et al., contribution to the parallel sessions of the Saint-Malo ELFE Workshop, Saclay Report DAPNIA-SPhN-96-35, 1996.
16. K.S. Egiyan et al., Cebaf Experiment 94-019.
17. J.-M. Laget, Phys. Rep. 69 (1981) 1.
18. P.E. Argan et al., Phys. Rev. Lett. 41 (1978) 86.
19. M. Guidal, J.-M. Laget and M. Vanderhaeghen, Phys. Lett. B400 (1997) 6; Nucl. Phys. A627 (1997) 645; Phys. Rev. C57 (1998) 1454.
20. J. Napolitano et al., Phys. Rev. Lett. 61 (1988) 2530.
21. R. Holt et al., Cebaf Experiment 89-012.
22. O. Bing et al., Conference Proceeding 44 "The ELFE Project"; J. Arvieux and E. De Sanctis Eds., SIF, Bologna 1993, p. 475.
23. E. Keil and G. Geschonke, Report CERN SL-060(RF).

COHERENT NUCLEAR DIFFRACTIVE PRODUCTION OF MINI-JETS — SIGNATURE OF COLOR TRANSPARENCY

G. A. MILLER
Department of Physics, Box 351560
University of Washington, Seattle, Washington 98195-1560, USA
E-mail: miller@phys.washington.edu

The process $\pi + A$ leading to a pair of mini-jets at high relative transverse momentum ($k_t \gtrsim 2$ GeV) while leaving the nucleus in its ground state was selected as a definitive test of the existence of color transparency by Frankfurt, Miller and Strikman in 1993. The preliminary results of Fermilab experiment E791, led by Ashery and Weiss–Babai, show a strong A-dependence, consistent with the notion of color transparency.

1 Introduction

The process we discuss is $\pi + A$ going to a $q\bar{q}$ pair of high relative transverse momentum ($k_t \gtrsim 2$ GeV) and leaving the nucleus in its ground state. The q and $\bar{q}$ each decay into a mini-jet; I shall not discuss this aspect much. Observing the nuclear (A) dependence of this process was pointed out as a good way to study color transparency by us[1] in 1993. This was a long time ago. The reason for discussing this topic now is the recent apparent confirmation of the color transparency idea in Fermilab E791 by Ashery and Weiss–Babai[2]. The use of the very high energy $E_\pi = 500$ GeV beam makes these observations possible – a small-sized wave packet can travel through the nucleus without expanding[3].

2 The Process

Why should color transparency occur in π + A $\rightarrow q\bar{q}$+ A (ground state)? The selection of the final state as a $q\bar{q}$ pair moving at high relative momentum ($k_t \gtrsim 2$ GeV) plus the nuclear ground state causes the $q\bar{q}$ component of the pion to be the dominant component[4]. At 500 GeV, the pion breaks up into a $q\bar{q}$ pair well before hitting the target. Since k_t is large, the transverse separation b between the q and $\bar{q}$ is small. The gluon emission from a color singlet $q\bar{q}$ pair (or any color–singlet, quark–gluon system) of small spatial extent is suppressed[5]. The forward scattering amplitude $f \sim ib^2 \sim i\sigma_0 \frac{b^2}{<b^2>}$, where σ_0 is a normal size cross section ($\sim 20 - 30$ mb) and $< b^2 >$ is the pionic average of the squared transverse of the $q\bar{q}$ pair. The origin of f is the exchange of two dipole gluons with the nuclear target. Two gluons are needed to maintain the color neutrality of the projectile and target. For large values of $k_t, f \sim -i\sigma_0/(< b^2 > k_t^2)$. The

$q\bar{q}$ production process (which simultaneously leaves the nucleon in its ground state) occurs rarely. Thus the dominant nuclear amplitude occurs via only one interaction of f. There are no other initial or final state interactions because of the small size of the $q\bar{q}$ pair. This is color transparency. The single interaction of f can occur on any nucleon and, because the final state is the nuclear ground state, is a coherent process. The nuclear amplitude $\mathcal{M}_A \propto Af$. This is a huge nuclear enhancement.

In the following, I'll discuss our 1993 calculation, the experiment and their relationship.

3 Nucleon Amplitude

The amplitude, $\mathcal{M}(N)$, for the process $\pi N \to q\bar{q}N$ to occur on a nucleon N is given, for large enough values of k_t by

$$\mathcal{M}(N) = \int d^2b\, \psi_\pi(x,b)\, i\sigma_0 \frac{b^2}{< b^2 >} e^{-i\vec{k}_t\cdot\vec{b}}\,, \tag{1}$$

where $\psi_\pi(x,b)$ is the pion wave function and x is the fraction of the pion longitudinal momentum carried by the final state quark. The anti-quark carries a fraction $1-x$. The b^2 operator can be replaced by $-\nabla^2_{k_t}$ acting on the Fourier transform, $\tilde{\psi}_\pi(x,k_t)$, of the pion wave function. For wave functions with a power fall–off in k_t, the term b^2 then acts as $\sim -1/k_t^2$. Note that $< b^2 >\approx 0.24$ fm^2, while $1/k_t^2 \sim 0.01$fm^2, so the interaction is weak indeed. For the asympotic pion wave function we find [1]

$$\mathcal{M}(N) \sim \frac{i}{k_t^2}\frac{x(1-x)}{k_t^2}\alpha_s(k_t^2)\,. \tag{2}$$

In this case $\mathcal{M}(N) \propto \tilde{\psi}_\pi(x,k_t) = \alpha_s(k_t^2)\frac{x(1-x)}{k_t^2}$. This especially simple relation occurs only if the final $q\bar{q}$ pair do not interact with each other by the exchange of a high momentum gluon [6].

4 Nuclear Amplitude

The picture we have is that the pion becomes a $q\bar{q}$ pair of essentially zero transverse extent well before hitting the nuclear target. This point like configuration PLC can move through the entire nucleus without expanding. The $q\bar{q}$ can interact with one nucleon and can pass undisturbed through any other nucleon. For zero momentum transfer, q_t, to the nucleus, the amplitude $\mathcal{M}(A)$ takes the form

$$\mathcal{M}(\mathrm{A}) = \mathrm{A}\mathcal{M}(N)\left(1 + \frac{\epsilon}{< b^2 > k_t^2}\mathrm{A}^{1/3}\right) \equiv \mathrm{A}\mathcal{M}(N)\gamma\,, \tag{3}$$

where the real number $\epsilon > 0$ and $\gamma > 1$. Observe the factor A which is the dominant effect here. The ϵ correction term arises from a soft rescattering which can occur as the PLC moves through the nuclear length ($R_A \propto \mathrm{A}^{1/3}$). The action of f produces the $1/(< b^2 > k_t^2)$ factor.

The differential cross section takes the form

$$\frac{d\sigma(A)}{dq_t^2} = A^2\gamma^2\frac{d\sigma}{dq_\perp^2}(N)e^{-q_t^2R_A^2/3}\,. \tag{4}$$

One measures [2] the integral

$$\sigma(A) = \int dq_\perp^2\frac{d\sigma(A)}{dq_t^2} = \frac{3}{R_A^2}A^2\gamma^2\sigma(N)\,. \tag{5}$$

A typical procedure is to parametrize $\sigma(A)$ as

$$\sigma(A) = \sigma_1 A^\alpha \tag{6}$$

in which σ_1 is a constant independent of A. For the R_A corresponding to the two targets Pt (A = 195) and C(A = 12) of E791, one finds $\alpha \approx 1.45$, if γ is taken to be unity. Including the value of γ using the results of [1] leads to

$$\alpha \approx 1.58\,. \tag{7}$$

That $\gamma > 1$ was a somewhat surprising feature of our 1993 calculation because the usual second order scattering reduces cross sections. The key features of the usual first order term are: $f = i\sigma_0$, and those of the second order term are $if^2 = -i\sigma_0^2$. Note the opposite signs. For us here $f = i\sigma_0 b^2/ < b^2 >$, which for very large values of k_t^2 becomes $f = -i\sigma_0/(< b^2 > k_t^2)$. The second order term depends on $if^2 = i[-i\sigma_0/(< b^2 > k_t^2)^2]^2$, which now has the same sign as the first–order term.

5 Requirements for Color Transparency

These were discussed in Ref. [1]. The two jets should have total transverse momentum to be very small. The relative transverse momentum should be $\gtrsim 2$ GeV/c. One must identify the final nucleus as the target ground state. This is done [2] by isolating the q_t^2-dependence of the elastic diffractive peak. We need substantial A-dependence, $\sigma \equiv \sigma_1 A^{1.6}$, for large enough values of

k_t. The background processes involving nuclear excitation vary as A, so an unwanted counting of such would actually weaken the signal we seek. The amplitude varies as $\mathcal{M}(A) \sim \alpha_s/k_t^4$ and $\sigma(A) \sim \alpha_s^2/k_t^8$; this should be checked experimentally. For the amplitude discussed here, $\sigma(A) \sim (x(1-x))^2$. See however [6].

6 The experiment – ELAB E791

All of the information I have is from Ashery's webpage[2]. This discusses the excellent resolution of the transverse momentum, shows the identification of the di-jet using the Jade algorithm, and displays the identification of the diffractive peak by the q_t^2 dependence for very low q_t^2. This dependence is consistent with that obtained from the previously measured radii $R_c = 2.44$ fm, and $R_{Pt} = 5.27$ fm.

Their preliminary result is

$$\alpha \approx 1.55 \pm 0.05 \,, \tag{8}$$

which is remarkably close to the theoretical value shown in Eq.(9). The k_t^2 dependence of the scattering amplitude has not been checked, but we are told that this is both feasible and in progress.

The measured cross section does have an x-dependence which is reasonably well described by $[x(1-x)]^2$. This lends support to the notion that the asymptotic behavior is manifest in nature. However, $\mathcal{M}(A)$ is only approximately proportional to the quark distribution amplitude [6].

7 Electromagnetic Background

Because very low values of q_t^2 are involved, one could ask if the process occurs by one photon exchange (Primakoff effect) instead of two gluon exhange. Even if the Primakoff process were dominant, the unusual A-dependence would be a consequence of color transparency. However, we can estimate the relative importance of the two effects. We have

$$\mathcal{M}_P(N) = \frac{e^2 \tilde{\psi}_\pi(x, k_t) Z}{q_t^2} \,, \tag{9}$$

which should be compared with the amplitude of Eq. (2) written as

$$\mathcal{M}(N) = \tilde{\psi}_\pi(x, k_t) A \frac{\sigma_0}{< b^2 >} \frac{i}{k_t^2} \tag{10}$$

using $q_t^2 \approx 0.02$ GeV^2, Z = 78, $e^2 = 1/137, k_t = 2$ GeV, and $\sigma_0/ < b^2 > \approx 10$ gives

$$\frac{\mathcal{M}_P(N)}{\mathcal{M}(N)} = -0.06i\ . \tag{11}$$

The Primakoff term is small and, because of its real nature, does not interfere with the larger stong amplitude.

8 Meaning of $\alpha = 1.55$

The use of $\alpha = 1.55$ in Eq. (6) leads to

$$\frac{\sigma(Pt)}{\sigma(C)} = 75\ . \tag{12}$$

The usual dependence of a diffractive process is $\approx A^{2/3}$ or $\alpha = 2/3$. This would give

$$\frac{\sigma_{\text{USUAL}}(Pt)}{\sigma_{\text{USUAL}}(C)} = 7. \tag{13}$$

Thus color transparency causes a factor of 10 enhancement! This seems to be the huge effect of color transparency that many of us have been hoping to find. A word of caution must be sounded. The k_t^2 dependence of the cross section must be verified at least roughly to be consistent with the $(\alpha_s/k_t^4)^2$ behavior discussed above.

9 Implications

Suppose color transparency has been correctly observed in $\pi + A \rightarrow q\bar{q} + A$ (ground state). So what?

There are many implications. The spectacular enhancement of the cross section would be a new novel effect. The point like configurations PLC would be proved to exist. This is one more verification of the concept and implications of the idea of color. Previous experiments [7] showing hints [8] of color transparency (for a review see Ref [3]) probably do show color transparency. Efforts [9] to observe color transparency at Jefferson Lab should be increased.

The idea that a nucleon is a composite object is emphasized by these findings. If there are PLC, there must also exist blob–like configurations of Huskyons [10]. These different configurations have wildly different interactions with a nucleus [11], so that the nucleon in the nucleus can be very different from a free nucleon. This leads to an entirely new view of the nucleus, one in which the nucleus is made out of oscillating, pulsating, vibrating, color singlet, composite objects.

Acknowledgments

This work is partially supported by the USDOE, and is based on a collaboration with L. Frankfurt and M. Strikman.

References

1. L. Frankfurt, G.A. Miller and M. Strikman, *Phys. Lett.* B **304**, 1 (1993).
2. Fermilab experiment E791, R. Weiss-Babai, talk at Hadron '97 and D. Ashery, talk at International Workshop of Diffractive Physics, Rio de Janeiro, Feb. 1998.
3. See the review - L. Frankfurt, G.A. Miller and M. Strikman, *Ann. Rev. Nucl. Part. Sci.* **44**, 501 (1994).
4. The pion has a component of $q\bar{q}$ plus wee partons. The wee partons have the size of a normal hadron and break up when traveling through the nucleus. The irrelevance of this component is one aspect of the filtering idea of P. Jain, B. Pire and J.P. Ralston, *Phys. Rept.* **271**, 67 (1996).
5. G.F. Bertsch, S.J. Brodsky, A.S. Goldhaber and J.F. Gunion, *Phys. Rev. Lett.* **47**, 297 (1981).
6. There are other contributions to $\mathcal{M}(N)$, also suppressed by color coherence, which lead to terms of the same form as Eq.(2). These will be discussed in forthcoming publications by L. Frankfurt, G.A. Miller and M. Strikman including: L. Frankfurt, G.A. Miller and M. Strikman, "Perturbative pion wave function in coherent pion nucleon dijet production," hep-ph/9907214, to appear in Foundations of Physics.
7. A.S. Carroll *et al.*, *Phys. Rev. Lett.* **61**, 1698 (1988).
8. B.K. Jennings and G.A. Miller, *Phys. Rev. Lett.* **69**, 3 (6)19 1992; B.K. Jennings and G.A. Miller, *Phys. Lett.* B **236**, 209 (1990); B.K. Jennings and G.A. Miller, *Phys. Rev.* D **44**, 692 (1991); B.K. Jennings and G.A. Miller, *Phys. Lett.* B **318**, 7 (1993).
9. K. Egiyan, L. Frankfurt, W.R. Greenberg, G.A. Miller, M. Sargsian and M. Strikman, *Nucl. Phys.* A **580**, 365 (1994); L.L. Frankfurt, W.R. Greenberg, G.A. Miller, M.M. Sargsian and M.I. Strikman, *Z. Phys.* A **352**, 97 (1995); L. Frankfurt, W.R. Greenberg, G.A. Miller, M.M. Sargsian and M.I. Strikman, *Phys. Lett.* B **369**, 201 (1996).
10. A. Bulgac and L. Frankfurt, *Phys. Rev.* D **48**, ((1)993) 1894.
11. L.L. Frankfurt and M.I. Strikman, *Nucl. Phys.* B **250**, 1 (4)3 (1985). M.R. Frank, B.K. Jennings and G.A. Miller, *Phys. Rev.* C **54**, 9 (2)0 (1996).

SPIN DEPENDENCE OF DIFFRACTIVE VECTOR MESON PRODUCTION

N.N. Nikolaev

Institut f. Kernphysik, Forschungszentrum Jülich, D-52450 Jülich, Germany
L.D.Landau Institute for Theoretical Physics, 142432 Chernogolovka, Russia
E-mail: N.Nikolaev@fz-juelich.de

I review the recent progress in the theory of s-channel helicity nonconservation (SCHNC) effects in diffractive DIS. SCHNC in diffractive vector meson production is an unique probe of the spin-orbit coupling and Fermi motion of quarks in vector mesons. Photo- and electroproduction of the ϕ and its radial and angular excitations at Jlab is perfectly posed to probe SCHNC in QCD pomeron exchange. I also discuss the unitarity driven demise of the Burkhardt-Cottingham sum rule and large & scaling departure from the Wandzura-Wilczek relation.

1 Introduction

The common wisdom is that spin-dependence vanishes in the high energy limit, as an example recall the rapidly decreasing longitudinal spin asymmetry , A_1, in polarized DIS. In high energy QCD the well known quark helicity conservation was believed universally to entail the s-channel helicity conservation (SCHC) at small x, i.e. the QCD pomeron was supposed to decouple from helicity flip, i.e., spin-dependence of diffractive DIS was supposed to vanish st small x. Here I review the recent work [1,2,3,4,5,6,7], which shows this belief was groundless, and expose an extremely rich SCHNC physics in diffractive DIS at small x. Furthermore, by the virtue of unitarity diffractive SCHNC is found to change dramatically the small-x behaviour of transverse spin asymmetry A_2 and lead to the demise of the Burkhardt-Cottingham sum rule and the departure from the Wandzura-Wilczek relation.

2 Is SCHNC compatible with quark helicity conservation?

The backbone of DIS is the Compton scattering (CS) $\gamma^*_\mu p \to \gamma^{*'}_\nu p'$, which at small-$x$ can be viewed as a (i) dissociation $\gamma^* \to q\bar{q}$ followed by (ii) elastic scattering $q\bar{q}p \to q\bar{q}p'$ with **exact** conservation of quark helicity and (iii) fusion $q\bar{q} \to \gamma^{*'}$. The CS amplitude $A_{\nu\mu}$ can be written as $A_{\nu\mu} = \Psi^*_{\nu,\lambda\bar{\lambda}} \otimes A_{q\bar{q}} \otimes \Psi_{\mu,\lambda\bar{\lambda}}$ where $\lambda, \bar{\lambda}$ stands for $q, \bar{q}$ helicities, $\Psi_{\mu,\lambda\bar{\lambda}}$ is the wave function of the $q\bar{q}$ Fock state of the photon. The $q\bar{q}$-proton scattering kernel $A_{q\bar{q}}$ does not depend on, and conserves exactly, the $q, \bar{q}$ helicities. For nonrelativistic massive quarks, $m_f^2 \gg Q^2$, one only has transitions $\gamma^*_\mu \to q_\lambda + \bar{q}_{\bar{\lambda}}$ with $\lambda + \bar{\lambda} = \mu$. However, the

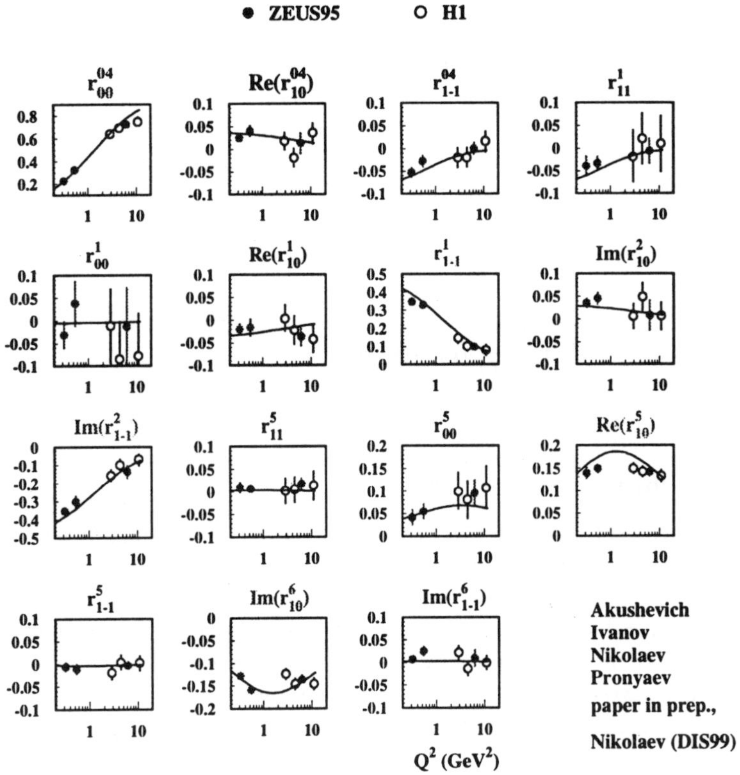

Figure 1: Our prediction[5] for the spin density matrix r^n_{ik} of diffractive ρ^0-meson vs. the experimental data from ZEUS[9] and H1[10].

relativistic P-waves give rise to transitions of transverse photons $\gamma^*_\pm$ into the $q\bar{q}$ state with $\lambda+\bar{\lambda}=0$ in which the helicity of the photon is transferred to the $q\bar{q}$ orbital angular momentum. Consequently, the SCHNC transitions $\gamma^*_\pm \to (q\bar{q})_{\lambda+\bar{\lambda}=0} \to \gamma^*_L$ and $\gamma^*_\pm \to (q\bar{q})_{\lambda+\bar{\lambda}=0} \to \gamma^*_\mp$ are allowed[1,2]. Furthermore, the SCHNC and SCHC amplitudes have similar x-dependence, i.e., the QCD pomeron exchange contributes to the both SCHC and SCHNC transitions. We emphasize that the above argument for SCHNC does not require applicability of pQCD.

By conservation of the angular momentum the single-flip $T \to L$ and double-flip $\pm \to \mp$ transition amplitudes are $\propto \Delta$ and $\propto \Delta^2$, respectively. Here $\mathbf{\Delta}$ is the (γ, γ') and/or (p', p) momentum transfer.

3 LT interference in diffractive DIS into continuum $\gamma^* p \to p'X$

The first ever direct evaluation[1] of SCHNC effect in QCD - the LT-interference of transitions $\gamma^*_L p \to p'X$ and $\gamma^*_\pm p \to p'X$ into the same continuum diffractive states X - has been reported in 1997 and went unnoticed. Experimentally, it can be measured at HERA by both H1 and ZEUS via azimuthal correlation between the (e,e') and (p,p') scattering planes. The detailed discussion of this asymmetry A_{LT} and its use for the determination of the otherwise elusive $R = \sigma_L/\sigma_T$ for diffractive DIS is found in[2]. Here I only recall that azimuthal asymmetry is the twist-3 effect,

$$A_{LT} \propto \frac{\Delta}{Q} g^D_{LT}(x_{\mathbb{P}}, \beta, Q^2)\,, \tag{1}$$

where g^D_{LT} is the scaling structure function. It does not decrease at $x \to 0$!

4 SCHNC in diffractive $\gamma^* p \to Vp'$

Evidently, diffractive vector meson production $\gamma^* p \to Vp'$ is obtained from CS by continuation from spacelike γ^* to timelike V. The azimuthal correlation of the (e,e') and (p,V) planes with the vector meson decay plane, and the in-decay-plane angular distributions of decay products, allow the experimental determination of all helicity amplitudes $A_{\mu\nu}$. The consistent analysis of the S-wave and D-wave states of the vector mesons is presented only in[4]. At small x and whithin the diffraction cone helicity amplitudes for $\gamma^*_\mu \to V_\nu$ have the following form[3,4]

$$A_{\nu\mu}(\bar{x}, Q^2, \mathbf{\Delta}) = is\frac{c_V\sqrt{4\pi\alpha_{em}}}{2\pi^2}\exp(-\frac{1}{2}B_{3\mathbb{P}}\Delta^2)\int_0^1 \frac{dz}{z(1-z)}$$
$$\int d^2k_\perp \psi(z,k_\perp)\int \frac{d^2\kappa_\perp}{\kappa^4}\alpha_S \cdot I_{\nu\mu}(\gamma^* \to V)\frac{\partial G(\bar{x},\kappa^2)}{\partial \log \kappa^2} \tag{2}$$

and are calculable in terms of the gluon structure function of the target proton $G(x,\kappa^2)$ taken at $\bar{x} = \frac{1}{2}x_{\mathbb{P}} = \frac{1}{2}(Q^2+m_V^2)/(Q^2+W^2)$. Here $z,(1-z)$ and $\mathbf{k}$ are are the Sudakov lightcone variables of the q and $\bar{q}$ in the vector meson. The typical QCD hard scale is $\overline{Q}^2 = z(1-z)Q^2 + m^2$, where m is the quark mass. To the LL$\overline{Q}^2$, i.e., for $\kappa^2 \lesssim \overline{Q}^2$, one finds for pure S-wave mesons

$$I^S_{0L} = -4QMz^2(1-z)^2\frac{2\kappa^2}{\overline{Q}^4}\left[1+\frac{(1-2z)^2}{4z(1-z)}\frac{2m}{M+2m}\right],$$
$$I^S_{\pm\pm} = \frac{2\kappa^2}{\overline{Q}^4}\left[m^2 + 2k^2(z^2+(1-z)^2) + \frac{m}{M+2m}k^2(1-2(1-2z)^2)\right]$$

$$I^S_{\mp\pm} = 4z(1-z)\Delta^2 \frac{k^2}{2\overline{Q}^4}\left(1+\frac{6\kappa^2(1-2z)^2}{\overline{Q}^2}\right)\left[1+\frac{(1-2z)^2}{4z(1-z)}\frac{2m}{M+2m}\right]$$

$$I^S_{0\pm} = -2Mz(1-z)(1-2z)^2\Delta\frac{2\kappa^2}{\overline{Q}^4}\left[1+\frac{(1-2z)^2}{4z(1-z)}\frac{2m}{M+2m}\right],$$

$$I^S_{\pm L} = -2Qz(1-z)(1-2z)^2\Delta\frac{2\kappa^2}{\overline{Q}^4}\frac{2k^2}{\overline{Q}^2}\frac{M}{M+2m}, \tag{3}$$

where M is an invariant mass of the $q\bar{q}$ pair. No separation of the S and D-wave has been done in [7].

First, notice how the transverse and longitudinal Fermi motion of quarks in vector mesons are necessary for helicity flip and SCHNC, which would suppress SCHNC in production of heavy quarkonia in which quarks are nonrelativistic. Second, apart from the double-helicity flip, $I^S_{\nu\mu} \propto \kappa^2$ which after the κ^2 integration leads to pQCD calculable $A(\bar{x}, Q^2, \mathbf{\Delta}) \propto G(\bar{x}, \sim \frac{1}{4}(Q^2+m_V^2))$. Third, for the double-helicity flip $A_{\mp\pm}(\bar{x}, Q^2, \mathbf{\Delta}) \propto G(\bar{x}, \mu_G^2)$ where $\mu_G^2 \sim$(0.5-1)GeV^2 and it is not pQCD calculable at any large Q^2. Finally, by exclusive-inclusive duality [8] the above results for $I^S_{\mu\nu}$ can be related SCHNC LT interference in diffractive DIS into continuum [1,2] and indeed the dominant SCHNC effect in vector meson production is the interference of production of longitudinal vector mesons by (SCHC) longitudinal and (SCHNC) transverse photons, i.e., the element r^5_{00} of the vector meson polarization density matrix. The overall agreement between our theoretical estimates [5] of the spin density matrix r^n_{ik} for diffractive production of the ρ^0 and the ZEUS [9] and H1 [10] experimental data is very good. More theoretical analysis of the sensitivity to the wave function of vector mesons is called upon.

5 Sensitivity of SCHNC to spin-orbit coupling in vector mesons

Production of D-wave vector mesons nicely demonstrates a unique sensitivity of helicity flip in $\gamma^* p \to V p'$ to spin-orbit coupling [3,4]:

$$I^D_{0L} = -\frac{Q}{M}\cdot\frac{32\mathbf{r}^4}{15(M^2+Q^2)^2}\cdot\left(1-8\frac{M^2}{M^2+Q^2}\right)\kappa^2,$$

$$I^D_{\pm\pm} = \frac{32\mathbf{r}^4}{15(M^2+Q^2)^2}\cdot\left(15+4\frac{M^2}{M^2+Q^2}\right)\kappa^2,$$

$$I^D_{\pm L} = -\frac{24\Delta Q}{M^2+Q^2}\frac{32\mathbf{r}^4}{15(M^2+Q^2)^2}\cdot\frac{24Q}{M^2+Q^2}\kappa^2,$$

$$I^D_{L\pm} = \frac{8\Delta}{M}\frac{32\mathbf{r}^4}{15(M^2+Q^2)^2}\cdot\left(1+3\frac{M^2}{M^2+Q^2}\right)\kappa^2,$$

$$I^D_{\pm\mp} = \Delta^2 \cdot \frac{32\mathbf{r}^4}{15(M^2+Q^2)^2} \cdot \left(1 - \frac{96}{7}\frac{\kappa^2\mathbf{r}^2}{M^2(M^2+Q^2)}\right), \qquad (4)$$

where $4r^2 = M^2 - 4m^2$. In the D-wave state the total spin of $q\bar{q}$ pair is predominantly opposite to the spin of the D-wave vector meson. As a results, in contrast to S-wave states there is no nonrelativistic suppression of helicity flip. Such an enhancement of SCHNC may facilitate the D-wave vs. $2S$-wave assignment of the $\rho'(1480)$ and $\rho'(1700)$ and of the $\omega'(1420)$ and $\omega'(1600)$, which remains one of hot issues in the spectroscopy of vector mesons. Notice abnormally large higher twist corrections. For instance, A_{0L}, and LT interference thereof, changes the sign at $Q^2 \sim 7m_V^2$. The ratio $R^D = \sigma_L/\sigma_T$ has thus a non-monotonous Q^2 behavior and $R^D \ll R^S$. Furthermore, $R^D \lesssim 1$ in a broad range of $Q^2 \lesssim 225m_V^2$.

6 SCHNC physics at Jefferson Lab

Above we focused on SCHNC in QCD pomeron exchange which is described in pQCD by a generalized two-gluon ladder in the t-channel. There will be a substantial secondary reggeon contribution to diffractive ρ^0 and ω^0 production at CEBAF energies even after energy upgrade. In pQCD the reggeon exchange is modeled by generalized quark-antiquark ladder in the t-channel [11]. An exhaustive analysis of SCHNC for secondary reggeons has not been carried out yet. Still, the above predictions of SCHNC are fully applicable to, and can be tested in, the hidden-strangeness ϕ^0 and radial and angular-excited ϕ' diffractive photo- and electroproduction at CEBAF, because secondary reggeons do not contribute to the ϕ^0 production. Recall, for instance, the Zweig rule.

7 Dramatic impact of SCHNC diffraction upon the small-x behaviour of transverse spin structure function g_2

The transverse spin asymmetry in polarized DIS is proportional to the amplitude of forward CS $\gamma_L^* p\uparrow \rightarrow \gamma_T^* p\downarrow$. This CS amplitude and the transverse spin asymmetry are proportional to $g_{LT} = g_1 + g_2$. Because the photon helicity flip is compensated by the target proton helicity flip, the familiar forward zero of this helicity amplitude is lifted. However, in the standard two-gluon t-channel tower approximation the cross-talk of the target and beam helicity flip is only possible if in the Gribov-Lipatov decomposition of the gluon propagator only one of the gluons is having the nonsense polarization whereas the second one has the transverse polarization. The price one pays for such a combination of polarizations of gluons is the suppression of small-x behaviour of g_{LT} by the

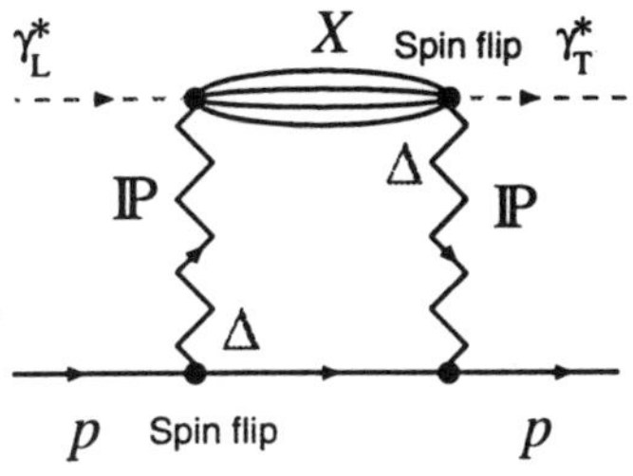

Figure 2: The unitarity diagram for diffractive contribution to g_{LT}.

extra factor $\sim x$ compared to the pomeron exchange in which the both gluons have the nonsense polarization .

The more familiar argument for the vanishing A_2 has been the parton model Wandzura-Wilczek relation between g_{LT} and g_1 [12]:

$$g_{LT}(x,Q^2) = \int_x^1 \frac{dy}{y} g_1(x,Q^2)\,,$$

which entails that $A_2 \sim A_1$. Because the diffractive pomeron exchange does not contribute to g_1, the WW relation can be reinterpreted as a vanishing pomeron exchange contribution to g_{LT}. This vanishing of the pomeron contribution has been the principal motivation behind the much discussed Burkhardt-Cottingham [13] (BC) sum rule $\int_0^1 dx g_2(x,Q^2) = 0$. Our recent discovery is that diffractive SCHNC destroys the both WW relation and BC sum rule [6].

The opening of diffractive DIS channels affects, via unitarity, the Compton scattering amplitudes. In [6] we have shown how diffractive LT interference in conjunction with spin-flip pomeron-nucleon coupling r_5 give rise to the transverse spin asymmetry $A_2 \propto x^2 g_{LT}$ which does not vanish at small x. The building blocks of the unitarity diagram shown in fig. 2 are the diffractive amplitudes $\gamma^* p \to p'X$ in which there is a helicity flip sequence, $\gamma^*_{\rightarrow} X_L \gamma^*_T$ in the top blob and helicity flip sequences either $p\uparrow \to p'\uparrow \to p\downarrow$ or $p\uparrow \to p'\downarrow \to p\downarrow$ in the bottom blob. The both amplitudes are proportional to $\mathbf{\Delta}$ and vanish in the forward direction, but upon the integration over the phase space of $p'X$ one finds the nonvanishing $\int d^2\mathbf{\Delta}\Delta_i\Delta_k$ and unitarity driven transition $\gamma^*_L p\uparrow \to \gamma^*_T p\downarrow$ which does not vanish in the forward direction. In the old Regge theory language it can be reinterpreted as the two-pomeron cut contribution which for the QCD pomeron has about the same small-x behaviour as the pomeron exchange. The principal difference from the single pomeron exchange is that the

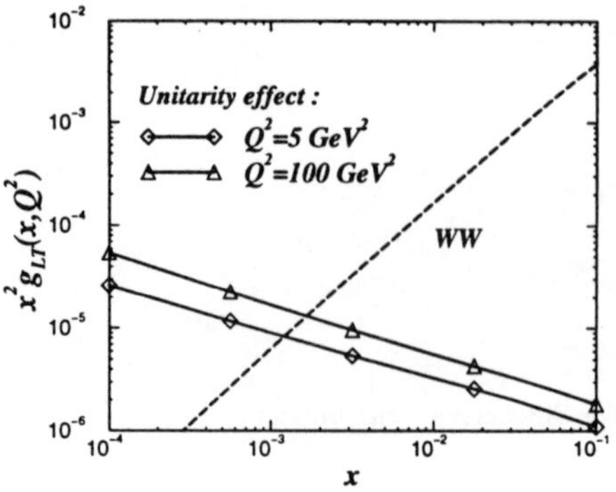

Figure 3: The unitarity correction to, and WW relation based evaluation of, g_{LT} [6].

unitarity diagram starts with the four-gluon state in the t-channel and four gluons can furnish the cross-talk of the beam and target helicity flip with pure nonsense polarizations of all the four exchanged gluons.

Our result [6] for the diffraction-driven g_{LT} reads

$$g_{LT}(x,Q^2) \propto \frac{1}{x^2} r_5 \int_x^1 \frac{d\beta}{\beta} g_{LT}^D(x_{\mathbb{P}} = \frac{x}{\beta}, Q^2)\,. \tag{5}$$

It rises steeply at small x. It is the scaling function of Q^2 because the diffractive LT structure function $g_{LT}^D(x_{\mathbb{P}}, Q^2)$ is the scaling one. The corresponding transverse spin asymmetry $A_2 \propto x g_{LT}/F_1$ does not vanish at small x, furthermore, at a moderately small x it even rises because $g_{LT}^D(x_{\mathbb{P}}, Q^2) \propto G^2(x_{\mathbb{P}}, \overline{Q}^2)$ where $\overline{Q}^2 \sim 0.5$-1 GeV^2.

In fig. 3 we show how the steeply rising unitarity correction overtakes at small x the standard g_{LT} evaluated from the Wandzura-Wilczek (WW) relation starting with fits to the world data on g_1. As such our unitarity effect is the first nontrivial scaling departure from the WW relation

Finally, the above breaking of the WW relation implies $g_{LT} \gg g_1$ and $g_2 = g_{LT}$ at very small x. Consequently, the unitarity-driven rise of g_2 destroys the BC sum rule because the BC integral would diverge severely. Incidentally, the BC sum rule has always been suspect.

Conclusions

The QCD pomeron exchange does not conserve the s-channel helicity. The mechanism of SCHNC is well understood. SCHNC offers an unique window at the spin-orbit coupling in vector mesons. SCHNC in diffractive DIS drives, via unitarity relation, a dramatic small-x rise of the transverse spin structure function g_2 which breaks the Wandzura-Wilczek relation and invalidates the Burkhardt-Cottingam sum rule. Jlab is perfectly posed to study SCHNC QCD pomeron exchange in photo- and electroproduction of the ϕ and its radial and orbital excitations.

Acknowledgments. Thanks are due to C.Carlson and A.Radyushkin for the invitation to the Workshop.

References

1. N.N. Nikolaev and B.G. Zakharov, Deep Inelastic Scattering and QCD, Proceedings of the 5th International Workshop on Deep Inelastic Scattering and QCD (DIS'97), Chicago, IL, 14-18 April 1997, AIP Conference Proceedings No.407, editors J. Repond and D. Krakauer, pp. 445-455.
2. N.N. Nikolaev, A.V. Pronyaev and B.G. Zakharov, *Phys. Rev.* **D59** (1999) 091501.
3. E.V. Kuraev, N.N. Nikolaev and B.G. Zakharov, *JETP Lett.* **68** (1998) 667.
4. I.P.Ivanov and N.N.Nikolaev, *JETP Letters* **69** (1999) 268.
5. I. Akushevich, I. Ivanov, N.N. Nikolaev and A. Pronyaev, paper in preparation.
6. I. Ivanov, N.N. Nikolaev, A. Pronyaev and W. Schaefer, *Phys. Lett.* **B457** (1999) 218.
7. D. Ivanov and R. Kirschner, *Phys. Rev.* **D58** (1998) 114026.
8. M.Genovese, N.N.Nikolaev and B.G.Zakharov, *Phys. Lett.* **B380** (1996) 213..
9. ZEUS Collab: A. Savin, DIS'99 and **DESY-99-102**.
10. H1 Collab: B. Clerbaux, DIS'99 and **DESY-99-010**.
11. W. Schäfer, Deep inelastic scattering and QCD, Proceedings of DIS'98, World Scientific, eds. Gh.Corenmans and R.Roosen, pp. 404-407.
12. S.Wandzura and F.Wilczek, *Phys. Lett.* **B72** (1977) 195.
13. H.Burkhardt and W.N.Cottingham, *Ann. of Phys. (USA)*, **56** (1970) 453.

FLAVOR DECOMPOSITION OF THE NUCLEON

W. MELNITCHOUK

Jefferson Lab, 12000 Jefferson Avenue, Newport News, VA 23606, and Special Research Centre for the Subatomic Structure of Matter, University of Adelaide, Adelaide 5005, Australia

I review some recent developments in the study of quark flavor distributions in the nucleon, including (i) valence quark distributions and the quark–hadron duality prediction for the $x \to 1$ F_2^n/F_2^p ratio, and (ii) sea quark asymmetries and electromagnetic form factors.

1 Introduction

While the problem of how the proton's *spin* is distributed among its constituents continues to captivate the attention of a large segment of the hadron physics community[1] and stimulate development of new approaches to the problem [2], the question of how the proton's momentum is distributed among its various *flavors* is far from being satisfactorily answered. Recent experiments and refined data analyses have indeed forced us to go far beyond the naive view of a nucleon in terms of three non-relativistic valence quarks in a sea of perturbatively generated $q\bar{q}$ pairs and gluons.

A classic example of this is the asymmetry of the light quark sea, dramatically confirmed in the recent Drell-Yan experiment at Fermilab [3], whose interpretation defies any perturbative understanding[4]. In a self-consistent representation of the nucleon, the dynamics responsible for any non-perturbative effects visible in quark distributions should also leave traces in other flavor-sensitive observables such as electromagnetic form factors. One can quantitatively study the connection between the quark distributions and form factors in the context of non-forward parton distributions, which is one of the offshoots of the recent proton spin decomposition studies[2]. An important test of any realistic model of nucleon structure is that it be able to account not only for the asymmetries in sea quark distributions such as the $\bar{d}/\bar{u}$ ratio, but also observables such as the electric form factor of neutron, which is particularly sensitive to the spin-flavor dynamics of quarks.

At the same time as the proton continues to reveal a rich substructure of its sea, some important details of quark distributions in the valence region still remain elusive. Most conspicuous of these is the valence d quark distribution, or the d/u ratio, whose $x \to 1$ limit remains controversial [5,6]. A number of proposals [7] have been made recently for determining the large-x behavior of d/u in semi-inclusive deep-inelastic scattering and other high-energy processes.

In the following I recall an old prediction for the d/u ratio based on empirical observations of quark–hadron duality first made nearly 3 decades ago.

2 Valence Quarks and Duality

The valence d/u ratio contains important information about the spin-flavor structure of the proton [8], and its asymptotic $x \to 1$ behavior reflects the mechanism(s) responsible for the breaking of $SU(2)_{\rm spin} \times SU(2)_{\rm flavor}$ symmetry. There are a number of predictions for this ratio, ranging from 1/2 in the non-relativistic SU(6) quark model, to 0 in broken SU(6) with scalar-isoscalar spectator quark dominance [9], to 1/5 in perturbative QCD [10]. Indeed, this is one of the very few predictions for the x-dependence of parton distributions which can be drawn from perturbative QCD, and its verification or failure would be an important indicator of the appropriate kinematics at which QCD can be treated perturbatively.

The biggest obstacle to an unambiguous determination of d/u at large x is the fact that this ratio is extracted from the ratio of inclusive neutron and proton structure functions, where the former is never measured directly but rather inferred from proton and deuteron cross sections. The deuteron cross sections must however be corrected for nuclear effects in the structure function, which can become quite significant [11,12] at large x. In particular, whether one corrects for Fermi motion only, or in addition for binding and nucleon off-shell effects, the extracted neutron structure function for $x > 0.7$ can differ dramatically [5]. The original observation [5] that the d/u ratio, when corrected for nuclear effects in deuterium, is larger than the value assumed in global data parameterizations was confirmed recently in a subsequent reanalysis [13] based on the assumption that the nuclear corrections scale with nuclear density [11]. While this may be a reasonable assumption for heavy nuclei such as ^{56}Fe, whether it can be applied to the deuteron, whose density is qualitatively very different from that in ^{56}Fe, is somewhat dubious. Nevertheless, the conclusions of both analyses do suggest that the neutron structure function may be significantly underestimated through the neglect of nuclear effects.

While a number of suggestions have been made how to avoid the nuclear contamination problem [7], one of the more direct ways is to measure relative yields of π^+ and π^- mesons in semi-inclusive deep-inelastic scattering from protons [14]. At large z (z being the energy of the pion relative to the photon) the u quark fragments primarily into a π^+, while a d fragments into a π^-, so that at large x the ratio $R^\pi = \sigma(\pi^-)/\sigma(\pi^+)$ is given by the ratio d/u weighted by $q \to \pi$ fragmentation functions. Indeed, in the limit $z \to 1$, where the $u \to \pi^+$ fragmentation function dominates [14], the ratio $R^\pi \to d/4u$.

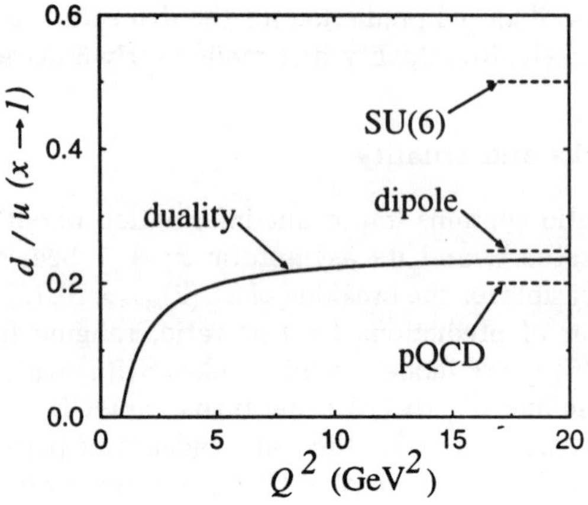

Figure 1: Quark–hadron duality prediction for the $x \to 1$ behavior of the d/u ratio.

A direct semi-inclusive measurement of fast pion production will require relatively large Q^2 and W^2, which may not be feasible until a facility such as the upgraded 12 GeV electron beam at Jefferson Lab becomes available. In the meantime, one must look for clues from other sources for information about d/u at large x, and one of the more obscure ones is the phenomenon of quark–hadron duality in inelastic structure functions.

As observed originally by Bloom and Gilman [15], when averaged over some interval of $\omega' = (2M\nu + M^2)/Q^2$, or more precisely [16] the Nachtmann scaling variable $\xi = 2x/(1+\sqrt{1+4M^2x^2/Q^2})$, the inclusive structure function in the resonance region at low W is approximately equal to the scaling structure function at much larger Q^2. This was later reinterpreted by de Rújula, Georgi and Politzer [16] in terms of an operator product expansion of the Nachtmann moments of F_2, in which the equality of the low moments was understood to arise from the relatively small size of higher twist ($1/Q^2$ suppressed) contributions compared with the leading twist.

Recent experiments at Jefferson Lab [17] confirm that this observation is reasonably accurate for each of the low-lying resonances, including the extreme case of elastic scattering. If one takes the latter seriously, then the first moment of the elastic structure function of the proton is given by the electromagnetic form factors [16]:

$$\int d\xi\ F_2^{\rm el}(\xi, Q^2) = \frac{\xi_0^2}{2-\xi_0}\frac{1}{1+\tau}\left(G_E^2(Q^2) + \tau G_M^2(Q^2)\right), \qquad (1)$$

with $\tau = Q^2/4M^2$ and $\xi_0 = 2/(1+\sqrt{1+1/\tau})$ is the value of ξ at $x = 1$.

Differentiating both sides of Eq.(1) with respect to Q^2 gives the $x \to 1$ F_2^n/F_2^p ratio in terms of the Q^2-derivatives of G_E and G_M. Using empirical values for the form factors and inverting F_2^n/F_2^p gives the local duality prediction for the d/u ratio in the $x \to 1$ limit in Fig.1. If one further assumes that both the proton and neutron magnetic form factors have the same dipole form at large Q^2, then $d/u \to (\mu_p^2 - 4\mu_n^2)/(\mu_n^2 - 4\mu_p^2) \approx 0.25$ in the limit $Q^2 \to \infty$.

It's interesting to observe that above $Q^2 \approx 5$ GeV^2 the duality prediction is similar numerically to the expectation from perturbative QCD. This is consistent with the operator product expansion interpretation of de Rújula et al. [16] in which duality should be a better approximation with increasing Q^2. Whether this is a coincidence or an indicator of common underlying physics remains to be settled by future experiments.

3 Role of the Light Quark Sea

The $\bar{d}/\bar{u}$ ratio is an important testing ground for our ideas about the long-range structure of the nucleon, and the origin of the non-perturbative spin-flavor interaction. The latest round of discussion about the proton's $\bar{d}$ and $\bar{u}$ distributions has been spurred on by the recent measurement by the E866 Collaboration [3] at Fermilab of the x-dependence of the pd to pp cross section ratio for Drell-Yan production, which is sensitive to the $\bar{d}/\bar{u}$ ratio at small x. As pointed out originally by Field and Feynman [18], the Pauli exclusion principle implies that $\bar{u}u$ pair creation is suppressed in the proton relative to $\bar{d}d$. Later, Thomas [19] observed that an excess of $\bar{d}$ quarks in the proton also arises naturally from the chiral structure of QCD, in the form of a pion cloud. In simple terms, if part of the proton's wave function has overlap with a virtual $\pi^+ n$ state, a deep-inelastic probe scattering from the virtual π^+, which contains a valence $\bar{d}$ quark, will automatically lead to $\bar{d} > \bar{u}$ in the proton.

Whatever the ultimate origin of the asymmetry, it is likely to involve some non-perturbative spin-flavors interaction between quarks. In order to identify the different possible origins of the asymmetry, consider a model of the nucleon in which the nucleon core consists of valence quarks, possibly interacting via gluon exchange, with sea quark effects introduced through the coupling of the core to $q\bar{q}$ states with light meson quantum numbers (many variants of such a model exist — see for example Refs.[20,21]). In practice, because of their pseudo-Goldstone nature and their anomalously light mass, the pseudoscalar pions (and for strange observables, kaons) are the most important $q\bar{q}$ states. The effects of the Pauli exclusion principle on the $\bar{d}/\bar{u}$ ratio can then be associated with antisymmetrization of $q\bar{q}$ pairs created inside the core [22].

Figure 2 shows the $\bar{d}/\bar{u}$ ratio in the proton including flavor symmetry

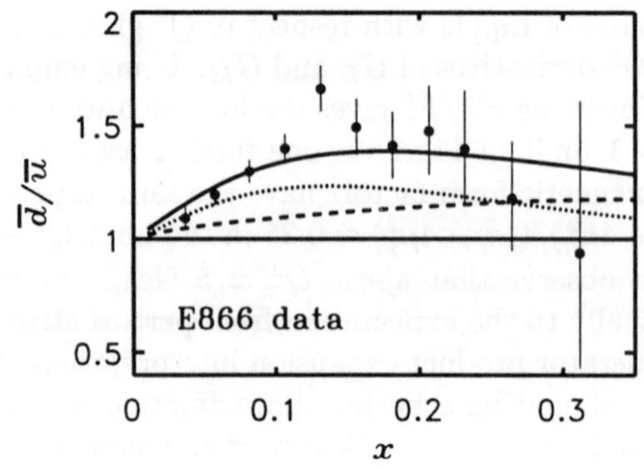

Figure 2: Flavor asymmetry of the light antiquark sea, including pion cloud (dashed) and Pauli blocking effects (dotted), and the total (solid) [4].

breaking effects generated by a pion cloud, with both N and Δ recoil states (for soft hadronic form factors the contributions from heavier mesons and baryons are small [23,24]), and an admixture of the Pauli blocking effect [4]. The pion cloud parameters, namely the hadronic πNN and $\pi N\Delta$ vertex form factors, are taken from the measured values of the axial elastic N and $N\Delta$ transition form factors[25], and give an overall pion probability in the proton of $\approx 10-15\%$.

If a pseudoscalar cloud of $q\bar{q}$ states plays an important role in the $\bar{d}/\bar{u}$ asymmetry, its effects should also be visible in other flavor-sensitive observables such as electromagnetic form factors. An excellent example is the electric form factor of the neutron, a non-zero value for which can arise from a pion cloud, $n \to p\pi^-$. Although in practice other effects [26,27] such as spin-dependent interactions due to one gluon exchange between core quarks may also contribute to G_E^n, it is important nevertheless to test the consistency of the above model by evaluating its consequences for all observables that may carry its signature.

To illustrate the sole effect of the pion cloud, all residual interactions between quarks in the core are switched off, so that G_E^n has only two contributions: one in which the photon couples to the virtual π^- (labeled "$\pi^- p$" in Fig.3) and one where the photon couples to the recoil proton ("$p\pi^-$" in Fig.3). Both contributions are large in magnitude but opposite in sign, so that the combined effects cancel to give a small positive G_E^n, consistent with the data.

One should stress that *the same* pion cloud parameters are used in the calculation of $G_E^n(Q^2)$ as for the $\bar{d}/\bar{u}$ asymmetry in Fig.2, with the exception that since here one is not creating $q\bar{q}$ pairs inside the nucleon, the Pauli block-

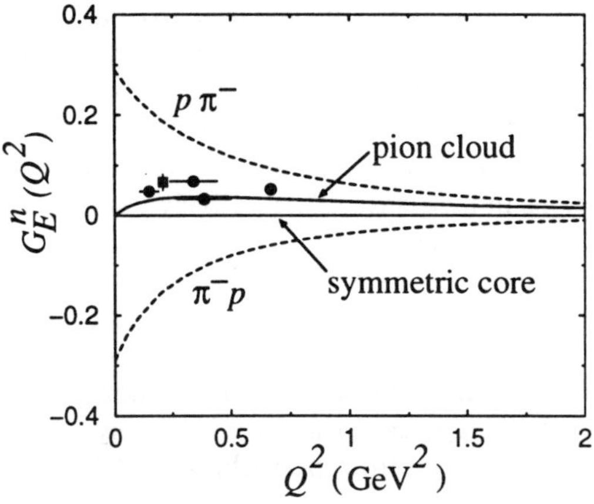

Figure 3: Neutron electric form factor in the pion cloud model.

ing plays no direct role for form factors. The fact that the model prediction slightly underestimates the strength of the observed G_E^n suggests that other mechanisms, such as one gluon exchange between quarks in the core [26], could be responsible for some of the difference.

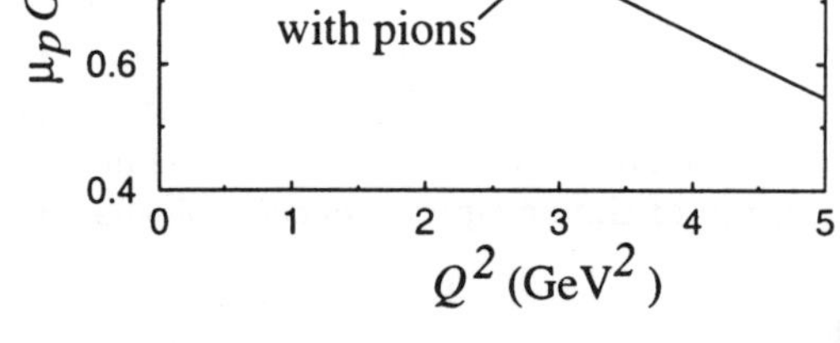

Figure 4: Electric to magnetic proton form factor ratio, with and without pseudoscalar $q\bar{q}$ clouds.

Another quantity sensitive to details of the flavor distributions in the proton is the electric to magnetic form factor ratio, which was recently measured at Jefferson Lab [28]. Perturbative QCD predicts that asymptotically this ratio should be Q^2-independent[29], so that any deviation from a constant ratio would be due to the influence of non-perturbative dynamics. Again assuming a symmetric core which leaves $\mu_p G_E^p = G_M^p$ for all Q^2, the pion cloud contribution to the Pauli form factor is suppressed in G_E^p relative to that in G_M^p, resulting in the softening of the G_E^p/G_M^p ratio. Compared with a dipole parameterization, the pion corrections in fact leave G_M^p relatively unaffected, but make G_E^p softer, leading to the ratio in Fig.4.

More quantitative comparisons with data would require one to consider more sophisticated models of the nucleon, incorporating explicitly the dynamics of core quarks as well as $q\bar{q}$ pairs. These simple examples, however, serve to illustrate the point that the structure and interactions of light quarks and antiquarks in both high-energy scattering and low energy form factors is far richer than could ever be inferred from perturbative QCD.

4 Future

One can anticipate progress to be made on each of the issues addressed here in the near future, as better quality data from high energy, high luminosity experiments, capable of accessing extreme kinematic regions, become available. The semi-inclusive π production is an example of a straightforward way to cleanly extract the d/u ratio at large x, free of the nuclear contamination inherent in earlier analyses. Once this is achieved, the origin of the SU(6) symmetry breaking responsible for the softening of the d quark distribution may be within reach.

For the $\bar{d}/\bar{u}$ ratio, it is important experimentally to confirm the downward trend of the ratio at large x, where pion cloud models generally predict a flattening out rather than any dramatic decrease in $\bar{d}/\bar{u}$. Issues concerning the size of the gluon contribution at large x may need to be resolved, however, before definitive conclusions from the present data can be reached. Future measurements of the neutron's electric form factor at Jefferson Lab and elsewhere, as well as the Q^2-dependence of the proton's form factors at large Q^2, should provide critical tests of our understanding of the dynamics of light quarks in the nucleon and the origin of the non-perturbative spin-flavor interaction.

Acknowledgments

This work was supported by the Australian Research Council.

References

1. B. Lampe and E. Reya, Report MPI-PHT-98-23, hep-ph/9810270.

2. X. Ji, *J. Phys.* G **24**, 1181 (1998); A.V. Radyushkin, *Phys. Rev.* D **56**, 5524 (1997).
3. E.A. Hawker *et al*, *Phys. Rev. Lett.* **80**, 3715 (1998).
4. W. Melnitchouk, J. Speth and A.W. Thomas, *Phys. Rev.* D **59**, 014033 (1999).
5. W. Melnitchouk and A.W. Thomas, *Phys. Lett.* B **377**, 11 (1996).
6. A.D. Martin, R.G. Roberts, W.J. Stirling and R.S. Thorne, Report DTP-99-64, hep-ph/9907231.
7. W. Melnitchouk and J.C. Peng, *Phys. Lett.* B **400**, 220 (1997). W. Melnitchouk, M. Sargsian and M.I. Strikman, *Z. Phys.* A **359**, 99 (1997).
8. N. Isgur, *Phys. Rev.* D **59**, 034013 (1999).
9. R.P. Feynman, *Photon Hadron Interactions* (Benjamin, Reading, Massachusetts, 1972); F.E. Close, *Phys. Lett.* B **43**, 422 (1973); F.E. Close and A.W. Thomas, *Phys. Lett.* B **212**, 227 (1988).
10. G.R. Farrar and D.R. Jackson, *Phys. Rev. Lett.* **35**, 1416 (1975).
11. L.L. Frankfurt and M.I. Strikman, *Phys. Rep.* **160**, 235 (1988).
12. W. Melnitchouk, A.W. Schreiber and A.W. Thomas, *Phys. Lett.* B **335**, 11 (1994); *Phys. Rev.* D **49**, 1183 (1994).
13. U.K. Yang and A. Bodek, *Phys. Rev. Lett.* **82**, 2467 (1999).
14. W. Melnitchouk, J. Speth and A.W. Thomas, *Phys. Lett.* B **435**, 420 (1998).
15. E.D. Bloom and F.J. Gilman, *Phys. Rev. Lett.* **16**, 1140 (1970).
16. A. De Rújula, H. Georgi and H.D. Politzer, *Ann. Phys.* **103**, 315 (1975).
17. I. Niculescu, Ph.D. thesis, Hampton University, 1999; C. Keppel, talk presented at 7th International Workshop on Deep Inelastic Scattering and QCD (DIS 99), Zeuthen, Germany, Apr. 1999.
18. R.D. Field and R.P. Feynman, *Phys. Rev.* D **15**, 2590 (1977).
19. A.W. Thomas, *Phys. Lett.* B **126**, 97 (1983).
20. A.W. Thomas, *Adv. Nucl. Phys.* **13**, 1 (1984); D.H. Lu, A.W. Thomas and A.G. Williams, *Phys. Rev.* C **57**, 2628 (1998).
21. P. Geiger and N. Isgur, *Phys. Rev.* D **55**, 299 (1997).
22. F.M. Steffens and A.W. Thomas, *Phys. Rev.* C **55**, 900 (1997).
23. W. Melnitchouk and A.W. Thomas, *Phys. Rev.* D **47**, 3794 (1993).
24. J. Speth and A.W. Thomas, *Adv. Nucl. Phys.* **24**, 83 (1998).
25. T. Kitagaki *et al*, *Phys. Rev.* D **42**, 1331 (1990).
26. N. Isgur, G. Karl and D.W.L. Sprung, *Phys. Rev.* D **23**, 163 (1981).
27. N. Isgur, *Phys. Rev. Lett.* **83**, 272 (1999).
28. M.K. Jones *et al*, in preparation for Physical Review Letters.
29. G.P. Lepage and S.J. Brodsky, *Phys. Rev. Lett.* **43**, 545 (1979).

SEMI-INCLUSIVE ELECTRON SCATTERING OFF THE DEUTERON AND THE NEUTRON STRUCTURE FUNCTIONS

Silvano SIMULA
Istituto Nazionale di Fisica Nucleare - Sezione Roma III
Via della Vasca Navale 84, I-00146 Roma, Italy

The detection of slow nucleons in semi-inclusive electron scattering processes off the deuteron, $^2H(e,e'N)X$, is a very promising tool for the extraction of *model-independent* information on the neutron structure function $F_2^n(x,Q^2)$. The key point is the occurrence of a peculiar scaling property of the semi-inclusive cross section, which holds in the deep-inelastic regime as well as in the nucleon-resonance production regions. Moreover, the possibility to get unique information on the neutron asymmetry $A_1^n(x,Q^2)$ from polarised $^2\vec{H}(\vec{e},e'N)X$ processes is illustrated.

1 INTRODUCTION

Till now the investigation of the neutron structure function $F_2^n(x,Q^2)$ has been carried out by means of inclusive lepton scattering experiments off nuclear (typically deuteron) targets [1,2]. However, the unfolding of the neutron contribution from the inclusive cross section is known to lead to non-trivial ambiguities, which are mainly related to our non-precise knowledge of the nucleon-nucleon (NN) interaction and of the reaction mechanism (see Ref. [2] and references therein quoted). Moreover, the neutron asymmetry $A_1^n(x,Q^2)$ is still a poor-known quantity, particularly at large Bjorken x, where nucleon structure models might be tested (see Ref. [3]).

A new way to get information on the neutron structure function has been recently proposed in Refs. [4,5]. There it has been shown that the detection of slow nucleons in semi-inclusive electron scattering processes off the deuteron, $^2H(e,e'N)X$, is a very promising tool for the extraction of the neutron structure function in a *model-independent* way. In this contribution we recall the main results obtained in Refs. [4,5] in case of unpolarised scattering and we briefly extend the analysis to the case of polarised $^2\vec{H}(\vec{e},e'N)X$ processes.

2 UNPOLARISED $^2H(e,e'N)X$ SCATTERING AND THE SPECTATOR SCALING

Let us consider the so-called spectator mechanism in which, after virtual photon absorption by a nucleon N_1 in the deuteron, the recoiling nucleon N_2 is emitted and detected in coincidence with the scattered electron. The corresponding semi-inclusive cross section can be cast in the following form [4,5]

$$\frac{d^4\sigma}{dE_{e'}\,d\Omega_{e'}\,dE_2\,d\Omega_2} = K M p_2 n^{(D)}(p_2)\,\frac{F_2^{N_1}(x^*,Q^2)}{x^*}\,D^{N_1}(x,Q^2,\vec{p}_2) \quad (1)$$

where $n^{(D)}(p_2)$ is the (non-relativistic) nucleon momentum distribution in the deuteron, $\vec{p}_2$ (E_2) is the momentum (energy) of the detected nucleon, $p_2 \equiv |\vec{p}_2|$, $Q^2 \equiv |\vec{q}|^2 - \nu^2$ is the squared four-momentum transfer, $x \equiv Q^2/2M\nu$ is the Bjorken scaling variable, $x^* \equiv Q^2/(Q^2 + M_1^{*2} - M^2)$ and M_1^* is the final invariant mass of the struck nucleon N_1, given through energy and momentum conservation by: $M_1^* = \sqrt{(\nu + M_D - E_2)^2 - (\vec{q} - \vec{p}_2)^2}$, with M_D (M) being the deuteron (nucleon) mass. In Eq. (1) $K \equiv (2Mx^2 E_e E_{e'}/\pi Q^2)\,(4\pi\alpha^2/Q^4)$ $\left[1 - y + (y^2/2) + (Q^2/4E_e^2)\right]$, with $y \equiv \nu/E_e$, and the quantity $D^{N_1}(x,Q^2,\vec{p}_2)$ depends both upon kinematical factors and the ratio $R_{L/T}^{N_1}$ of the longitudinal to transverse cross section off the nucleon (see Refs. [4,5] for more details).

Let us introduce the following semi-inclusive response function [5]

$$F^{(s.i.)}(x,Q^2,\vec{p}_2) = \frac{1}{\tilde{K}}\frac{d^4\sigma}{dE_{e'}d\Omega_{e'}dE_2 d\Omega_2} = M p_2 n^{(D)}(p_2)\frac{F_2^{N_1}(x^*,Q^2)}{x^*}\tilde{D}^{N_1}(x,Q^2,\vec{p}_2) \quad (2)$$

where $\tilde{K}$ is a pure kinematical factor given by $\tilde{K} \equiv K\,[D^{N_1}(x,Q^2,\vec{p}_2)]_{R_{L/T}^{N_1}=0}$ and $\tilde{D}^{N_1}(x,Q^2,\vec{p}_2) = D^{N_1}(x,Q^2,\vec{p}_2)\,/\,[D^{N_1}(x,Q^2,\vec{p}_2)]_{R_{L/T}^{N_1}=0}$. In the Bjorken limit one should have $R_{L/T}^{N_1} \to_{Bj} 0$ apart from logarithmic QCD corrections; thus, $\tilde{D}^{N_1} \to_{Bj} 1$ and

$$F^{(s.i.)}(x,Q^2,\vec{p}_2) \to_{Bj} M p_2 n^{(D)}(p_2)\,\frac{F_2^{N_1}(x^*)}{x^*} \quad (3)$$

where $F_2^{N_1}(x^*) \to_{Bj} F_2^{N_1}(x^*,Q^2)$. Therefore, in the Bjorken limit and at fixed values of $p_2 (\equiv |\vec{p}_2|)$ the semi-inclusive response function (2) does not depend separately upon x and the nucleon detection angle θ_2 ($cos\theta_2 \equiv \vec{p}_2 \cdot \vec{q}/p_2 q$), but it depends only upon the variable $x^* \to_{Bj} x/(2 - z_2)$, with $z_2 = [E_2 - p_2 cos\theta_2]/M$ being the light-cone momentum fraction of the detected nucleon. In what follows, we will refer to the variable x^* and the quantity $F^{(sp)}(x^*,Q^2,p_2)$, given explicitly by

$$x^* = \frac{Q^2}{Q^2 + (\nu + M_D - E_2)^2 - (\vec{q} - \vec{p}_2)^2 - M^2} \quad (4)$$

$$F^{(sp)}(x^*,Q^2,p_2) = M p_2 n^{(D)}(p_2)\,F_2^{N_1}(x^*,Q^2)/x^* \quad (5)$$

as the *spectator scaling* variable and function, respectively.

A natural question is whether and to what extent the spectator scaling can hold as well at finite values of Q^2. To this end, Eq. (2) has been calculated for various kinematical conditions corresponding to $Q^2 = 1$ and 10 $(GeV/c)^2$; the Bjorken variable x and the nucleon detection angle θ_2 have been varied in the range $0.20 \div 0.95$ and $10^o \div 170^o$, respectively (for sake of simplicity, the polar angle ϕ_2 has been chosen equal to 0). As for the nucleon structure function, the parameterization of the *SLAC* data of Ref. [2], containing $R_{L/T}^{N_1} \simeq 0.18$, has been adopted. The results of the calculations of Eq. (2), performed at $p_2 = 0.3\ GeV/c$, are reported in Fig. 1 and compared with the corresponding values of the spectator-scaling function (5). It can clearly be seen that the spectator scaling is fulfilled within few % in the *deep inelastic scattering* (*DIS*) regime as well as in *nucleon-resonance production* regions. This result is due to the fact that the spectator variable x^* collects final electron and nucleon kinematics corresponding to the same value of M_1^*; therefore, the spectator scaling does not depend upon the occurrence of the Bjorken scaling of the nucleon structure function.

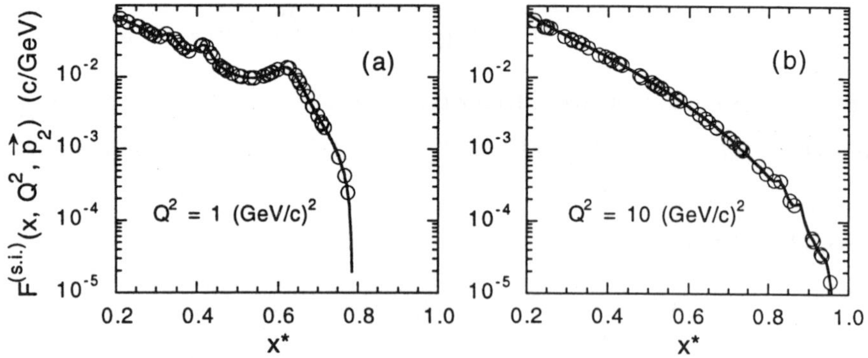

Figure 1. The response function $F^{(s.i.)}(x, Q^2, \vec{p}_2)$ (Eq. (2)) for the process $^2H(e, e'p)X$ versus the spectator-scaling variable x^* (Eq. (4)) calculated at $Q^2 = 1\ (GeV/c)^2$ (a) and $Q^2 = 10\ (GeV/c)^2$ (b) for $p_2 = 0.3\ GeV/c$. The solid line is the spectator-scaling function (5).

In the spectator-scaling function (5) all the nuclear effects are contained in the nucleon momentum distribution $n^{(D)}(p_2)$. At $p_2 \lesssim 0.3\ GeV/c$ the dependence of $n^{(D)}(p_2)$ upon the nuclear force model is quite negligible, because at large internucleon distances the NN interaction is almost totally governed by the one-pion-exchange contribution. Therefore, in the spectator-scaling regime it is possible to extract the nucleon structure function with small uncertainty due to nuclear effects. In particular, the process $^2H(e,e'p)X$ can provide information on the neutron structure function, while the reaction $^2H(e,e'n)X$, involving the proton structure function, can be used for a consistency check.

Within the spectator scaling regime the measurement of the semi-inclusive cross section both for the $^2H(e,e'p)X$ and $^2H(e,e'n)X$ processes allows to investigate two spectator-scaling functions, involving the same nuclear part, $Mp_2n^{(D)}(p_2)$, and the neutron and proton structure functions, respectively. Since moreover $R^n_{L/T} \simeq R^p_{L/T}$ (as suggested by recent *SLAC* data analyses [6]), at fixed values of p_2 both the nuclear part and the factor $\tilde{D}^{N_1}(x,Q^2,\vec{p}_2)$ cancel out in the ratio of semi-inclusive cross sections (1), yielding

$$R^{(s.i.)}(x,Q^2,\vec{p}_2) \equiv \frac{d^4\sigma[^2H(e,e'p)X]}{d^4\sigma[^2H(e,e'n)X]} \to \frac{F_2^n(x^*,Q^2)}{F_2^p(x^*,Q^2)} \quad (6)$$

and, therefore, the neutron to proton structure function ratio can directly be inferred from the ratio of semi-inclusive cross sections. Note that with respect to the response function $F^{(s.i.)}(x,Q^2,\vec{p}_2)$ the ratio $R^{(s.i.)}(x,Q^2,\vec{p}_2)$ exhibits a more general scaling property, for at fixed Q^2 it does not depend separately upon x and $\vec{p}_2$, but only on x^*. This implies that any p_2-dependence of the semi-inclusive ratio (6) would be a signature for off-shell deformations of the nucleon structure function (see Ref. [4]). Finally, note that in the quasi-elastic scattering regime the spectator scaling analysis can be applied to extract *model-independent* information on the elastic-peak contribution to the nucleon structure functions (see Ref. [5]).

Since the spectator scaling is a peculiar feature of the spectator mechanism, it can be in principle violated by different reaction mechanisms. In Refs. [4,5] the nucleon emission arising from the target fragmentation of the struck nucleon, which is thought to be responsible for the production of slow hadrons in *DIS* processes, has been investigated at high Q^2 adopting the factorization approach described in Ref. [7]. The results of the calculations of the semi-inclusive ratio (6), performed at $Q^2 = 10\ (GeV/c)^2$, are reported in Fig. 2 and compared with the free neutron to proton structure function ratio given in Ref. [2]. It can clearly be seen that: i) only at $x^* \lesssim 0.4$ fragmentation processes can produce relevant violations of the spectator scaling, as it is expected from the fact that the diquark remnant has a light-cone momentum fraction which decreases as x increases; ii) *backward kinematics* appear to be the most appropriate conditions for the extraction of the neutron to proton structure function ratio.

Violations of the spectator scaling can be expected also from the distortions of the energy and angular distributions of the detected nucleon due to

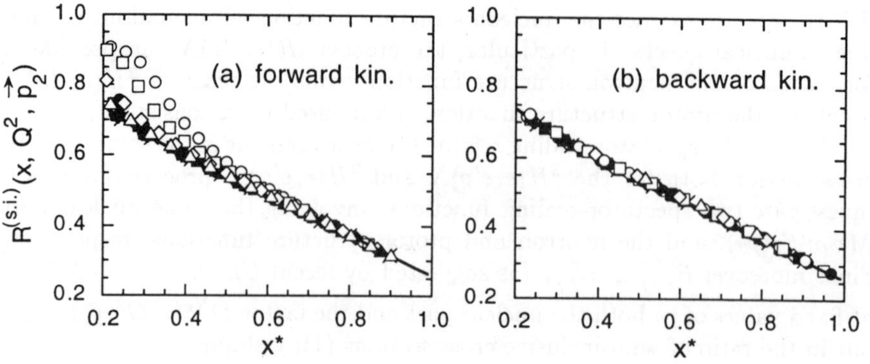

Figure 2. The semi-inclusive ratio $R^{(s.i.)}(x, Q^2, \vec{p}_2)$ (Eq. (6)) versus the spectator-scaling variable x^* (Eq. (4)), calculated at $Q^2 = 10$ $(GeV/c)^2$ and $p_2 = 0.1, 0.3, 0.5$ GeV/c by adding to the spectator mechanism the contribution arising from the target fragmentation of the struck nucleon. The various markers correspond to different values of the nucleon detection angle θ_2; forward ($\theta_2 < 90^o$) and backward ($\theta_2 > 90^o$) nucleon emissions are shown in (a) and (b), respectively. The solid line is the free neutron to proton structure function ratio of Ref. [2].

the reinteraction of the debris produced by the fragmentation of the struck nucleon with the recoiling spectator nucleon. Estimates of these final state interactions in the DIS regime have been discussed in Refs. [4,5], where it has been argued that backward nucleon emission is not significantly affected by forward-produced hadrons. However, at low Q^2 ($\sim$ few $(GeV/c)^2$) pion production may lead to reinteraction effects with the recoiling spectator nucleon, if both particles are emitted backward in the lab frame. For sake of simplicity let us consider the process of virtual photon absorption on a nucleon initially at rest in the lab frame, producing a $\Delta(1232)$ resonance which subsequently decays into a nucleon plus a pion. The momentum of the nucleon and the pion in the center-of-mass (CM) of the resonance is given from energy-momentum conservation by $p_{CM} = \{(M_\Delta^2 - M^2 + m_\pi^2)^2/4M_\Delta^2 - m_\pi^2\}^{1/2} \simeq 0.23\ GeV/c$. The boost velocity from the CM to the lab frame is $V_q = q/\sqrt{M_\Delta^2 + q^2}$, with $q = |\vec{q}|$ being the three-momentum transfer. The component of the pion momentum in the lab frame along the direction of $\vec{q}$ can be easily evaluated and shown to be proportional to $(V_q - V_\pi)$, where $V_\pi = p_{CM}/\sqrt{m_\pi^2 + p_{CM}^2} \simeq 0.85$. Therefore, the condition for having pions emitted only forward in the lab frame is $V_q \geq V_\pi$, which implies $q \gtrsim 2\ GeV/c$. Since in the lab frame $M + \nu = \sqrt{M_\Delta^2 + q^2}$, one gets $Q^2 \gtrsim 2\ (GeV/c)^2$. Thus, it can be safely argued that for $Q^2 \gtrsim 1\ (GeV/c)^2$ the final state interactions of the recoiling nucleon emitted in the backward hemisphere should drop quickly with increasing Q^2 and become almost totally negligible for $Q^2 \gtrsim$ few $(GeV/c)^2$.

3 POLARISED $^2\vec{H}(\vec{e}, e'N)X$ SCATTERING

Let us now extend our analysis to the case of $^2\vec{H}(\vec{e}, e'N)X$ processes, in which both the target and the incident electron beam are polarised. In what follows we will limit ourselves to the Bjorken limit, since this suffices to illustrate the relevant features of the slow nucleon production. We start by defining the semi-inclusive deuteron asymmetry $A_1^{D(s.i.)}$ in terms of the helicity components of the semi-inclusive cross section, viz. $A_1^{D(s.i.)} = (\sigma_{3/2}^{(s.i.)} - \sigma_{1/2}^{(s.i.)})/(\sigma_{3/2}^{(s.i.)} + \sigma_{1/2}^{(s.i.)})$, which in the Bjorken limit can be cast in the following form

$$A_1^{D(s.i.)}(x, \vec{p}_2) = 2x \frac{g_1^{D(s.i.)}(x, \vec{p}_2)}{F_2^{D(s.i.)}(x, \vec{p}_2)} \tag{7}$$

where $g_1^{D(s.i.)}(x, \vec{p}_2)$ and $F_2^{D(s.i.)}(x, \vec{p}_2)$ are polarised and unpolarised semi-inclusive deuteron structure functions, respectively. Within the spectator mechanism one easily gets

$$g_1^{D(s.i.)}(x, \vec{p}_2) = \frac{1}{z_1} g_1^{N_1}(\frac{x}{z_1}) \Delta n^{(^2H)}(\vec{p}_2)[1 + \frac{p_2}{M} cos(\theta_2)]$$
$$F_2^{D(s.i.)}(x, \vec{p}_2) = z_1 F_2^{N_1}(\frac{x}{z_1}) n^{(^2H)}(p_2) \tag{8}$$

where $z_1 = 2 - z_2 \simeq 1 + p_2 cos(\theta_2)/M + O(p_2^2/M^2)$ is the light-cone momentum fraction of the struck nucleon N_1. In Eq. (8) the deuteron structure is encoded in the functions $n^{(^2H)}(p_2)$ and $\Delta n^{(^2H)}(\vec{p}_2)$. In terms of the non-relativistic deuteron S- and D-wave functions $u(p)$ and $w(p)$, one has (cf. Ref. [8])

$$n^{(^2H)}(p) = u^2(p) + w^2(p)$$
$$\Delta n^{(^2H)}(\vec{p}) = u^2(p) + \frac{3cos^2(\theta) - 1}{\sqrt{2}} u(p)w(p) + [\frac{3}{2} cos^2(\theta) - 1]w^2(p) \tag{9}$$

Using Eqs. (7-9), the semi-inclusive deuteron asymmetry $A_1^{D(s.i.)}(x, \vec{p}_2)$ acquires in the spectator mechanism a simple factorized form, viz.

$$A_1^{D(s.i.)}(x, \vec{p}_2) = A_1^{N_1}(x^*) D(\vec{p}_2) \tag{10}$$

where $A_1^{N_1}(x^*) = 2x^* g_1^{N_1}(x^*)/F_2^{N_1}(x^*)$ is the asymmetry of the struck nucleon, $x^* = x/(2 - z_2)$ is the spectator variable and $D(\vec{p}_2)$ is the depolarisation factor arising from the D-wave of the deuteron, namely

$$D(\vec{p}_2) = \frac{1}{1+\frac{w^2(p_2)}{u^2(p_2)}} \left\{ 1 + \frac{3cos^2(\theta_2) - 1}{\sqrt{2}} \frac{w(p_2)}{u(p_2)} + [\frac{3}{2}cos^2(\theta_2) - 1] \frac{w^2(p_2)}{u^2(p_2)} \right\} \quad (11)$$

Thus, the semi-inclusive deuteron asymmetry (10) does not exhibit the spectator scaling property in the variable x^*, because of the explicit dependence of the depolarisation factor (11) on the nucleon detection angle θ_2. A remarkable sensitivity of $D(\vec{p}_2)$ to the specific value of θ_2 and to the NN interaction model (at least for $p_2 \gtrsim 0.2\ GeV/c$) is illustrated in Fig. 3.

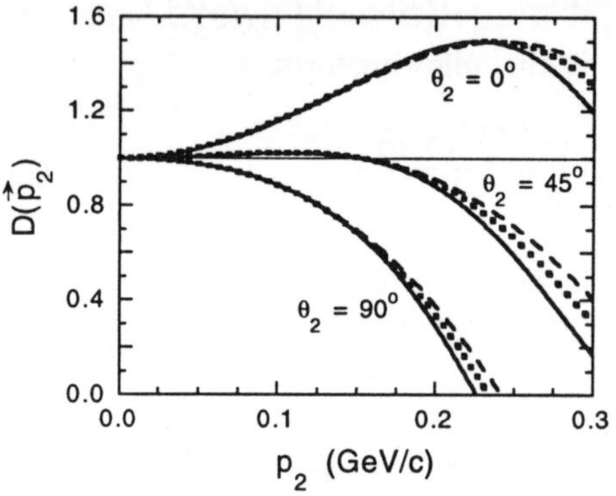

Figure 3. The depolarisation factor (11) versus the nucleon momentum p_2 for various values of the nucleon detection angle θ_2. The solid, dashed and dotted lines correspond to the use of the *RSC* [9(a)], Paris [9(b)] and Bonn [9(c)] models of the NN interaction, respectively.

However, from Eq. (10) the spectator scaling is fulfilled by the ratio of semi-inclusive deuteron asymmetries, viz.

$$R_A^{D(s.i.)}(x, \vec{p}_2) \equiv \frac{A_1^{D(s.i.)}[^2\vec{H}(\vec{e}, e'p)X]}{A_1^{D(s.i.)}[^2\vec{H}(\vec{e}, e'n)X]} \to \frac{A_1^n(x^*)}{A_1^p(x^*)} \quad (12)$$

so that the ratio $R_A^{D(s.i.)}(x, \vec{p}_2)$ does not depend separately on x and $\vec{p}_2$, but only on x^*. The value of the scaling function is now directly the neutron to proton asymmetry ratio $A_1^n(x^*)/A_1^p(x^*)$, which can be therefore extracted from deuteron data in a *model-independent* way.

Before closing this Section we point out that in the process $^2\vec{H}(\vec{e}, e'n)X$ the deuteron asymmetry $A_1^{D(s.i.)}(x, \vec{p}_2)$ involves the proton asymmetry $A_1^p(x^*)$ and the depolarization factor $D(\vec{p}_2)$. Since the former can be obtained using hydrogen targets, the depolarisation factor (11) can be extracted at low momenta from the reaction $^2\vec{H}(\vec{e}, e'n)X$. In this way unique information on the D- to S-wave ratio in the deuteron can be obtained and compared with theoretical predictions based on different NN interaction models, provided $p_2 \simeq 0.2 \div 0.3\ GeV/c$ (see Fig. 3).

4 CONCLUSIONS

The production of slow nucleons in semi-inclusive (polarised and unpolarised) electron scattering off the deuteron may exhibit a peculiar scaling property, the spectator-scaling, which can be used to extract *model-independent* information on the neutron structure functions. Namely, the neutron structure function F_2^n, the neutron to proton structure function ratio F_2^n/F_2^p and the neutron to proton asymmetry ratio A_1^n/A_1^p can be obtained directly from the semi-inclusive cross section data in the kinematical regions corresponding to deep inelastic scattering ($HERA$-like kinematics) as well as to nucleon-resonance production ($JLAB$-like kinematics), provided slow nucleons (i.e., with momentum of the order of $0.1 \div 0.2 \ GeV/c$) are detected in the backward hemisphere.

Acknowledgments

The author gratefully acknowledges Gunther Piller for supplying him with the parameterizations of the deuteron wave function used in this work.

References

1. J.J. Aubert et al., Nucl. Phys. B **293**, 740 (1987). A.C. Benvenuti et al., Phys. Lett. B **237**, 599 (1990). P. Amaudruz et al., Nucl. Phys. B **371**, 3 (1992).
2. A. Bodek and J.L. Ritchie, Phys. Rev. D **23**, 1070 (1981). L.W. Whitlow et al., Phys. Lett. B **282**, 745 (1992).
3. N. Isgur, Phys. Rev. D **59**, 341033 (1999).
4. S. Simula, Phys. Lett. B **387**, 245 (1996); e-print archive nucl-th 9608053, Proc. of the Workshop on *Future Physics at HERA* (DESY, September '95 - May '96), eds. G. Ingelman, A. de Roeck and R.K. Klanner, DESY (Hamburg, 1996), p. 1058.
5. S. Simula, Nucl. Phys. A **631**, 602c (1998).
6. S. Dasu et al., Phys. Rev. D **49**, 5641 (1994).
7. C. Ciofi degli Atti and S. Simula, Few Body Systems **18**, 55 (1995). S. Simula, Few Body Systems Suppl. **9**, 466 (1995).
8. S.A. Kulagin et al., Phys. Rev. C **52**, 932 (1995). For a recent review see: G. Piller and W. Weise, preprint TUM/T39-99-11, e-print archive hep-ph9908230.
9. (a) V. Reid, Ann. Phys. (N.Y.) **50**, 411 (1968). (b) M. Lacombe et al., Phys. Rev. C **21**, 861 (1980). (c) R. Machleidt et al., Phys. Rep. **149**, 1 (1987).

PERTURBATIVE COLOR TRANSPARENCY IN ELECTRON BEAM EXPERIMENTS VIA IMPACT-PARAMETER FACTORIZATION

JOHN P. RALSTON

Department of Physics and Astronomy, University of Kansas, Lawrence, KS 66045, USA

Email:ralston@ukans.edu

PANKAJ JAIN, BIJOY KUNDU

Department of Physics, IIT Kanpur, Kanpur-208 016, India

E-mail: pkjain@iitk.ac.in, bijoyk@iopb.res.in

JIM SAMUELSSON

Department of Theoretical Physics, University of Lund , Lund S-22362, Sweden

E-mail: jim@thep.lu.se

Color transparency is a prediction of perturbative QCD. Yet detailed calculations have been lacking, and aspects of the required factorization have been controversial. We report on the first complete calculations entirely within a perturbative QCD framework. We also comment on the underlying factorization method and assumptions.

1 Introduction

Color transparency is an important tool to investigate exclusive processes using nuclear targets. The earliest predictions of the phenomenon, going back to 1982,[1] were based on having asymptotically large Q^2 select short distance. Given the difficulties over short-distance dominance, the transition to color transparency then multiplied the controversy. However the discovery of *nuclear filtering* shifted the emphasis.[2,3,4] Components of a hadronic quark wave function with large transverse spatial separations cannot propagate elastically in nuclear matter. This fact of QCD, noted earlier [5] for diffractive processes, was re-discovered for color transparency. It was realized that asymptotically large Q^2 is unnecessary to motivate perturbative color transparency.

Several experiments indicate that color transparency and nuclear filtering have been observed at large nuclear number A. Perhaps the most spectacular is the first color transparency experiment of Carroll *et al*,[6] which convincingly showed that interference effects in proton-proton scattering were filtered away in nuclear targets. The FNAL E-665 experiment [7] also proved consistent with filtering effects,[8] especially in the observation of longitudinal final state polarization in $\gamma^* A \to \rho A$. Electron beam experiments have been more difficult, partly because of low rates and reduced resolving power.

There are important theoretical reasons not to over-rely on the Q^2 dependence of the transparency ratio.[9] The ratio compares the object of study (an exclusive process in a nuclear target) with something not well-understood (an analogous free-space process). The implicit assumption of a universal hard-scattering form factor with fixed normalization and Q^2 dependence (to be followed by some model of propagation) is theoretically unsupported. Hybrid models, in which free-space form factors are explicitly used, run into the difficulty that exclusive processes are not self-consistently described at laboratory energies by the asymptotic formalism.

Better methods exist to characterize the data empirically. The A dependence is particularly powerful. O'Neill *et al*[10] showed that effective attenuation cross sections extracted from $A(e, e'p)$ SLAC data were smaller than Glauber theory calculations by a statistically significant amount. However, choice of normalization and uncertainties in the nuclear spectral distributions complicated the interpretation, and the precision of the data[11] was insufficient to establish a large effect. New $(e, e'p)$ beam experiments are underway at Jefferson Laboratory.

The theory to describe color transparency in pQCD actually forced a revision of basic factorization methods.[3] It was found that the asymptotic factorization of Lepage and Brodsky[12] (LB) was inadequate. To go beyond this, it was necessary to introduce an integration over the transverse separation of quarks into the description. An antecedent was an important paper by Botts and Sterman.[13] The promising new factorization methods, which might be called "impact parameter factorization", have been highly refined in applications by Li and Sterman[14,15] to free-space form factors. Kundu *et al*[16] developed the method further, and responded to criticisms.[17]

The purpose of the transverse integrations is to incorporate regions of finite quark spatial separation. In free-space the Sudakov form factor tends to suppress these regions somewhat; in nuclei a stronger suppression comes from the filtering effects of nuclear matter.[18] The new method also enables calculation of helicity-flip form factors,[19] which cannot be described in the old formalism, and which would be very interesting to measure in nuclear matter, if somehow it could be done.

Happily one does not give up the impulse approximation, which has been the cornerstone of all successful pQCD calculations. Often misconstrued to be the same as the "frozen approximation" of high-energy diffractive processes, the impulse approximation allows one to start a clock allowing time evolution of the outgoing system, while separating the fast time scale of the scattering from the slow time scale of hadron formation. Controversies in the literature still rage over whether "expansion" in one form or the other has been incorporated properly.[4] Our calculations in perturbation theory integrate over light-cone "minus" components of wavefunctions. This is the step which separates the hadron time scale from the rest. The rest of the calculation is Feynman diagrams, which have the time evolution of quarks built in. Feynman diagrams faithfully reproduce time-evolution of the quark and gluon degrees of freedom order-by-order, creating the same kind of amplitude (a Green function) that the experiment can measure. The physical picture of the impact-parameter factorization, then, is somewhat different from the picture of the asymptotic limit. In the asymptotic limit, quarks participate at zero distance, and move so fast that perfect transparency always occurs. In impact-parameter factorization, quarks with all sideways and longitudinal separations are superposed coherently over the whole nucleus. The hard scattering and the propagation remain coupled during propagation.

2 The Method

It is worthwhile to review the different frameworks of exclusive processes in free space before introducing nuclear targets. Lepage and Brodsky [12] (LB) calculate a meson electromagnetic form factor with a factorization written as

$$F_\pi(Q^2) = \int dx_1 dx_2 \phi(x_2, Q) H(x_1, x_2, Q) \phi(x_1, Q). \tag{1}$$

Here $\phi(x, Q)$ are the distribution amplitudes, which can be expressed in terms of the pion wave function $\psi(x, \vec{k}_T)$ as $\phi(x, Q) = \int^Q d^2 k_T \psi(x, \vec{k}_T)$. We use x for the longitudinal momentum fraction and $\vec{k}_T$ for the transverse momentum carried by the quark. The factorization is justified provided the external photon momentum Q^2 is *asymptotically large*. Then the k_T integrals of perturbation theory decouple, and can be applied to make the distribution amplitudes. The k_T dependence of the hard scattering H can be expanded in a power series, retaining the trivial, constant term. One directly obtains the power-law scaling of the quark-counting method,[20] with logarithmic corrections.

Note that the asymptotic limit is taken right away, in fact prematurely. All distances separating quarks then become asymptotically short. For the purposes of color transparency, taking Q^2 arbitrarily large (but fixed), one might think all targets become perfectly transparent. But then taking $A \to \infty$ at fixed, arbitrarily large Q^2, we know that all targets must become opaque. Thus there is a limit interchange problem in the LB factorization, because the limit of large Q^2 and large A do not commute. The scheme is fundamentally limited to asymptotic Q^2, and there is no way to fix it to describe the phenomena of color transparency at laboratory energies.

A proper description of the phenomena follows from a impact-parameter factorization scheme incorporating the transverse degrees of freedom.[3,13] By including a broader integration region, impact-parameter factorization is more general than the LB method. Li and Sterman [14] simplify the calculation of form-factors by dropping the weak k_T dependence of quark propagators in a hard scattering kernel H. Working in configuration (impact-parameter b) space the expression for a form-factor becomes:

$$F_\pi(Q^2) = \int dx_1 dx_2 \frac{d^2\vec{b}}{(2\pi)^2} \mathcal{P}(x_2, b, P_2, \mu) \tilde{H}(x_1, x_2, Q^2, \vec{b}, \mu) \mathcal{P}(x_1, b, P_1, \mu), \quad (2)$$

where $\mathcal{P}(x, b, P, \mu)$ and $\tilde{H}(x_1, x_2, Q^2, \vec{b}, \mu)$ are the Fourier transforms of the wave function, including Sudakov factors, and hard scattering respectively; $\vec{b}$ is conjugate to $\vec{k}_{T1} - \vec{k}_{T2}$, μ is the renormalization scale and P_1, P_2 are the initial and final momenta of the meson.

We now discuss filtering. When a fast hadron traverse the nuclear medium, each quark interacts primarily through exchange of transverse momentum. While longitudinal momentum can also be exchanged, this degree of freedom does not affect the overall structure, and is in fact lost in the present uncertainties of Feynman x wave-functions. Nuclear targets provide a much better transverse filter than free space, and so it was thought at first that the LB method might apply. The profound differences between impact-parameter factorization and the LB factorization was not understood at first. It was thought that the Ralston and Pire 1990 [3] method was a way to generate distribution amplitudes within the LB framework. Eventually we realized that the large Q^2 dependence of the impact-parameter method is vastly more flexible than just logarithmic behavior, while LB factorization cannot be otherwise.

Transverse momentum integrations conserving overall momentum, and at small momentum transfer, turn into products in b space. Thus the nuclear medium modifies the quark wave function such that [3] $\mathcal{P}_A(x, b, P, \mu) = f_A(b; B)\mathcal{P}(x, b, P, \mu)$, where $\mathcal{P}_A$ is the wave function probed inside the medium and f_A is the nuclear filtering amplitude. An eikonal form [3] appropriate for f_A is: $f_A(b; B) = exp(-\int_z^\infty dz'\sigma(b)\rho(B, z')/2)$. Here $\rho(B, z')$ is the nuclear number density at longitudinal distance z' and impact parameter B relative to the nuclear center. We parametrize $\sigma(b)$ as kb^2 for our calculations. Finally, we must include the probability to find a target at position B, z inside the nucleus. Putting together the factors, the process of knocking out a hadron from inside a nuclear target has an amplitude M given by

$$M = \int_0^\infty d^2B \int_{-\infty}^{+\infty} dz\rho(B,z) \times F_\pi(x_1, x_2, b, Q^2) \times f_A(b, B) \qquad (3)$$

For the proton the important transverse scale is the maximum of the three quark separation distances, $b_{max} = max(b_1, b_2, b_3)$. The calculation of the process in the nuclear target needs a 9 dimensional integration, which is performed by Monte Carlo. The calculation of proton targets, then, follows the same basic rules as knocking out pions, but with more degrees of freedom. In our calculations we found that uncertainties on the nuclear correlations at the 10% level were a major concern, in some cases exceeding the theoretical uncertainties from the rest of the calculation.[18] We also find that the physics cannot reasonably be captured by a free-space hard scattering, followed by some model of propagation with or without "expansion". This is because the integrations over the transverse quark variable extend over the whole volume of the nucleus. Color transparency is something probing the internal structure of hadrons.

3 Results and Discussions

We calculated color transparency and nuclear filtering for populations of both pions and nucleons (protons) in the nucleus. We explored the Q^2 dependence of the proton transparency ratio using popular distribution amplitude models, and later (after the meeting) with the asymptotic and other models. Realizing belatedly that wave functions more central in Feynman x would possibly be more purely short-distance, we explored such cases and include them for comparison. To make our trial transparency ratio, we divided the rate in the nuclear target by a cross section including the free-space form factor.

Some results are shown in Fig. 1. It is noteworthy that the *slope* of the ratio versus Q^2 depends on the wave-functions. The ratio for a model having a more-central wave function rises faster than that of model with large endpoint contributions. This is consistent with the known tendency for endpoint regions of the CZ and KS models[21] to enhance soft regions of integration, or "big-fat" protons. The asymptotic model, for example, is seen to look like a "smaller" proton in the calculation with the quarks sitting in the central regions.

We are not concerned here with controversies over different trial wave functions, and use the asymptotic one merely to illustrate a point. It is quite

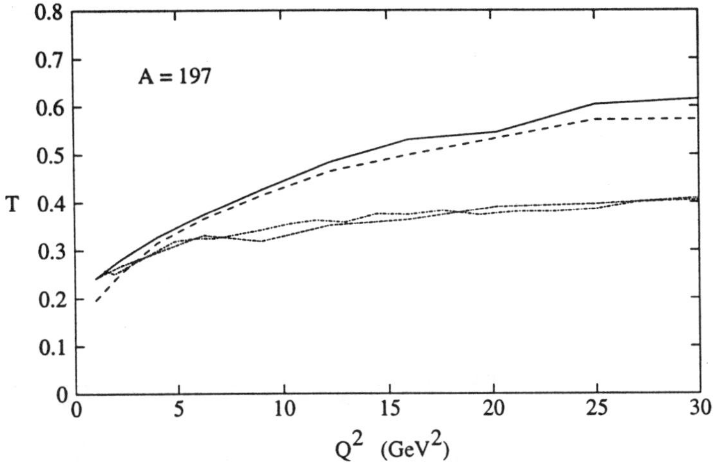

Figure 1: Color transparency ratio for different proton trial wave functions. Solid curve: asymptotic wave function. Dashed curve: a more centrally peaked wave function of the type $x_1^2 * x_2^2 * x_3^2$. Dash-dot curve: KS wave function. Dash-dot-dot curve: CZ wave function.

striking, and possibly important, that the Q^2 dependence of the transparency ratio measures something as fundamental as the basic x depedence of the wave functions. The calculations indicate that the slope can distinguish centrally peaked from endpoint-peaked models.

For the endpoint-dominated models, the quark transverse separation cut-off showed a significant reduction in sensitivity inside the nucleus compared to free space. However, it must be remembered that the effects of filtering in electron beam experiments are comparatively modest compared to hadron-hadron reactions, because the hadron tends to be knocked out the "back-side" of the nucleus. This accounts for an intrinsic reduced sensitivity of electron beam reactions, which hopefully is compensated by their high intrinsic precision.

Finally, following,[9] we extracted the effective attenuation cross section $\sigma_{eff}(Q^2)$, which serve as a litmus test of whether "color transparency" has actually been achieved. The results (Fig. 2) show a significant decrease of $\sigma_{eff}(Q^2)$ with increasing Q^2 to values well below the Glauber model attenuation cross section. This indicates color transparency. Consistently, the transparency of the asymptotic model is more dramatic than the cases of the CZ and KS models.

Uncertainties in nuclear spectral functions remain, and are becoming problematic for calculation of absolute normalizations. This strongly affects the Q^2

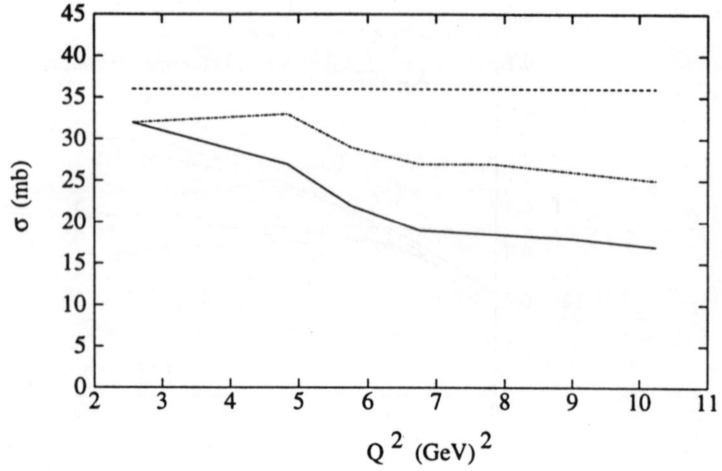

Figure 2: Extracted effective attenuation cross sections $\sigma_{eff}(Q^2)$ as a function of Q^2 exhibit color transparency. The decrease of $\sigma_{eff}(Q^2)$ with Q^2 is sufficiently large that conventional nuclear physics might be ruled out with large Q^2 or sufficient precision. Solid curve: asymptotic wavefunction. Dash-dot curve: KS wave function. The dashed line is the Glauber value of 36 mb.

dependence, but is somewhat ameliorated by studying the nuclear matter limit of large A. Several conclusions are nevertheless well supported by the investigations, reported in more detail elsewhere.[18] These first quantitative perturbative calculations support the claim that exclusive processes are theoretically under better control than the free-space analogues. Next, observables such as the slope of the transparency ratio, sometimes calculated in terms of an ad-hoc "expansion time scale" in hadronic models, probe the interplay of the transverse and longitudinal development of the amplitudes in perturbation theory. The slope depends directly on the x-dependence of wave functions. Next, the calculations show significant reduction in attenuation, even in circumstances where *no rise is seen with Q^2 in transparency ratios*. The explanation, we believe, is that purification to short-distance tends to deplete amplitude normalizations. Division by a free-space cross section, where uncontrolled amplitudes dominate, overestimates the magnitude of hard scattering inside a nucleus, and can give a misleading ratio. This competition is unlikely to be unraveled with Q^2 dependence alone, and calls for another experimetal handle. The A dependence separates the competition between process normalization and attenuation, as revealed by σ_{eff}.

Acknowledgments

This work was supported by BRNS grant No. DAE/PHY/96152, the Crafoord Foundation, the Helge Ax:son Johnson Foundation, DOE Grant number DE-FGO2-98ER41079, the KU General Research Fund and NSF-K*STAR Program under the Kansas Institute for Theoretical and Computational Science.

References

1. S. J. Brodsky and A. H. Mueller, *Phys. Lett.* B **206**, 685 (1988).
2. J. P. Ralston and B. Pire, *Phys. Rev. Lett.* **61**, 1823 (1988).
3. J. P. Ralston and B. Pire, *Phys. Rev. Lett.* **65**, 2343 (1990).
4. P. Jain, B. Pire and J. P. Ralston, *Phys. Rep.* **271**, 67 (1996).
5. G. Bertsch, S. J. Brodsky, A.S. Goldhaber, and J.F. Gunion, *Phys. Rev. Lett.* **47**, 297 (1981).
6. A. S. Carroll *et. al.*, *Phys. Rev. Lett.* **61**, 1698 (1988).
7. M. R. Adams *et al Phys. Rev. Lett.* **74**, 1525 (1995).
8. B. Z. Kopeliovich, J. Nemchick, N. N. Nikolaev, and B. G. Zakharov, *Phys. Lett.* B **309**, 179 (1993); *Phys. Lett.* B **324**, 469 (1994).
9. P. Jain and J. P. Ralston, *Phys. Rev.* D **48**, 1104 (1993).
10. T. G. O ' Neill *et. al*, *Phys. Lett.* B **351**, 87 (1995).
11. N. Makins *et. al.*, *Phys. Rev.* D **12**, 163 (1994).
12. S. J. Brodsky and G. P. Lepage, *Phys. Rev.* D **24**, 2848 (1981).
13. J. Botts and G. Sterman, *Nucl. Phys.* **B325**, 62 (1989).
14. H-n. Li and G. Sterman, *Nucl. Phys.* **B381**,129 (1992).
15. H-n. Li, *Phys. Rev.* D **48**, 4243 (1993).
16. B. Kundu, H. N. Li, J. Samuelsson and P. Jain, *Eur. Phys. J.* **C8**, 637 (1999), hep-ph/9806419.
17. J. Bolz, R. Jakob, P. Kroll, M. Bergmann, and N.G. Stefanis, *Z. Phys. C* **66**, 267 (1995).
18. B. Kundu, J. Samuelsson, P. Jain and J. P. Ralston, (under preparation), hep-ph/9812506.
19. P. Jain and J. P. Ralston, in *Future Directions in Particle and Nuclear Physics at Multi-GeV Hadron Beam Facilities*, edited by D. Geesaman (BNL 1993).
20. S. J. Brodsky and G. R. Farrar, *Phys. Rev.* D **11**, 1309 (1975); V. A. Matveev, R. M. Muradyan and V. A. Tavkhelidze, *Nuovo Cimento* **7**, 719 (1973).
21. V. L. Chernyak and A. R. Zitnitsky, *Nucl. Phys.* **B246**, 52 (1984); I.D. King and C.T. Sachrajda, *Nucl. Phys.* **B279**, 785 (1987).

ASYMPTOTIC GLUON SHADOWING

S. LIUTI

Institute of Nuclear and Particle Physics, University of Virginia, McCormick Road, Charlottesville, Virginia 22901, USA

and

INFN-Sezione Roma Tre, Dipartimento di Fisica E. Amaldi, Via Vasca Navale, 84. 00146 Roma, Italy.

E-mail: sl4y@virginia.edu

F. CANO

Dipartmento di Fisica, Universitá di Trento, 038050 Povo, Trento, Italy

E-mail: cano@science.unitn.it

We examine the low Bjorken x gluon distribution in nuclei in the asymptotic region.

1 Introduction

Nuclear shadowing or the depletion at low Bjorken x of the nuclear Deep Inelastic (DI) structure function, F_2^A, with respect to the nucleon one, F_2^N, has been observed in a number of experiments (see [1] and references therein). Assuming the universality of parton distributions in nuclei, one expects nuclear shadowing to be present in other high energy processes as well, such as Drell-Yan pair, J/ψ and Υ production in lepton-nucleus, hadron-nucleus and nucleus-nucleus collisions. In particular nuclear shadowing might be concurring with the other mechanisms among which quark-gluon plasma formation, that result in a depletion of the observed cross sections for these processes [2,3]. A quantitative understanding of both the x and Q^2 dependences of the nuclear parton distributions at low and very low x is therefore a necessary step for interpreting the outcome of future experiments at RHIC and at the LHC. Recent calculations rely on non-perturbative models for the nuclear parton distributions at a given (low) scale, Q_o^2, combined with DGLAP [4] perturbative evolution. They are all therefore affected by the uncertainty in the initial parton distributions and, in particular, in the gluon distribution which governs evolution at low x, and which is poorly known experimentally. Moreover, evolution to moderate values of Q^2 ($Q^2 \approx 10\,\mathrm{GeV}^2$) depends even more strongly, by the value of Q_o^2 itself which, if varied within the range, $Q_o^2 = 0.8 - few\,\mathrm{GeV}^2$ can lead to completely different evolution patterns for the shadowing of both the structure functions and the gluon distribution. These simple facts hamper the possibility of predicting the Q^2 dependence of nuclear shadowing although its behavior within pQCD is quite well understood and it can be predicted accurately for each choice of initial conditions.

In this paper we suggest an avenue for dealing with the model dependence associated with the initial conditions of perturbative evolution, which is based on a study of the low x and high Q^2 asymptotic behavior of the shadowing ratios $R_G = G_A/G_N$ and $R_F = F_2^A/F_2^N$, $G_{N(A)}$ and $F_2^{N(A)}$ being the gluon distributions and the Deep Inelastic (DI) structure function in a nucleon, (N), and in an isoscalar nucleus, (A), respectively. In the asymptotic regime defined by $Q^2 \gg Q_o^2$ and $x \ll x_o$, $x_o \leq 0.1$, the Double Logarithmic Approximation (DLA) to pQCD evolution applies [5]. The proton structure function data analyzed recently at HERA [6,7] have been proven [6] after the suggestion of Ref.[8], to evolve according to DLA. The key test is to show that the data obey Double Asymtptotic Scaling (DAS) in the variables $\rho = \gamma((Y - Y_o)/\xi)^{1/2}$ and $\sigma = \gamma^{-1}((Y - Y_o)\xi)^{1/2}$, $\gamma = 6/(33 - 2N_f)^{1/2}$, $Y = \ln 1/x$, $\xi = \gamma^2 \ln(\ln Q^2/\Lambda_{QCD}^2/ \ln Q_o^2/\Lambda_{QCD}^2)$. Violations from DAS (other than due to the fact that the data lie in a pre-asymptotic region or to NLO DGLAP corrections [8]b) would signal either the onset of other approximations such as leading-(next-to-leading)-log in $1/x$, LL_x (NL_x) resummation, or the beginning of parton saturation. Here we wish to apply a DAS type of analysis to nuclear DI scattering.

As a first step we show that if evolution proceeds through ordinary asymptotic pQCD evolution equations in nuclei as well as in a free nucleon, the ratios R_F and R_G become a function of ρ only, which is *per se* a model independent result. We use this result as a basis for exploring the origin of scaling violations in nuclei. In nuclei in fact, the asymptotic regime is entered in principle at different values of x_o^A and $Q_{o,A}^2$ than in the proton. In particular Unitarity Shadowing Corrections (USC) [9,10,11,12] are expected to affect evolution at $x_o^A > x_o$ because of the increase of the gluon density in a nucleus due to the overlapping of nucleons in the longitudinal direction. On a more speculative basis one might also expect the transition to the $\ln(1/x)$ resummation to appear in a different regime. In our approach such questions can be addressed systematically as they introduce specific ***scaling violations*** from the DLA result, appearing as a σ dependence in the ratios R_G and R_F.

In summary, although it is technically predictable that in a proton at very low values of x and sufficiently large Q^2, *i.e.* deeply in the asymptotic region, DGLAP evolution and the DLA should break down and give way to $\ln(1/x)$ summation and to USC, it is still a major task to be able to pinpoint where and if the transition from the different regimes is going to take place in the kinematical regimes currently under exploration. Our goal is to obtain some new insight by using nuclear targets where the asymptotic regime can in principle be reached at larger x. As a by-product we obtain quantitative predictions for RHIC and the LHC.

2 DGLAP Evolution in Nuclei

We first summarize results for ordinary DGLAP evolution applied to the nuclear ratios at low x, assuming that the proton and the nuclear distributions evolve along similar paths. As it is well known evolution is driven by the gluon distribution which dominates over the sea quarks one and one can predict the behavior of the shadowing ratios, R_G and R_F with Q^2:

$$\begin{aligned}\frac{\partial R_G}{\partial \ln Q^2} &\simeq \int_x^1 P_{GG}\left(\frac{x}{y},\alpha_S(Q^2)\right)\frac{G_N(y,Q^2)}{G_N(x,Q^2)} \\ &\qquad \times \left[R_G(y,Q^2)-R_G(x,Q^2)\right]\frac{dy}{y} \\ &\propto \frac{\partial G_A(x,Q^2)/\partial \ln Q^2}{\partial G_N(x,Q^2)/\partial \ln Q^2} - R_G(x,Q^2)\ , \qquad (1)\end{aligned}$$

$$\begin{aligned}\frac{\partial R_F}{\partial \ln Q^2} &\simeq \int_x^1 P_{qG}\left(\frac{x}{y},\alpha_S(Q^2)\right)\frac{G_N(y,Q^2)}{\Sigma_N(x,Q^2)} \\ &\qquad \times \left[R_G(y,Q^2)-R_F(x,Q^2)\right]\frac{dy}{y}\ , \qquad (2)\end{aligned}$$

where we have disregarded the sea quarks distribution on the *r.h.s.* of the coupled DGLAP evolution equations; P_{qG} and P_{GG} are the splitting functions evaluated at NLO; and we used the approximation $F_2^{N(A)} \approx 5/18\Sigma_{N(A)}$. Eqs.(1) and (2) show that the Q^2 dependence of R_G and R_F is regulated by a subtle balance involving the parton distributions and the ratios themselves [13]. The following predictions can be made for the ratios R_G and R_F: If, as predicted by current non-perturbative shadowing models R_G is a growing function of x, then it also grows with Q^2, the *r.h.s.* of Eq.(1) being positive $(y \geq x)$. On the other side, defining $R_G(x,Q_o^2) \equiv R_G^o$ and $R_F(x,Q_o^2) \equiv R_F^o$, one obtains two opposite behaviors for R_F namely: *i)* if $R_F^o < R_G^o$, then R_F grows with Q^2; *ii)* if instead $R_F^o > R_G^o$, then R_F initially decreases with Q^2 until it reaches the value of R_G^o and it subsequently starts increasing along with R_G. [a] The "rate" of change with Q^2 is governed by the ratios $G_N(y)/G_N(x)$ and $G_N(y)/\Sigma_N(x)$, at $Q^2 = Q_o^2$, respectively. If $Q_o^2 \leq 1\,\mathrm{GeV}^2$, then a rapid

[a] Although the form of Eq.(1) might seem suggestive of a fixed point behavior [14], this is *a priori* not the case, since the quantity $\partial G_A(x,Q^2)/\partial \ln Q^2/\partial G_N(x,Q^2)/\partial \ln Q^2$ depends on Q^2.

evolution strongly reduces the shadowing in both R_G and R_F, between Q_o^2 and $Q^2 \approx 2-3\,\mathrm{GeV}^2$. If on the contrary Q_o^2 ranges from 2 to $5\,\mathrm{GeV}^2$ then the evolution is slower and shadowing remains large even at $Q^2 \approx 10-100\,\mathrm{GeV}^2$ (these results affect quarkonia production where $Q^2 \equiv M_{J/\psi}^2 \approx 10\,\mathrm{GeV}^2$ and $Q^2 \equiv M_{\Upsilon}^2 \approx 100\,\mathrm{GeV}^2$).

3 Double Asymptotic Scaling

We now examine the nuclear DI structure function and gluon distribution in the asymptotic regime. The derivation of the equations of the DLA in a nucleus parallels the one for the proton, namely one first writes the LO DGLAP evolution equation for the gluon distribution in the limit $x \to 0$, in moments space (we have omitted the subscripts $N(A)$ unless necessary):

$$\frac{\partial g(n,Q^2)}{\partial \ln Q^2} = \frac{\alpha_s(Q^2)}{2\pi}\gamma_{GG}^0(n) g(n,Q^2), \tag{3}$$

$\gamma_{gg}^0(n) \approx 2C_A/(n-1)$ being the anomalous dimension in the limit $n \to 1$. Solutions in (x,Q^2) are found by evaluating the anti-Mellin transform,

$$G(x,Q^2) = \frac{1}{2\pi i}\int_C dn\, g(n,Q_0^2) \exp\{(n-1)Y + \xi/(n-1)\}, \tag{4}$$

with the saddle point method. In Eq.(4) $g(n,Q_0^2) = \int_0^1 dx\, x^{(n-1)} g(x,Q_0^2)$, $g(x,Q_o^2)$ being the inital gluon distribution, and $G(x,Q^2) = xg(x,Q^2)$. If one takes a "soft" initial condition such as, $G(x,Q_0^2) \approx A_N x^{-\lambda}$, $\lambda \approx 0$, and Y and ξ are both similarly large, then $g(n_0,Q_0^2) \approx A_N/(n-1)$, and the saddle point is: $n_0 = 1/(2\Delta Y) + \sqrt{1/(4\Delta Y)^2 + \xi/\Delta Y} \approx \sqrt{1+\xi/\Delta Y}$, $\Delta Y = Y - Y_o$, $Y_o = \ln(1/x_o)$, yielding:

$$G(x,Q^2) = \sqrt{2\pi}\left(\frac{\tilde{g}(n_o,Q_o^2)}{\tilde{g}''(n_o,Q_o^2)}\right)^{1/2} \tilde{g}(n_o,Q^2), \tag{5}$$

with $\tilde{g}(n,Q^2) = g(n,Q^2)\exp[(n-1)Y]$. [b] In terms of the DAS variables, ρ and σ, $n_0 = 1 + \rho/\gamma^2$ and:

$$G \equiv G^{DLA}(x,Q^2) = f(\rho,\sigma)\exp(2\gamma\sigma), \tag{6}$$

where $f(\rho,\sigma)$ is a function that depends on the explicit form of the initial conditions. The main result is that $\ln(G^{DLA})/f(\rho,\sigma))$ is predicted to be linear

[b] We have omitted sub-leading corrections for simplicity.

in σ, the slope being fixed by γ. A similar behavior is found for F_2^N by solving the equation:

$$\frac{\partial F_2^N(x,Q^2)}{\partial \ln Q^2} = \frac{2}{9}\frac{\alpha_S}{\pi} G(x,Q^2). \tag{7}$$

Deviations from DAS due to hard initial conditions *i.e.* when $\lambda \geq 0.2$ are technically obtained as follows: $g(n_0, Q_0^2) \approx A_N/(n-(\lambda+1))$ and a physical solution in the limit $Y \gg \xi$ is

$$G(x,Q^2) = C/\lambda e^{\lambda(Y-Y_o)+\xi/\lambda}, \tag{8}$$

corresponding to the saddle point value, $n_0(\lambda) = 1/\Delta Y + \lambda + 1$ (see also [8,11]). Physically this behavior supports the appearance of USC which become more important just because of the steep rise at small x. The data of Ref.[6] favor the DAS scenario, whereas the more recent data at lower x and Q^2 [7] are best interpreted in terms of USC[16]. In this context it is therefore important to study nuclear shadowing because, due to the A-dependent increase in gluon density, the physics we are interested in is expected to show up at larger x values and because one can tune the variable A in order to discriminate among models.

In a nucleus we make the two following assumptions for the asymptotic regime: *i)* the initial distributions are shadowed because of some non-perturbative mechanism, the evolution equations are not affected by the medium; *ii)* besides the non-perturbative shadowing the evolution equations have screening corrections at a larger x_A. The derivation of the ratio R_G in the hypothesis *i)* yields:

$$\begin{aligned} R_G^{Asym} &= \left[\frac{G_A(n_o,Q^2)}{G_N(n_o,Q^2)}\right]^{3/2} \left[\frac{G''_N(n_o,Q^2)}{G''_A(n_o,Q^2)}\right]^{1/2} \\ &= f_A(\rho)\frac{\rho h_N^{(1)}(\rho) + \gamma^2\sigma h_N^{(2)}(\rho)}{\rho h_A^{(1)}(\rho) + \gamma^2\sigma h_A^{(2)}(\rho)} . \end{aligned} \tag{9}$$

where $f_A(\rho) = G_A(n_o,Q_o^2)/G_N(n_o,Q_o^2)$, and the functions $h_{N(A)}^{(1,2)}$ are listed explicitely elsewhere. The main point is that the exponential terms appearing in G^{DLA}, Eq.(6) calculated for a nucleus and for a nucleon respectively, cancel exactly and the remaining dependence on σ is small asymptotically. This scaling result (shown in Fig.1) will be our reference point.

The onset of a different evolution mechanism in the nucleus will appear as a ***σ-scaling violation*** modifying the exponential behavior of Eq.(5). In general one can write

$$R_G = R_G^{Asym}(\rho) \times \exp\left\{2\gamma(\sigma_A - \sigma)\right\}, \tag{10}$$

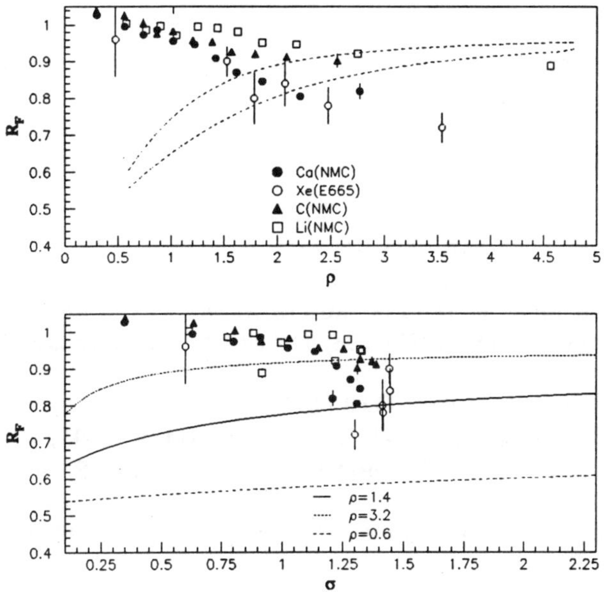

Figure 1: Nuclear structure functions ratios, R_F, vs. the DAS variables ρ (top) and σ (bottom). The theoretical curves [20] show the onset of scaling in σ. The non-perturbative shadowing model used in the calculation is from Ref.[15]. Experimental data from [17,18] (Li, C, Ca) and [19] (Xe).

where R_G^{Asym} is the large σ limit of Eq.(9). The form of $\sigma_A \equiv \sigma_A(x, Q^2)$ describes the approximation to pQCD evolution in a nucleus. In the case of a "hard pomeron" boundary conditions plus screening corrections [9] it is

$$\sigma_A = D_A(x, Q^2)\left(\lambda\sigma\rho + \frac{\gamma^2}{\lambda}\frac{\sigma}{\rho}\right), \tag{11}$$

D_A being a damping correction (more details are going to be found in [20]).

In conclusion, we have shown model independent predictions for gluon shadowing in nuclei in the asymptotic region. We have also outlined an approach for more detailed studies of the violations from DAS in nuclei as a signal of pQCD evolution mechanisms other than DGLAP or the DLA. Our calculations will be relevant in the regime accessible at future experiments at RHIC, LHC and possibly at the eA project at DESY [21].

Acknowledgments

The authors wish to thank the Institute of Nuclear and Particle Physics at the University of Virginia for financial support.

References

1. M. Arneodo, Phys. Rep. **240** (1994) 301.
2. Zi-wei Lin and M. Gyulassy, Phys.Rev.Lett. **77** (1996) 1222.
3. S. Liuti and R. Vogt, Phys. Rev. **C51** (1995) 2244.
4. L.V. Gribov and Lipatov, Yad.Fiz. **20** (1975) 181; G. Altarelli and G. Parisi, Nucl.Phys. **B126** (1977) 298; Dokshitzer, Sov.Phys.JETP **46** (1977) 641.
5. A. De Rujula et al., Phys. Rev. **D10** (1974) 1649.
6. H1 Collaboration, T.Ahmed et al., Nucl.Phys. **B439** (1995) 471; C. Adloff et al., Nucl.Phys. **B497** (1997) 3; S. Aid et al., Nucl.Phys. **B470** (1996), 3.
7. ZEUS Collaboration, M. Derrick et al., Z.Phys. **C69** (1996) 607; J. Breitweg et al., Phys.Lett. **B407** (1997) 432
8. R. Ball and S. Forte, Phys. Lett. **B335** (1994) 77; *ibid* **B336** (1994) 77; Nucl. Phys. Proc. Suppl. **54A** (1997) 163.
9. L.V. Gribov, E.M. Levin and M.G. Ryskin, Phys. Rep. **100** (1983) 1.
10. A.H. Mueller and J. Qiu, Nucl. Phys. **B268** (1986) 427. (1994) 44.
11. J.C. Collins and I. Kwiecinski, Nucl. Phys. **B335** (1990) 89.
12. K. Eskola, J. Qiu and X. Wang, Phys.Rev.Lett. **72** (1994) 36.
13. J. Qiu, Nucl. Phys. **B291** (1987) 746.
14. Z. Huang, H.J. Lu and I. Sarcevic, Nucl. Phys. **A637** (1998) 79.
15. L.L. Frankfurt and M.I. Strikman, Nucl. Phys. **B316** (1989) 340.
16. E. Gotsman, E.Levin, U.Maor and E. Naftali, Nucl. Phys. **B** (1999), hep-ph/9808257.
17. M. Arneodo et al., Nucl. Phys. **B441** (1995) 12.
18. P. Amaudruz et al., Z. Phys. **C53** (1992) 73.
19. M.R. Adams et al., Phys. Lett. **B287** (1992) 375; Phys. Rev. Lett. **68** (1992) 3266.
20. F. Cano and S. Liuti, *in preparation.*
21. *Workshop on Physics with HERA as eA Collider*, DESY, Hamburg, 25-26 May, 1999, C99/05/95.

NONPERTURBATIVE EFFECTS IN THE PROTON SEA

W. SCHÄFER, J.SPETH
Institut für Kernphysik, Forschungszentrum Jülich D-52425 Jülich, Germany
E-mail: Wo.Schaefer@fz-juelich.de

We revisit the evaluation of the pionic mechanism of the $\bar{u}-\bar{d}$-asymmetry in the proton structure function. Our analysis is based on the unitarity relation between contributions of different mechanisms to the inclusive particle production and the total photoabsorption cross-section (i.e. the proton structure function). We reanalyze the role of isovector reggeons in inclusive production of nucleons and Delta isobars in hadronic reactions. A rather large contribution of reggeon-exchange induced production of Delta isobars is found. This leaves much less room for the pion-exchange induced mechanism of Δ production and provides a constraint on the $\pi N\Delta$ form factor. The production of leading pions in proton-proton collisions puts additional constraints on the πNN vertex form factors. All these constraints are used then to estimate the pion content of the nucleon and allow to calculate parameter-free the x-dependence of $\bar{d}-\bar{u}$. We discuss the violation of the Gottfried Sum Rule and $\bar{d}$-$\bar{u}$ asymmetry and compare to the one obtained from the E866 experiment at Fermilab.

1 Introduction

Since the discovery of the Gottfried sum rule violation [1] there has been a long ongoing discussion on the $\bar{d}-\bar{u}$ asymmetry in the nucleon sea. Considerable atttraction has been received recently by the E866 experiment at Fermilab[2], which provided the first detailed measurement of the Bjorken-x dependence of the $\bar{d}-\bar{u}$-asymmetry from the comparison of *pp* and *pd* Drell–Yan production. Their striking finding is that the asymmetry tends to vanish at $x \sim 0.3$.

The strong observed asymmetry has no explanation in terms of the purely pQCD dynamics, and thus it gives important hints on the impact of the nonperturbative proton structure on the generation of the nucleons's sea quark content. Following the early work of Sullivan [3], a natural dynamical explanation emerges within the framework of the isovector meson cloud model of the nucleon (for a recent review see [4]). Here a special role is played by the pion, and it is intuitively appealing to account for the nonperturbative meson/baryon structure of the proton, by including the pion as the nonperturbative parton in the light cone wave function of the interacting nucleon:

$$|N\rangle_{phys} = |N\rangle_{bare} + |N\pi\rangle + |\Delta\pi\rangle + \dots . \qquad (1)$$

Here the pion, that is contained in the $N\pi, \Delta\pi$–Fock states can emerge as the target in the deep inelastic $\gamma^* N$ photoabsorption process, while the spectator baryonic constituent N, Δ appears in the final state, separated from the remnants of the $\gamma^*\pi$–interaction by a rapidity gap and carrying a large fraction z of the incoming proton's lightcone momentum. Hence, inclusive production of baryons, like $\gamma^* p \to Xn$ is naturally described in terms of the pionic 'partons' in the nucleon. But precisely *the same* dynamics is supposed to be at work, if we swap the virtual photon against the proton projectile. This opens the possibility to use a large body of experimental knowledge on inclusive particle production in hadronic reactions to constrain the meson/baryon dynamics relevant for deep inelastic scattering. In the following we shall demonstrate, that if one accounts consistently for the so–derived constraints, a satisfactory description of the Fermilab data emerges [5].

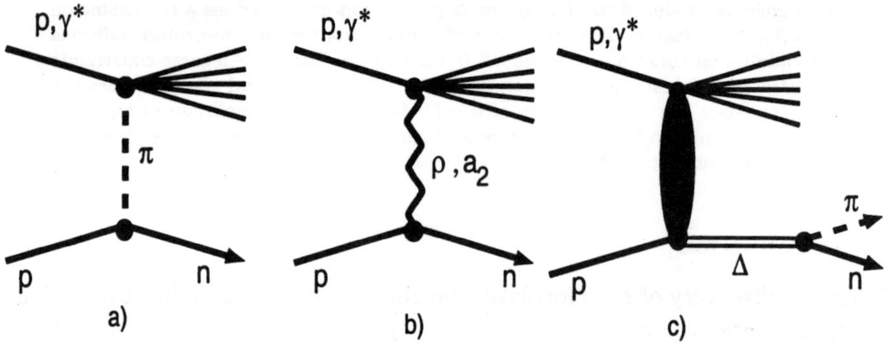

Figure 1: The mechanisms for inclusive production of forward neutrons. a) pion exchange, b) background from (reggeized) ρ, a_2 exchanges, c) neutrons originating from the decay of an intermediate Δ–resonance.

2 Inclusive production of baryons and mesons in hadronic reactions

2.1 Production of forward neutrons and Δ–isobars

Following the above described strategy, we first turn to the description of the forward neutron production in pp collisions. We take into account three production mechanisms: the dominant pion exchange, the background contribution from the isovector exchanges ρ, a_2, and, finally, the production of neutrons from the decay of an intermediate Δ–resonance (see fig 1). In addition, we have to account for the distortion of the incoming proton waves, employing standard methods of the generalized eikonal approximation [6]. Following the reasoning in [6], we apply the absorptive corrections only in the pp scattering, they can be neglected in the $\gamma^* p$–case. It is important to notice,that the phase space of the forward neutrons includes the kinematical boundary $z \sim 1$, which corresponds to a large Regge parameter $s/M_X^2 \gg 1$, where M_X^2 is the invariant mass squared of the inclusive system X. Hence the proper formalization of the ρ-exchange mechanism should be a Regge treatment [7].

In the Regge formulation, the contribution to the inclusive cross section from the pion exchange mechanism has the form:

$$z\frac{d\sigma(p\to n)}{dzdp_\perp^2} = \frac{g_{\pi pn}^2}{16\pi^2}\frac{(-t)}{[t-m_\pi^2]^2}\cdot F_{\pi NN}^2(t)(1-z)^{1-2\alpha_\pi(t)}\cdot\sigma_{tot}^{\pi p}(M_X^2)\,, \qquad (2)$$

with $-t = p_\perp^2/z + (1-z)^2 m_p^2/z$, z is the neutrons Feynman–variable, which for large z is equal to its lightcone momentum fraction, $\vec{p}_\perp$ is the neutron's transverse momentum. Due to the proximity of the pion pole, the departure of $\alpha_\pi(t) = \alpha'(t-m_\pi^2)\,, \alpha' \approx 0.7\,GeV^{-2}$ from $J_\pi = 0$ is small, and we can jusify taking the physical pion–nucleon cross section $\sigma_{tot}^{\pi p}$. The finite extension of the involved particles and/or the off shell effects are accounted for by the form factor $F_{\pi NN}^2(t)$, in the following we shall stick to the parametrization: $F_{\pi NN}^2(t) = \exp[(R^2(t-m_\pi^2))^2]\,, R^2 = 1.5GeV^{-2}$, for different choices of the functional form, see the discussion in [5].

In fig.2 we show our results for the invariant cross section against the experimental data. The data shown here are at relatively low $p_{LAB} \approx 24\,GeV/c$, hence the visible trace of the πN–resonance–region at large z. The $p_\perp$– dependence nicely illustrates the interplay of the different mechanisms. The pion exchange peaks in the region $0.7 \lesssim z \lesssim 0.9$, and dominates for $p_\perp \lesssim 0.3\,GeV$. The relative importance of the ρ, a_2 exchanges grows with $p_\perp$, which testifies to the dominant $\propto p_\perp^2$ spin–flip component of the ρN–coupling. The background contribution from the two–step process $p \to \Delta \to n$ turns out to be substantial only for $z \lesssim 0.8$.

We took care that the background from the Δ–decay is consistent with the experimental data on forward-Δ–production, to which we turn now. Unfortunately, experimental data are rather scarce, and often plagued by large uncertainties due to the nonresonant πN subtraction. We do not have any data differential in $p_\perp$ at hand, and obviously the shape of the $p_\perp$–integrated Δ–spectrum could be fitted by both extreme choices: the pure π–exchange as well as the pure ρ–exchange. Here valuable information comes from the two–body reactions at high energies. The cleanest manifestation of ρ–exchange is clearly the πN–charge exchange reaction $\pi^- p \to \pi^0 n$ and $\pi^+ p \to \pi^0 \Delta^{++}$, for which experimental data are available [8]. Assuming Regge factorization, we can now relate the ρ–exchange contribution in the $p \to n$ transition from our earlier considerations to the ρ–exchange in $p \to \Delta$. The result is a surprisingly large ρ–exchange contribution, which does not leave much room for the pionic contribution (see fig 3.).

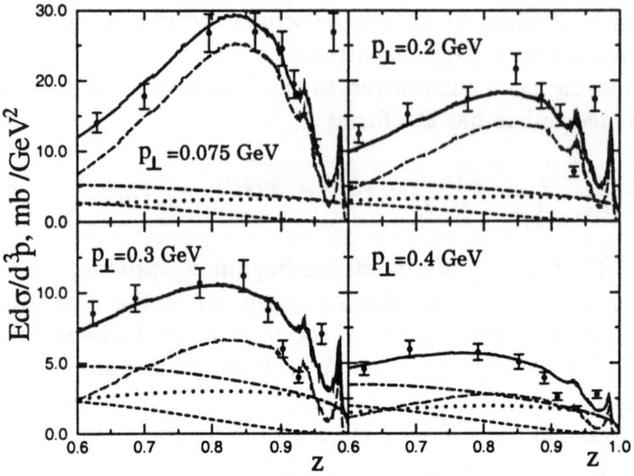

Figure 2: Invariant cross section for the reaction $pp \to nX$ at $p_{LAB} = 24\,\mathrm{GeV/c}$. The experimental data are taken from [9]. The long dashed curve shows the contribution from the pion exchange; the dotted curve is the ρ, a_2-exchange contribution, and the dashed curve shows the contribution from the two step process $p \to \Delta \to n$. In additionally we present the sum of the two background contributions as the dot-dashed line. Finally the solid curve represents the sum of all components.

2.2 Constraints from forward pion production

The good description of the forward baryon production can make us confident, that we have found a proper description of the $\pi N, \pi\Delta$–fock states in the region where the baryon carries a large momentum fraction $z \sim 0.6 \div 1$. However we should be aware, that the forward baryon production data do not put any constraint on the region $z \lesssim 0.6$; in such a kinematical region we cannot expect the meson/Reggeon exchanges to be the dominant reaction mechanisms. On the other hand, in the Fock–state picture a baryon carrying a small momentum fraction $z_B \ll 1$ is accompanied by the meson carrying large $z_M = 1 - z_B \sim 1$. Hence, the natural place to check the consistency of the Fock–state parameters is the production of forward, $z_\pi \sim 1$, pions. The important message from the experimental data (fig.4) is: there are *no pions carrying large momentum fractions* $0.6 \lesssim z \lesssim 1$ observed. This places a severe constraint on the cutoff in the employed πNN–formfactor. (Notice that we are not after a description of the forward pion production for $z \lesssim 0.5$). The specific formfactor choice cited

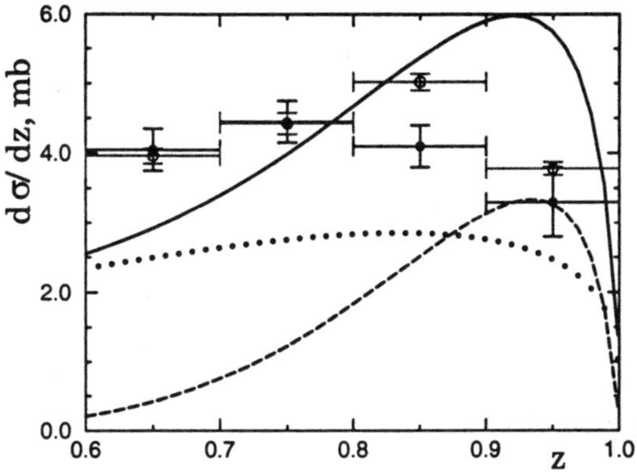

Figure 3: Differential cross section $d\sigma/dz$ for the reaction $pp \to \Delta^{++}X$ at p_{LAB} = 400 GeV/c. The dashed curve is the contribution from pion-exchange; the dotted curve shows the ρ, a_2 contribution. Shown by the solid curve is the sum of the two. Experimental data are taken from [10](open circles), and [11] (filled circles).

above leads to the middle solid curve in fig.4. Incidentally, the dotted line on fig.4 shows the result of a previously emloyed formfactor [12] that also did a reasonable job on forward neutrons, but does not respect the pion production data.

3 Regge mechanisms vs. Fock–state picture

In the introduction we stressed the relevance of the pion as the nonperturbative parton in the light–cone wave function of the interacting proton. Here the question arises what kind of Fock–state should then be associated with the ρ, a_2-Reggeon exchange production mechanisms, if any? In other words, knowing the reaction mechanisms that populate several inclusive channels, what can we say about their impact on the total cross section? We remind, that one of the firm predictions of Regge theory is a very specific phase of the amplitudes. While the inclusive cross section is calculated from the modulus of the amplitude squared, $|A|^2$, it is the product of two amplitudes, $A \cdot A$, that enters the evaluation of the total cross section. The result may be summarized by generalized AGK cutting rules [5,13]: for the purely real pion exchange amplitude, this implies that inelastic interactions of the projectile with the pions in the target hadrons enhances the total cross section, whereas the ρ, a_2–exchanges, which have a phase $\sim (1+i)$, happen to give a vanishing contribution to the total cross section. Hence, it does not make any sense, to identify the Reggeon ex-

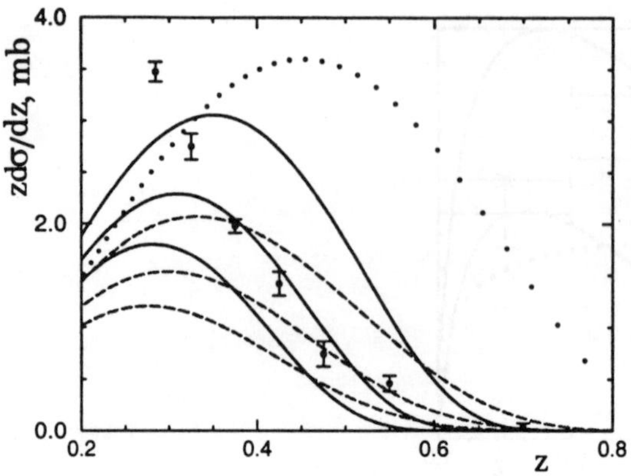

Figure 4: Differential cross section $zd\sigma/dz(pp \to \pi^0 X)$ at $p_{LAB} = 400$ GeV/c. The data are taken from [10]. The dotted curve shows a prediction of the model [12]. The solid curves were calculated with a 'Gaussian' form factor $F_{\pi NN}(t) = \exp(-\left[R_G^2(t-m_\pi^2)\right]^2)$. The curves in the figure correspond to $R_G^2 = 1.0\,, 1.5, and\, 2.0\ GeV^{-2}$ (from top to bottom). The dashed curves were calculated with an exponential form factor $F_{\pi NN}(t) = \exp(R_E^2(t-m_\pi^2))$. The curves in the figure are for $R_E^2 = 1.0\,, 1.5, and\, 2.0\ GeV^{-2}$ (from top to bottom).

change mechanism in inlusive reactions in terms of a meson/baryon–Fock state of the proton. We want to point out that there is no mystery connected with such a different impact of the phases on different observables. For instance in the inclusivex $pp \to pX$, diffractive channels (imaginary amplitude) dominate for $z \sim 1$, and it is well known that the opening of diffractive channels comes along with the *absorptive correction* to the total cross section, which is *negative* [13].

4 The $\bar{d} - \bar{u}$ asymmetry from pions in the nucleon

Following our reasoning above, we include the πN, $\pi\Delta-$ Fock states only. They lead to a contribution to the total $\gamma^* p$–cross section – i.e. the quark/antiquark–distributions in the proton given by

$$\Delta^\pi q(x, Q^2) = \int_x^1 \frac{dz_\pi}{z_\pi} f_\pi(z_\pi) \cdot q^\pi(x/z_\pi, Q^2)\,, \tag{3}$$

with the flux of pions which is determined through

$$f_\pi(z_\pi) = \int d^2\vec{p}_\perp E \frac{d\sigma(p \to n)}{\sigma_{tot}^{\pi p} d^3\vec{p}}\,, \tag{4}$$

and similarly for the $\pi\Delta$–Fock state. Besides the Fock–state parameters determined in our analysis above, we need as an input the quark distributions in the pion. Notice, that for the calculation of $\bar{d}(x) - \bar{u}(x)$ only the pion's valence distributions enter. These are reasonably well constrained down to $x \sim 0.2$ from the Drell–Yan experiments, for definiteness we take the GRV–parametrization[14]. We obtain a total multiplicity of pions in the proton associated with the πN–state of $n_{\pi N} \sim 0.21 \div 0.28$ and of $n_{\pi\Delta} \sim 0.03 \ll n_{\pi N}$ for the $\pi\Delta$–contribution. This translates to a Gottfried sum of $0.21 \lesssim S_G \lesssim 0.25$, compared to the NMC determination[1] $S_G = 0.235 \pm 0.026$ Our result for $\bar{d} - \bar{u}$ is shown in fig 5, we observe that the asymmetry is essentially driven by the πN–state. Notice that our good agreement in the region $x \gtrsim 0.2$ owes to the fact that we have no contributions from hard $z_\pi \gtrsim 0.5$ pions in the nucleon.

5 Summary

Our reanalysis of the pionic contribution to the $\bar{d} - \bar{u}$– asymmetry has been based on a unified treatment of the inclusive production of leading nucleons, Δ's and pions in hadronic high energy collisions. We paid special attention to the background contributions in the leading baryon production, and clarified the relation of Reggeon–exchange mechanisms and the Fock–state picture of the interacting nucleon. Severe constraints on our parameters were found to arise from the forward pion data. A good agreement with the recent E866 data on $\bar{d} - \bar{u}$ was found.

Finally we want to point at another spin–off of our analysis: namely one may use the pions in the nucleon as targets in deep inelastic scattering and hence obtain information on its structure function at small x, unexplored by the Drell-Yan experiments[15]. Our findings for the relevant background contributions show that this task is well feasible.

Acknowledgments

We wish to thank C. Carlson and A. Radyushkin for invitation. W.S. is indebted to INT/JLab for the financial support.

References

1. P. Amaudruz et al., Phys. Rev. Lett. **66** (1991) 2712.
2. E.A. Hawker et al., Phys. Rev. Lett. **80** (1998) 3715.
3. J.D. Sullivan, Phys. Rev. **D5** (1972) 1732.
4. J. Speth and A.W. Thomas, Adv. in Nuclear Physics Vol.**24** (1998) 84.

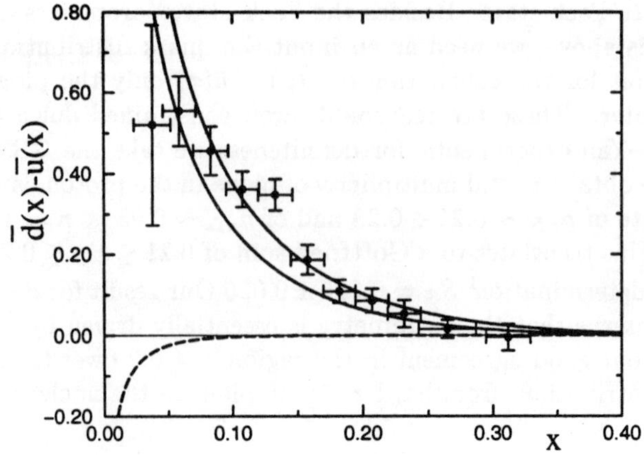

Figure 5: Flavour asymmetry $\bar{d}(x) - \bar{u}(x)$ at $Q^2 = 54\,\mathrm{GeV}^2$. Experimental data are from E866 [2]. The solid curves show the contribution from the πN-Fock state and were calculated for Gaussian form-factors; the upper curve belongs to $R_G^2 = 1\,\mathrm{GeV}^{-2}$, the lower one to $R_G^2 = 1.5\,\mathrm{GeV}^{-2}$ The dashed line shows the contribution of the $\pi\Delta$-Fock state.

5. N.N. Nikolaev, W. Schäfer, A. Szczurek, and J. Speth, Phys. Rev. **D60** (1999) 014004.
6. N.N. Nikolaev, J.Speth, and B.G. Zakharov, KFA-IKP(TH)-1997-17, unpublished, hep-ph/9708290.
7. A. Szczurek, N.N. Nikolaev and J. Speth, Phys. Lett. **B428** (1998) 383.
8. A.C. Irving and R.P. Worden, Phys. Rep. **34C** (1977) 117.
9. V. Blobel et al., Nucl. Phys. **B135** (1978) 379.
10. M. Aguilar-Benitez et al., Z. Phys. **C50** (1991) 405.
11. A. Breakstone et al., Z. Phys. **C21** (1984) 321.
12. H. Holtmann, A. Szczurek and J. Speth, Nucl. Phys. **A596** (1996) 631.
13. V.A. Abramovskii, V.N. Gribov and O.V. Kancheli, Sov. J. Nucl. Phys. **18** (1974) 308.
14. M. Glück, E. Reya and A. Vogt, Z. Phys. **C53** (1992) 651.
15. H. Holtmann, G. Levman, N.N. Nikolaev, A. Szczurek and J. Speth, Phys. Lett. **B338** (1994) 363.

LEADING BARYON PRODUCTION AT HERA

GARRY LEVMAN[a]
Department of Physics, University of Toronto, 60 St. George Street, Toronto ON M5S 1A7, Canada
E-mail: levman@physics.utoronto.ca

Recent results are presented on semi-inclusive leading proton and neutron production in e^+p collisions at HERA. The data, taken at a center-of-mass energy of $\sqrt{s} = 300$ GeV, are drawn from the neutral current data samples for both low ($0.1 < Q^2 < 0.8$ GeV^2) and high ($Q^2 > 2$ GeV^2) virtuality of the exchanged photon. The baryons are produced with small transverse momentum p_T, and carry a large fraction, x_L, of the initial proton's energy. The shape of the x_L spectra are independent of the leptonic kinematic variables Q^2 and Bjorken's x. The scaling violations in semi-inclusive interactions have the same behavior as in inclusive interactions. Outside the diffractive region, for $0.6 < x_L < 0.97$, proton production is approximately independent of x_L, Q^2 and x. In addition, the proton's p_T^2 distribution decreases exponentially with p_T^2 with a slope independent of the other kinematical variables. Neutron production is also independent of Q^2 and x, but shows a rapidly falling x_L distribution indicative of the dominance of pion exchange. The leading baryon data exhibit a strong factorization behavior, consistent with both limiting fragmentation and particle exchanges. Simple particle exchange models of leading baryon production are in good agreement with the data.

1 Introduction

The ZEUS and H1 collaborations have observed events with protons and neutrons carrying a large fraction x_L of the proton beam energy[1,2], but with small transverse momentum p_T. The data come from samples where the the photon's virtuality Q^2 spans a large range: from the transition region between photoproduction and deep inelastic scattering ($0.1 < Q^2 < 0.8$ GeV^2) to deep inelastic scattering $Q^2 > 2$ GeV^2. Leading baryon data cast considerable light on both the short and long range structure of the proton. Small p_T ensures that the baryon is produced at a soft scale where the quantum hadrodynamics of mesons and baryons dominates the production mechanism; at the same time, large Q^2 ensures that the quantum chromodynamics of quarks and gluons is very relevant.

In analogy with the inclusive structure function $F_2(x, Q^2)$, which is defined through the inclusive ep cross section,

$$\frac{d^2\sigma(ep \to eX)}{dxdQ^2} = \frac{4\pi\alpha^2}{dxQ^4}\left(1 - y + \frac{y^2}{2}\right) F_2(x, Q^2), \tag{1}$$

[a]representing the ZEUS and H1 Collaborations

the semi-inclusive structure function $F_2^{LB(4)}(x, Q^2, x_L, p_T^2)$ is defined through the four-fold differential cross section for leading baryon production,

$$\frac{d^4\sigma(ep \to eBX)}{dx dQ^2 dx_L dp_T^2} = \frac{4\pi\alpha^2}{xQ^4}\left(1 - y + \frac{y^2}{2}\right) F_2^{LB(4)}(x, Q^2, x_L, p_T^2), \qquad (2)$$

where $B = P$ or N for leading protons, or neutrons respectively, and $y = Q^2/(xs)$ is the positron's inelasticity in the proton's rest frame. The interdependence of the leptonic and baryonic variable is an important clue to the underlying production mechanisms. The hypothesis of limiting fragmentation[3] predicts that the leptonic variables Q^2 and x completely separate from the baryonic variable x_L and p_T^2 so that F_2^{LB} factorizes into a product of functions:

$$F_2^{LN} \sim f(x_L, p_T^2) \cdot \tilde{F}(x, Q^2). \qquad (3)$$

Particle exchange models[4] expect only a partial separation of the variables:

$$F_2^{LN} \sim f(x_L, p_T^2) \cdot \tilde{F}(x/(1 - x_L), Q^2). \qquad (4)$$

In this case, f is the flux factor and $\tilde{F}$ the structure function of the exchanged particle. From baryon number conservation we know that each bin in Q^2 and x contains a baryon; however, the conserved baryon can be arbitrarily distributed in species, p, n, Λ, and in the baryonic variables x_L, p_T^2. It is particularly useful, therefore, to study the yield of leading protons (and neutrons) as a fraction $r(x, Q^2)$ of the inclusive cross section in bins of x_L and p_T^2. The study of fractions is also advantageous experimentally since many systematic effects cancel in the ratio.

The Leading Proton Spectrometers and Forward Neutron Calorimeters of ZEUS and H1 will not be discussed here since they are described in detail elsewhere[2,5]. The data discussed are also more fully presented in the referenced papers[2,6,7].

2 Leading Proton Production

Fig. 1 shows the H1 measurement of F_2^{LP} for protons in the kinematical range $2 < Q^2 < 50$ GeV^2, $0.7 < x_L < 0.9$ and $p_T < 0.2$ GeV. In each bin of x and Q^2 the structure function is approximately flat in x_L; however, the level increases with increasing Q^2. The flatness of the x_L distribution for protons is also apparent in the ZEUS data[6] which shows that there is no difference in shape or rate between high ($Q^2 > 3$ GeV^2) and low ($0.1 < Q^2 < 0.8$ GeV^2) Q^2 data. ZEUS has also studied the angular distribution of leading protons as a function of x_L. The data are well described by the form $dN/dp_T^2 \propto \exp(-bp_T^2)$.

The values of the slope parameters b are largely independent of x_L and are the same in both the high and low Q^2 samples.

The fraction of events with a leading proton $0.6 < x_L < 0.97$, measured by ZEUS, is shown in Fig. 2. Within experimental errors r^{LP} is independent of the leptonic variables.

3 Leading Neutron Production

Fig. 3 shows the measurement of F_2^{LN} for neutrons in the kinematical range $2 < Q^2 < 50$ GeV^2, $0.2 < x_L < 1$ and $p_T < 0.2$ GeV. In each bin of x and Q^2 the structure function decreases quickly with increasing x_L; the fall off is roughly linear with a steepness increasing with increasing Q^2.

The observed (that is, uncorrected for acceptance of the FNC) fraction of events with a leading neutron $0.6 < x_L < 1$ and scattering angle $\theta < 0.8$ mrad is shown in Fig. 4. Within experimental errors r_{unc}^{LN} is independent of the leptonic variables.

4 Discussion

H1 data for F_2^{LB}, which are at relatively low values of Q^2, imply that the shape of the x_L distribution for leading baryons is independent of the leptonic variables x and Q^2. This observation is strengthened by the ZEUS comparisons of the x_L distributions at low and high Q^2. Moreover, ZEUS observes that the fraction of events with a leading baryon is approximately constant over a large kinematical range.

These results are in accord with the expectations of the hypothesis of limiting fragmentation; however, because of the kinematical condition $x_L < 1 - x$, limiting fragmentation is expected to hold only at low x. The rise of F_2^{LB} with Q^2 is an indication of scale breaking; the constancy of r^{LP} and r_{unc}^{LN} demonstrates that the behavior of scale breaking for leading baryon events is the same as for inclusive events. In fact, since $r_{(unc)}^{LB} \propto F_2^{LB}/F_2$, $dr_{(unc)}^{LB}/d\ln Q^2 = 0$ implies that

$$\frac{dF_2^{LB}/d\ln Q^2}{F_2^{LB}} = \frac{dF_2/d\ln Q^2}{F_2} \tag{5}$$

separately for $B = p$ and $B = n$ and so for $B = (p \text{ or } n)$.

The H1 and ZEUS collaborations have also compared to their data simple particle exchange models of leading baryon production[2,4]. H1 finds that the neutron data (at high x_L) is well described by π^+ exchange alone. The proton data requires Pomeron and Reggeon exchange as well as π^0 exchange, where the π^0 contribution is fixed by the neutron data and I-spin conservation. ZEUS has similar results[6] which show in addition that the angular distribution of leading protons is also well described. These studies indicate that particle exchange factorization is also exhibited by the data. It is remarkable that particle exchanges sum to give the extraordinary simplicity of behavior shown by the data.

Limiting fragmentation predicts that the shape of the x_L distribution is independent of Q^2 and x. The constancy of $r^{LB}_{(unc)}$ implies $\tilde{F} \propto F_2$. In order that limiting fragmentation and particle exchange factorization are consistent with each other, $\tilde{F}$ must satisfy a power law in x, $\tilde{F} \sim x^{\beta}$; the constancy of $r^{LB}_{(unc)}$, also $F_2 \sim x^{\beta}$. Eqn. 5 requires $(d\tilde{F}/d\ln Q^2)/\tilde{F} = (dF_2/d\ln Q^2)/F_2$.

5 Conclusions

For leading protons and neutrons produced at large x_L and small p_T the shape of the x_L spectra are independent of the leptonic kinematic variables Q^2 and x. Scaling violations are observed and have the same behavior as inclusive interactions. It is remarkable that outside the diffractive region ($0.6 < x_L < 0.97$), proton production is approximately independent of x_L, Q^2 and x. In addition, the proton's p_T^2 distribution decreases exponentially with p_T^2 with a slope independent of x_L, Q^2 and x. Neutron production is also independent of Q^2 and x, but has a rapidly falling x_L distribution indicative of the dominance of pion exchange. The leading baryon data exhibit a strong factorization behavior, consistent with both limiting fragmentation[3] and particle exchanges. Simple particle exchange models of leading baryon production[2,4] are in good agreement with the data.

References

1. ZEUS Collaboration, M. Derrick et al., *Phys. Lett.* B **384**, 388 (1996).
2. H1 Collaboration, C. Adoff et al., *Eur. Phys. J.* C **6**, 587 (1999).
3. J. Benecke, T. T. Chou, C. N. Yang, and E. Yen, *Phys. Rev.* **188**, 2159 (1969).
4. A. Szczurek, N. N. Nikolaev, and J. Speth, *Phys. Lett.* B **428**, 383 (1998).
5. S. Bhadra et al., *Nucl. Instrum. Methods* A **355**, 394 (121)1997.
6. ZEUS Collaboration, *Leading baryon production in ep scattering at HERA*, contributed paper 789 to the XXIX International Conference on High Energy Physics ICHEP98, Vancouver, Canada, 23-29 July 1998.
7. ZEUS Collaboration, *Leading baryon production in ep scattering at HERA*, contributed paper 3-501 to the International Europhysics Conference on High Energy Physics 99, Tampere, Finland, 15-21 July 1999.

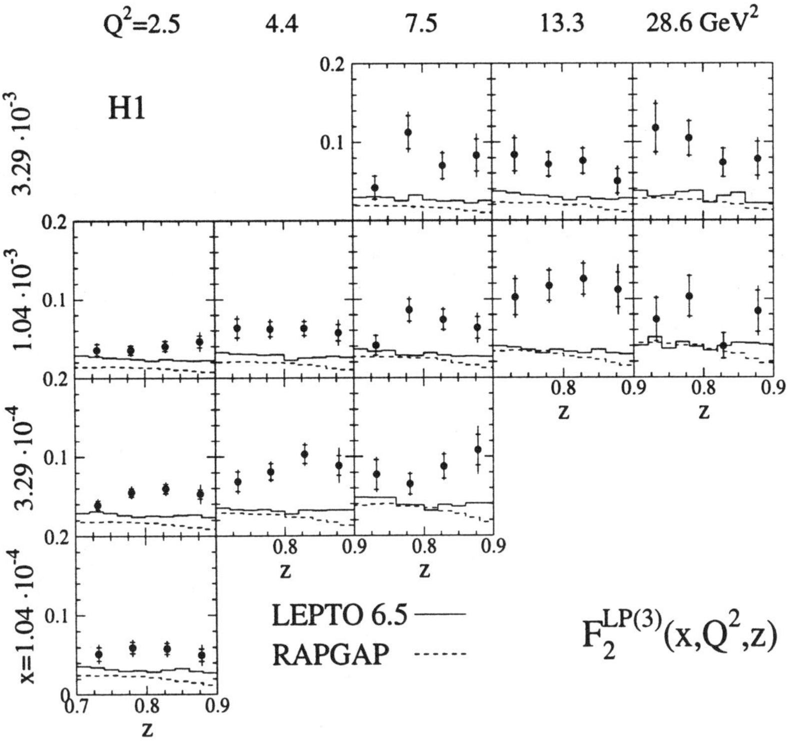

Figure 1: The H1 measurement of $F_2^{LP(3)} = \int F_2^{LP(4)} dp_T^2$ for protons with $p_T < 0.2$ GeV compared to the predictions of the LEPTO Monte Carlo with soft color interactions and RAPGAP Monte Carlo only with pion exchange. Note that $z \equiv x_L$.

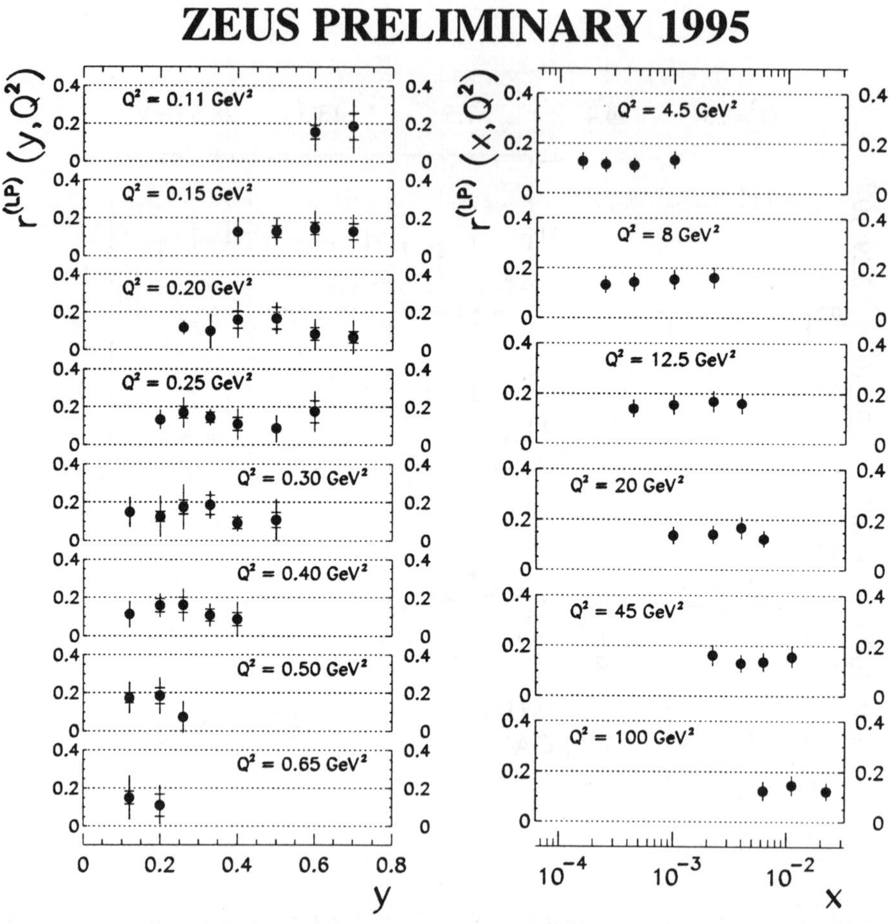

Figure 2: The fraction of events with a leading proton satisfying $0.6 < x_L < 0.97$ and $p_T^2 < 0.5$ GeV^2 in bins of Q^2 as a function of y (low Q^2 sample) or x (high Q^2 sample). In terms of structure functions $r^{LP} = (\int F_2^{LN(4)} dx_L dp_T^2)/F_2$. Within experimental errors r^{LP} does not depend on the leptonic variables x, y, Q^2.

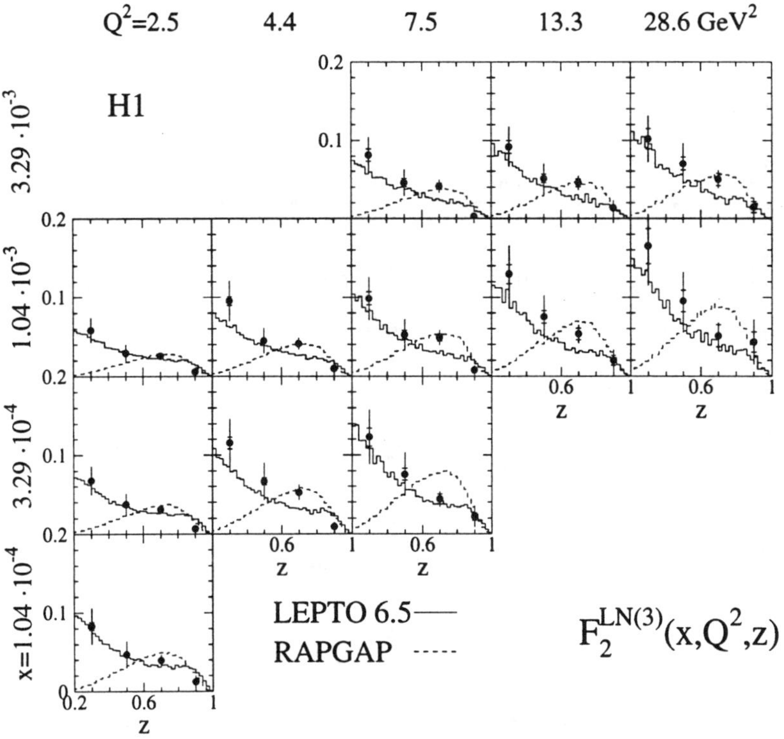

Figure 3: The H1 measurement of $F_2^{LN(3)} = \int F_2^{LN(4)} dp_T^2$ for neutrons with $p_T < 0.2$ GeV compared to the predictions of the LEPTO Monte Carlo with soft color interactions and RAPGAP Monte Carlo model only with pion exchange. Note that $z \equiv x_L$.

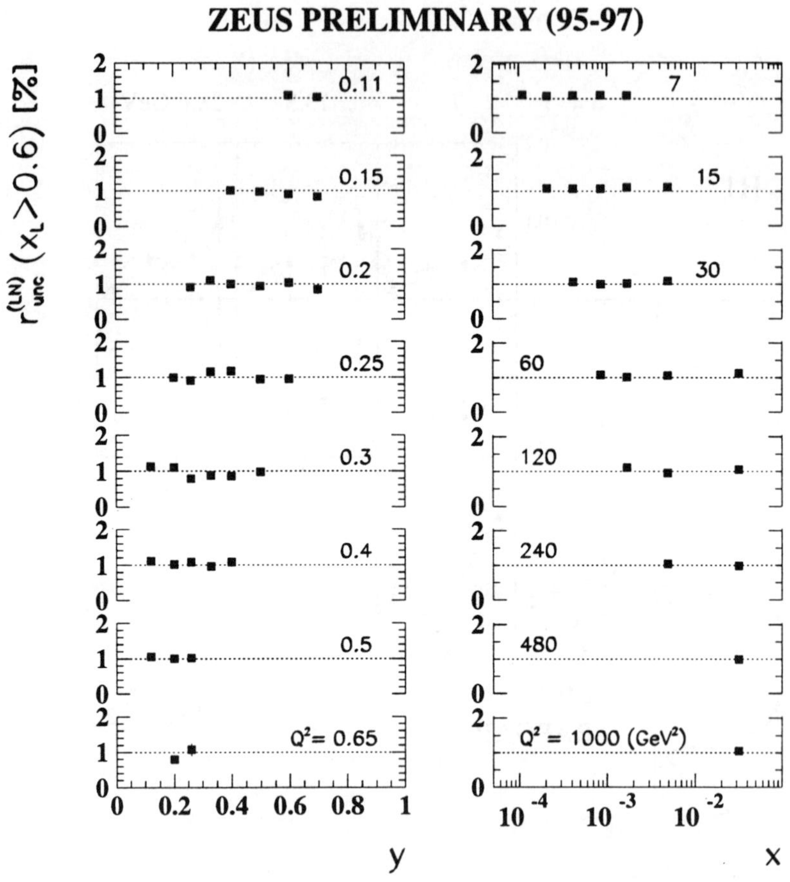

Figure 4: The observed fraction of events with a leading neutron satisfying $0.6 < x_L < 1$ and scattering angle $\theta < 0.8$ mrad in bins of Q^2 as a function of y (low Q^2 sample) or x (high Q^2 sample). The fraction is uncorrected for the acceptance of the FNC; in terms of structure functions $r^{LN}_{unc} \propto (\int F_2^{LN(4)} dx_L dp_T^2)/F_2$. Within experimental errors r^{LN}_{unc} does not depend on the leptonic variables x, y, Q^2.

TOPICAL RESULTS ON VECTOR-MESON PRODUCTION FROM THE HERA COLLIDER EXPERIMENTS[a]

James A. Crittenden
for the H1 and ZEUS collaborations

Physikalisches Institut, University of Bonn, Nußallee 12, 53115 Bonn, Germany

The HERA collider experiments H1 and ZEUS have established extensive measurement programs for diffractive vector-meson production processes during the first six years of their operation. The results provide stringent phenomenological tests of quantum chromodynamical descriptions of hard diffraction. We discuss recent topical results on Υ photoproduction, on decay-angle analyses of ρ^0, ϕ and J/ψ electroproduction and on ρ^0 and ϕ photoproduction at high momentum transfer.

1 Introduction

The HERA collider experiments H1 and ZEUS have developed measurement programs of diffractive vector-meson production in electron-proton collisions, extending the rich history of such investigations in the 1960s and 1970s to the modern era. Many of the topics addressed by the data have as sole precedent investigations performed before the development of quantum chromodynamics as the fundamental theory of the strong interaction. As a result, recent years have witnessed rapid expansion in a field of study in which the predictions of an asymptotically free field theory are tested for the first time in new domains of energy and photon virtuality. In light of the recent impressive theoretical successes in describing hard diffractive phenomena, in particular the factorization theorems proven for exclusive [1] and semi-exclusive [2] vector-meson production, these new measurements are providing an extensive phenomenology useful in testing the precepts of quantum chromodynamics in their application to diffractive processes.

In this report, we discuss three topics of particular interest to this workshop: the first measurements of Υ photoproduction, the results on the helicity structure of ρ^0, ϕ and J/ψ electroproduction and the semi-exclusive photoproduction of ρ^0 and ϕ mesons at high momentum transfer.

---

[a]Talk presented at the Joint INT/Jefferson Lab Workshop on Exclusive and Semi-Exclusive Processes at High Momentum Transfer, 20-22 May 1999, Jefferson Lab, Newport News, Virginia, USA

2 Υ Photoproduction

In stark contrast to the first published observation of photoproduction of the J/ψ meson a few months following its discovery in 1974, twenty years passed between the discovery of the Υ meson and the first report of its production in photon-proton interactions by the ZEUS experiment in 1997.[3] This measurement of exclusive Υ photoproduction at a photon-proton center-of-mass energy, W, of 120 GeV, has since been confirmed by the H1 collaboration, who have released a preliminary result[4] for the cross section at W=160 GeV. Each experiment identifies the Υ family via a resonant structure in the observed dimuon mass spectrum, which is dominated by the the continuum background from the Bethe-Heitler pair-production process. Figure 1 shows the preliminary measurement of this mass spectrum reported by H1. The H1 spectrum shows a peak which leads to an estimated signal of 8.3 ± 3.9 events. The mass resolution provided by the central tracking chambers and magnetic field does not allow the separation of the Υ, Υ' and Υ'' in either of the experiments, requiring the use of estimates of their relative production in order to extract

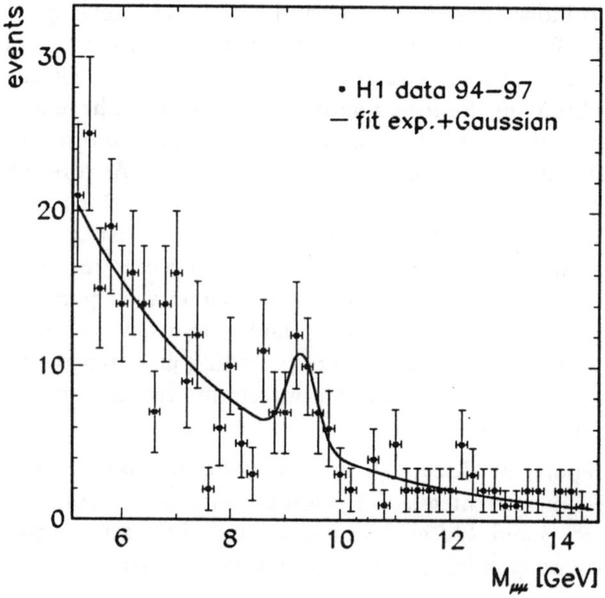

Figure 1: Invariant mass spectrum for diffractively photoproduced muon pairs reported by the H1 collaboration.[4] The curve shows the result of a fit to a sum of exponential and Gaussian functions

a production cross section for the Υ alone. The resulting determinations of the cross section for the exclusive process $\gamma p \to \Upsilon(1S)p$, shown in Fig. 2, exceed the leading-order QCD estimate[5] by a factor of 5-10. Detailed theoretical

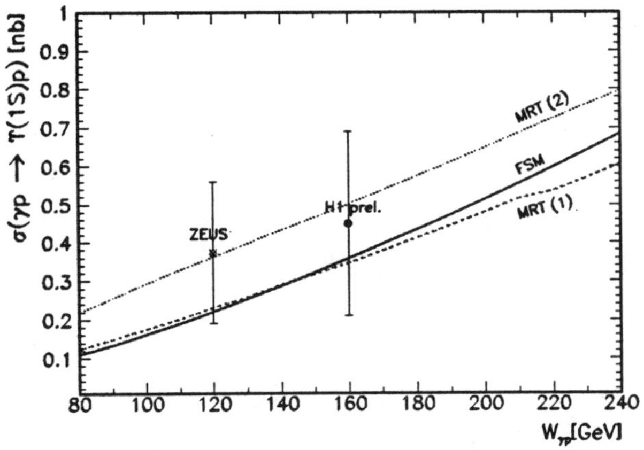

Figure 2: Measurements from the H1 and ZEUS collaborations of the elastic $\Upsilon(1S)$ photoproduction cross section. The error bars show the quadratic sum of statistical and systematic uncertainties. The curves show the results of QCD-based calculations which take into account a variety of effects beyond leading order[6]

investigations[6] into reasons for the large cross section have emphasized the importance of the evolution of the parton density functions (which are necessarily skewed by the kinematic lower bound on the momentum transferred to the proton resulting from the rest mass of the $\Upsilon(1S)$) and that of the contribution from the real part of the amplitude.

These initial measurements establishing the large cross section were based on an integrated luminosity of about 40 pb^{-1} for each experiment. The HERA luminosity upgrade program and operation schedule through 2005 will permit an increase in the Υ candidate event sample of at least a factor of twenty and is likely to enable decay-angle studies of the type described below for the lighter vector mesons.

3 Helicity Analyses of ρ^0, ϕ and J/ψ Electroproduction

The simplicity of the final state in the exclusive production of vector mesons permits detailed investigations of the decay-angle distributions, which have provided much information on the helicity structure of the diffractive production mechanism. These analyses of the decay-angle distributions have been employed as a means of measuring the ratio of the cross sections for longitudinal and transverse photons, R. This type of study provided another early success of the perturbative QCD models of exclusive diffractive processes by verifying the prediction that the longitudinal cross section exceed the transverse cross section at high Q^2.[2]

Three angles suffice to completely describe the exclusive electroproduction of vector mesons, as shown in Fig. 3 for the example of ρ^0 production: the azimuthal angle between the scattering plane and the production plane, Φ_h, and the two ρ^0 decay angles: ϕ_h, the azimuthal angle between the production

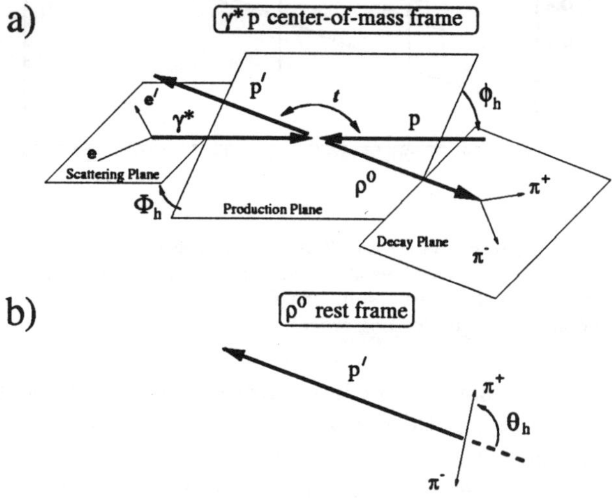

Figure 3: Schematic diagram of the process $ep \rightarrow e\rho^0 p$ (a) in the virtual-photon/proton center-of-mass system, and (b) in the rest frame of the ρ^0 meson

and decay planes, defined in either the virtual-photon/proton system or in the ρ^0 rest frame, and θ_h, which is the polar angle of the positively charged decay product, defined with respect to the direction of the ρ^0 momentum vector in the virtual-photon/proton system, which is opposite to the momentum vector of the final-state proton in the rest frame of the ρ^0 meson. This latter choice of spin-quantization axis defines the helicity frame, in which helicity conservation was found to hold approximately in exclusive ρ^0 photoproduction experiments at SLAC in the early 1970s.[7,8]

Following the work of Schilling and Wolf,[9] the three-dimensional angular distribution for this decay of a spin-1 state to two spinless particles has been parameterized as follows:

$$
\begin{aligned}
W(\cos\theta_h,\phi_h,\Phi_h) = \frac{3}{4\pi}\Bigg[&\frac{1}{2}(1-r_{00}^{04})+\frac{1}{2}(3r_{00}^{04}-1)\cos^2\theta_h \\
&-\sqrt{2}\,\mathrm{Re}\{r_{10}^{04}\}\sin 2\theta_h\cos\phi_h - r_{1-1}^{04}\sin^2\theta_h\cos 2\phi_h \\
&-\epsilon\cos 2\Phi_h(r_{11}^{1}\sin^2\theta_h + r_{00}^{1}\cos^2\theta_h - \sqrt{2}\,\mathrm{Re}\{r_{10}^{1}\}\sin 2\theta_h\cos\phi_h \\
&\qquad - r_{1-1}^{1}\sin^2\theta_h\cos 2\phi_h) \\
&-\epsilon\sin 2\Phi_h(\sqrt{2}\,\mathrm{Im}\{r_{10}^{2}\}\sin 2\theta_h\sin\phi_h + \mathrm{Im}\{r_{1-1}^{2}\}\sin^2\theta_h\sin 2\phi_h) \\
&+\sqrt{2\epsilon(1+\epsilon)}\cos\Phi_h(r_{11}^{5}\sin^2\theta_h + r_{00}^{5}\cos^2\theta_h \\
&\qquad -\sqrt{2}\,\mathrm{Re}\{r_{10}^{5}\}\sin 2\theta_h\cos\phi_h - r_{1-1}^{5}\sin^2\theta_h\cos 2\phi_h) \\
&+\sqrt{2\epsilon(1+\epsilon)}\sin\Phi_h(\sqrt{2}\,\mathrm{Im}\{r_{10}^{6}\}\sin 2\theta_h\sin\phi_h \\
&\qquad +\mathrm{Im}\{r_{1-1}^{6}\}\sin^2\theta_h\sin 2\phi_h)\Bigg], \qquad (1)
\end{aligned}
$$

where the superscripts of the combinations of spin-density matrix elements correspond to the helicity degrees of freedom of the virtual photon, and the subscripts to those of the dipion state. The fifteen coefficients r_{ik}^{04}, r_{ik}^{α} are related directly to various combinations of the helicity amplitudes, $T_{\lambda_\rho\lambda_\gamma}$, where λ_ρ and λ_γ are the helicities of the ρ^0 meson and of the photon, respectively. The assumption that helicity is conserved in the photon/vector-meson transition when the amplitudes are defined in the helicity frame ("s-channel helicity conservation" or "SCHC"), with the consequence that the degree of ρ^0 polarization is equal to the ratio of the longitudinal and transverse cross sections, allows the extraction of this ratio R from the distribution in polar angle alone. A compilation of determinations of R via this method is shown in Fig. 4. The data from the fixed-target muon experiment at FNAL, E665,[10] and the low-Q^2 data from the ZEUS collaboration identify a region of transition to increasing values of R, with the longitudinal cross section becoming dominant for $Q^2 \gtrsim 2$ GeV^2. Together with the latest results from the H1 collaboration,[11] these data indicate that the value of R reaches a plateau for $Q^2 \gtrsim 5$ GeV^2.

The H1 [11] and ZEUS [12] collaborations have recently completed analyses of the three-dimensional angular distribution for data samples of a few thousand events, extracting the fifteen coefficients r_{ik}^{04}, r_{ik}^{α}. Figure 5 shows the results from the ZEUS collaboration in the kinematic region $3 < Q^2 < 30$ GeV^2, $40 < W < 120$ GeV and $|t| < 0.6$ GeV^2, comparing them to the results from the H1 collaboration in a similar kinematic region, and to a calculation by Ivanov and Kirschner [13] (solid line). The inner error bars represent the statistical uncertainty; the outer error bars show the quadratic sum of statistical and

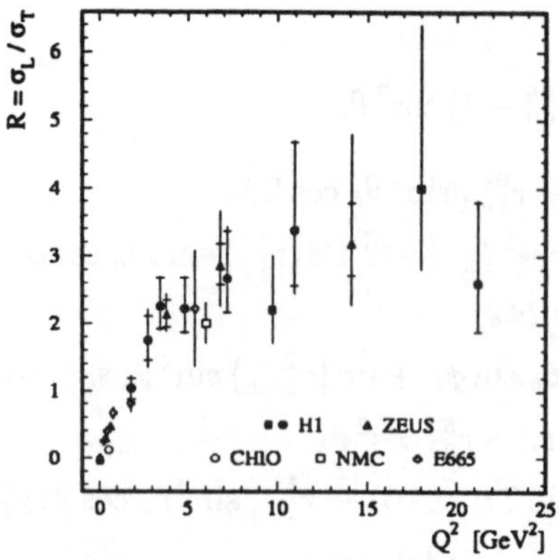

Figure 4: Measurements of the ratio of exclusive ρ^0 electroproduction cross sections for longitudinal and transverse photons, R, as a function of photon virtuality, Q^2

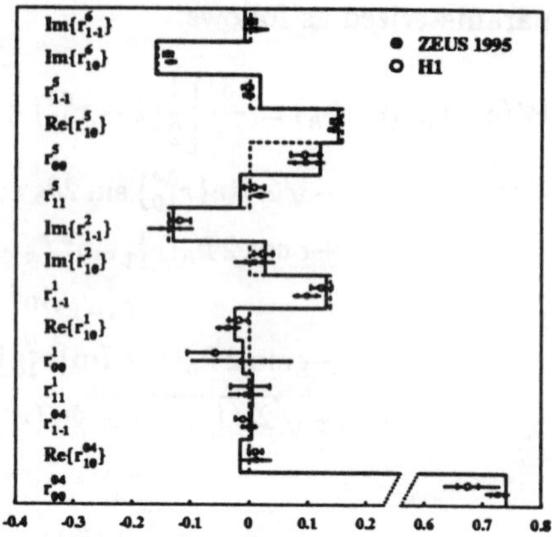

Figure 5: Combinations of spin-density matrix elements measured for ρ^0 electroproduction by the H1 [11] and ZEUS [12] collaborations. See text for full description

systematic uncertainties. Also shown are the values of the coefficients predicted according to the hypothesis of helicity conservation in the s-channel amplitudes [9] (dashed line). Of particular interest is the violation of SCHC evinced in the non-zero value for the coefficient r^5_{00}, which is manifested in the distribution in the azimuthal angle between the scattering and production planes (Φ_h). This observation implies a small non-zero single-flip amplitude for the production of ρ^0 mesons in helicity state 0 from transverse photons. Such a violation has now been reproduced in two further recent calculations based on differing assumptions.[14] These measurements show the level of violation of SCHC to be small enough that the effect of the assumption of SCHC in the earlier determinations of R is much smaller than the other sources of uncertainty in those determinations.

The H1 collaboration has recently extended their investigation to the proton-dissociative electroproduction of ρ^0 mesons;[15] verifying with similar significance that the violation of SCHC observed for the exclusive case holds also when the proton dissociates.

Preliminary results of a full decay-angle analysis with similar statistical power for exclusive ϕ electroproduction in the same kinematic region have been presented by the ZEUS collaboration last year.[16] They find values for the fifteen combinations of matrix elements consistent with those found for the ρ^0 meson, exhibiting a similar helicity violation and comparable values for R.

The ZEUS collaboration also performed the analysis [12] in the low-Q^2 region $0.25 < Q^2 < 0.85$ GeV^2, finding a value for r^5_{00} of similar magnitude, but also distinct indications for a more complex pattern of helicity violation, including double-flip contributions, as previously found in ρ^0 photoproduction (see Sect. 4).

Statistical limitations have precluded such full decay-angle analyses for J/ψ electroproduction; however, both the H1 [11] and ZEUS [17] collaborations have published measurements of R for the exclusive electroproduction of J/ψ mesons under the assumption of SCHC. The values are found to be significantly smaller than those found for the ρ^0 at similar Q^2.

4 Photoproduction of ρ^0, ϕ Mesons at High Momentum Transfer

The photoproduction of light vector mesons with high transverse momenta has raised interest recently [18,2] as a means of investigating the rôle of the momentum transferred to the proton in establishing the scale of the interaction, since the photoproduction of ρ^0 mesons at low $|t|$ has been shown to be governed by a soft diffractive production mechanism. However, diffractive photoproduction of the light vector mesons has been shown to be dominated by the proton-dissociative process for squared momentum transfers $|t| \gtrsim 0.5$ GeV^2, well below the perturbative regime.[19] The investigation of exclusive vector-meson production at high $|t|$ requires a trigger on the elastically scattered proton and such studies at HERA have lacked the integrated luminosity necessary to measure the small cross sections at such high values of $|t|$. However, the ZEUS collaboration [20] has recently employed a photoproduction-tagging method to investigate the proton-dissociative production of ρ^0 mesons for values of $|t|$ up to 11 GeV^2. The trigger conditions required the scattered positron to be detected in a special-purpose tungsten/scintillator calorimeter located 3 cm from the positron beam axis, 44 meters distant from the nominal e^+p interaction point in the positron-beam flight direction. The position of this photoproduction tagger determines the accepted range of energy lost by the positron to the photon which interacts with the proton, thus restricting the W range to the region $80 < W < 120$ GeV. Since the transverse momentum of the final-state positron is thus required to be small ($Q^2 < 0.01$ GeV^2), the transverse momentum of the ρ^0 (p_t) detected in the central detector via its dipion decay provides an accurate approximation for the square of the four-momentum transferred to the proton: $t \simeq -p_t^2$. Offline data-selection criteria include the reconstruction of exactly two tracks from the interaction vertex and reject events with calorimetric energy deposits in the rear and barrel sections of the calorimeter which are not associated with the extrapolation of either track. The selected events exhibit a substantial rapidity gap between the dissociated nucleonic system and the two tracks. Even for the highest values of $|t|$, the decay pions are in the rear half of the central detector.

Figure 6 shows the differential cross sections, $\frac{d\sigma}{dt}$, measured for ρ^0 and ϕ mesons. These exhibit extremely hard spectra; a fit to the form $\frac{d\sigma}{dt} \propto (-t)^{-n}$ results in a value for n of approximately 3. The results are compared to a QCD calculation by Ivanov and Ginzburg [2] which estimates both perturbative and non-perturbative contributions. In this model, the non-perturbative contribution is found to account for the hardness of the spectrum and to dominate in the region covered by the measurements for the ρ^0 meson. The calculation underestimates the ϕ cross section over most of the t range covered. The comparison of the data to the calculation of the purely perturbative contribution

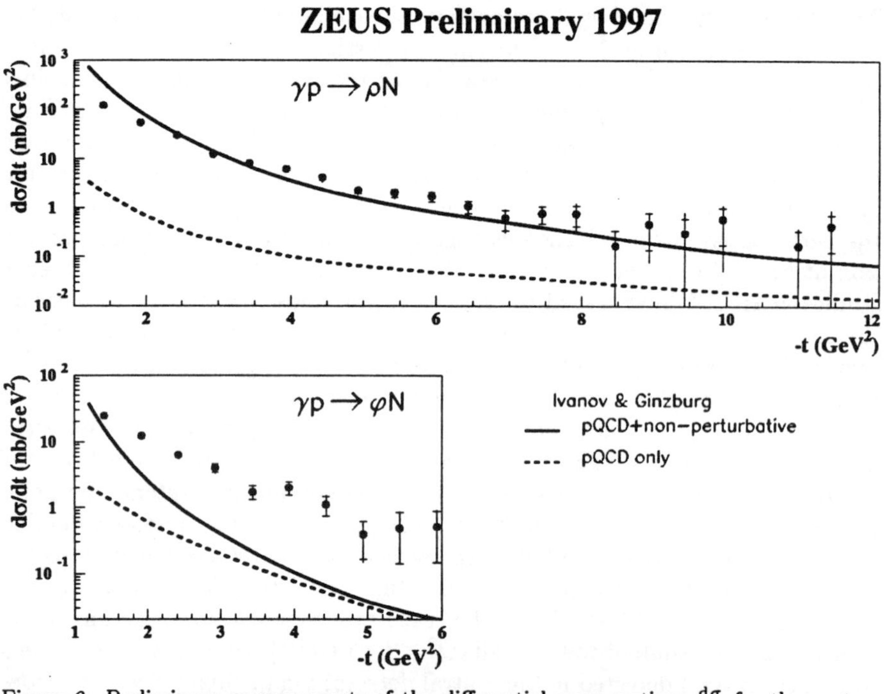

Figure 6: Preliminary measurements of the differential cross sections $\frac{d\sigma}{dt}$ for the proton-dissociative photoproduction of ρ^0 and ϕ mesons in the energy range $80 < W < 120$ GeV recently presented by the ZEUS collaboration.[20] The inner error bars represent the statistical uncertainties; the outer error bars show the quadratic sum of statistical and systematic uncertainties. The results of a QCD calculation by Ivanov and Ginzburg [2] (solid line) which distinguishes the purely perturbative contribution (dashed line) are shown for comparison

alone shows a good description of the shape of the t dependence, whereas the magnitude of the cross section is underestimated by more than one order of magnitude. The model thus fails to reproduce the ratio of ϕ production to that for the ρ^0 meson at high $|t|$, which is measured to be consistent with the value of 2/9 expected for a flavor-independent production mechanism.

The ZEUS collaboration has employed this photoproduction-tagging technique to perform decay-angle analyses of the diffractive photoproduction [21,20] of pion pairs in the ρ^0 mass region at values of $|t|$ up to 4 GeV^2. In this study, the small contribution from the elastic process was not subtracted. The decay-angle distribution was parameterized in terms of combinations of spin-density matrix elements in the Schilling-Wolf convention,[9] r_{ij}^{04}, as

$$\begin{aligned} W(\theta_h, \phi_h) = \frac{3}{4\pi}\Big[& \tfrac{1}{2}\left(1 - r_{00}^{04}\right) + \tfrac{1}{2}\left(3r_{00}^{04} - 1\right)\cos^2\theta_h \\ & -\sqrt{2}\mathrm{Re}(r_{10}^{04})\sin 2\theta_h \cos\phi_h - r_{1-1}^{04}\sin^2\theta_h \cos 2\phi_h \Big], \end{aligned} \quad (2)$$

where the three-dimensional distribution (Eq. 1) has been averaged over the unmeasured azimuthal angle between the positron scattering plane and the ρ^0 production plane, and thus no longer distinguishes the photon helicity states ± 1.

Under the assumption that the dipion final state is produced with one unit of angular momentum, and neglecting the contribution by longitudinal photons, these combinations of matrix elements are related to the helicity amplitudes, $T_{\lambda_\rho \lambda_\gamma}$, as follows[9,12]:

$$r_{00}^{04} \simeq \frac{|T_{01}|^2}{|T_{01}|^2 + |T_{11}|^2 + |T_{1-1}|^2}, \qquad r_{1-1}^{04} \simeq \frac{\mathrm{Re}(T_{11}T_{1-1}^*)}{|T_{01}|^2 + |T_{11}|^2 + |T_{1-1}|^2},$$

$$\mathrm{Re}(r_{10}^{04}) \simeq \frac{1}{2}\left[\frac{\mathrm{Re}(T_{11}T_{01}^*) + \mathrm{Re}(T_{1-1}T_{0-1}^*)}{|T_{01}|^2 + |T_{11}|^2 + |T_{1-1}|^2}\right]. \quad (3)$$

Figure 7 shows the results for the combinations of matrix elements obtained from a least-squares minimization procedure in which they served as fit parameters. The inner error bars represent the statistical uncertainty; the outer error bars show the quadratic sum of statistical and systematic uncertainties. The systematic uncertainties are dominated by the uncertainty in the acceptance corrections. The dipion mass range was restricted to the region $0.45 < M_{\pi\pi} < 1.1$ GeV. The results are compared to the results at lower $|t|$ for the exclusive reaction obtained with 9 GeV photons from a backscattered laser beam incident on a hydrogen bubble chamber at SLAC.[8] Also shown are the ZEUS 1994 results for exclusive ρ^0 photoproduction at low $|t|$.[22] The parameter

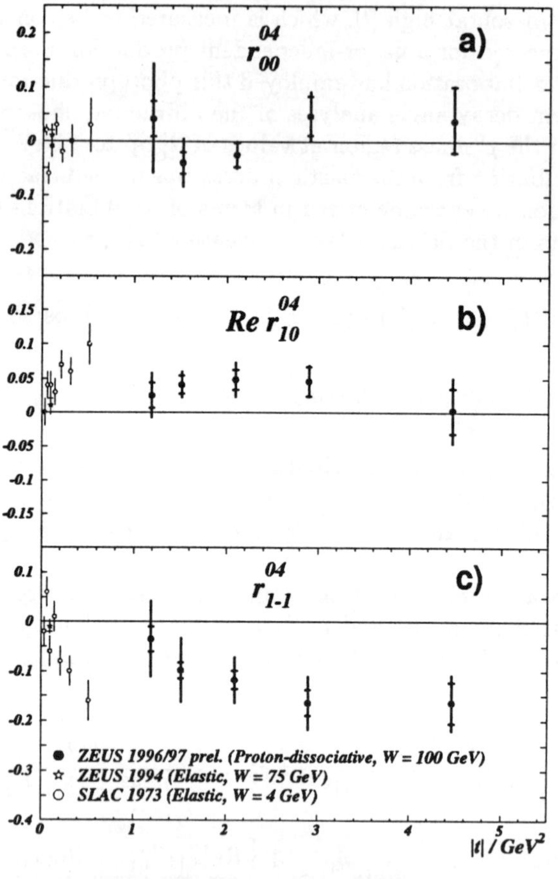

Figure 7: Recent measurements by the ZEUS collaboration [21,20] of the combinations of matrix elements (a) r_{00}^{04}, (b) Re r_{10}^{04}, (c) r_{1-1}^{04} for the diffractive photoproduction of pion pairs. The inner error bars represent the statistical uncertainties; the outer error bars show the statistical and systematic uncertainties added in quadrature. See text for full description

r_{00}^{04} is consistent with zero over the entire range in $|t|$, showing no evidence for the production of ρ^0 mesons in helicity state 0. The combination Re r_{10}^{04}, which is predominantly sensitive to the interference between the helicity-conserving amplitude and the single-flip amplitude, shows clear evidence for a small single-flip contribution in both the SLAC data and the high-$|t|$ ZEUS results. A clear indication of a sizeable double-flip contribution is shown by the measurements of r^{04}_{1-1} at high $|t|$, as was seen in the SLAC results at lower $|t|$ and lower energy.

In order to estimate the effect on the angular distributions of a hypothesized dipion background to ρ^0 decay, the decay-angle analysis was repeated for restricted dipion mass ranges above ($0.77 < M_{\pi\pi} < 1.0$ GeV) and below ($0.6 < M_{\pi\pi} < 0.77$ GeV) the nominal value for the ρ^0 mass. The observed value for each of the combinations of matrix elements was found to depend significantly on the mass range chosen, though the above conclusions concerning the lack of longitudinal polarization and the evidence for a double-flip contribution remain unchanged. This dependence on dipion invariant mass may suggest the presence of a non-resonant background. The extraction of the spin-density matrix elements for the ρ^0 meson alone from these dipion angular distributions thus awaits the understanding of this dependence.

The complexity of the helicity structure in pion-pair photoproduction is thus shown to persist at high values of $|t|$, where the hardness of the $|t|$ spectrum and the value of the ϕ/ρ^0 ratio encourage attempts to describe the semi-exclusive electroproduction of vector mesons in the framework of perturbative QCD.

5 Concluding Remarks

The selection of results on diffractive vector-meson production from the HERA collider experiments presented above exemplifies ongoing measurement programs undertaken by the H1 and ZEUS collaborations. The luminosity upgrade program scheduled for the year 2000 is expected to provide an increase in instantaneous luminosity of roughly a factor of five. This increase is of particular importance for investigations such as those covered above in the context of this workshop: Υ photoproduction, complete helicity analyses of vector-meson electroproduction, and semi-exclusive vector-meson photoproduction at high momentum transfer, since their accuracy is statistically limited. The future operation of HERA is therefore certain to continue to provide new results on exclusive and semi-exclusive processes at hard scales, ensuring a rigorous series of tests for new applications of quantum chromodynamics.

Acknowledgments

The author thanks the organizers for their hospitality during the workshop. Figure 2 was provided by P. Merkel. This work is supported by the Federal Ministry of Education and Research of Germany.

References

1. J. C. Collins, L. Frankfurt and M. Strikman, Phys. Rev. **D56** (1997) 2982.
2. D. Yu. Ivanov, Phys. Rev. **D53** (1996) 3564; I. F. Ginzburg and D. Yu. Ivanov, Phys. Rev. **D54** (1996) 5523; S. J. Brodsky, M. Diehl, P. Hoyer and S. Peigné, Phys. Lett. **B449** (1999) 306.
3. ZEUS Collaboration, J. Breitweg *et al.*, Phys. Lett. **B437** (1998) 432.
4. H1 Collaboration, C. Adloff *et al.*, Paper submitted to the XXIX International Conference on High-Energy Physics, 23-29 July 1998, Vancouver, Canada (Abstract 574).
5. M. F. McDermott, hep–ph/9905288 (1999).
6. L. L. Frankfurt, M. F. McDermott and M. Strikman, JHEP 9902 (1999) 002; A. D. Martin, M. G. Ryskin and T. Teubner, Phys. Lett. **B454** (1999) 339.
7. J. Ballam *et al.*, Phys. Rev. **D5** (1972) 545.
8. J. Ballam *et al.*, Phys. Rev. **D7** (1973) 3150.
9. K. Schilling and G. Wolf, Nucl. Phys. **B61** (1973) 381.
10. E665 Collaboration, M. R. Adams *et al.*, Z. Phys. **C74** (1997) 237.
11. H1 Collaboration, C. Adloff *et al.*, DESY Report 99-010 (1999), hep-ex/9902019, to appear in Eur. Phys. J.
12. ZEUS Collaboration, J. Breitweg *et al.*, DESY Report 99-102 (1999), hep-ex/9908026, to appear in Eur. Phys. J.
13. D. Yu. Ivanov and R. Kirschner, Phys. Rev. **D58** (1998) 114026.
14. Talks by I. Royen and N. Nikolaev, 7th International Conference on Deep Inelastic Scattering and QCD, 19-23 April, 1999, Zeuthen, Germany.
15. H1 Collaboration, C. Adloff *et al.*, Paper submitted to the International Europhysics Conference on High-Energy Physics, 15–21 July, 1999, Tampere, Finland (Paper 157n).
16. ZEUS Collaboration, J. Breitweg *et al.*, Paper submitted to the XXIX International Conference on High-Energy Physics, 23-29 July 1998, Vancouver, Canada (Abstract 792/3a).
17. ZEUS Collaboration, J. Breitweg *et al.*, Eur. Phys. J. **C6** (1999) 603.
18. J. Bartels, J. R. Forshaw, H. Lotter and M. Wüsthoff, Phys. Lett. **B375** (1996) 301.
19. ZEUS Collaboration, J. Breitweg *et al.*, Paper submitted to the XXIX International Conference on High-Energy Physics, 23-29 July 1998, Vancouver, Canada (Abstract 788).
20. ZEUS Collaboration, J. Breitweg *et al.*, Paper submitted to the International Europhysics Conference on High-Energy Physics, 15–21 July,

1999, Tampere, Finland (Abstract 499).
21. J. A. Crittenden, for the ZEUS Collaboration, contributed to the proceedings of the 7th International Conference on Deep Inelastic Scattering and QCD, 19-23 April, 1999, Zeuthen, Germany, hep-ex/9906005.
22. ZEUS Collaboration, J. Breitweg *et al.*, Eur. Phys. J. **C2** (1998) 247.

PARTON-HADRON DUALITY AND NUCLEON VALENCE STRUCTURE

C.E. Keppel
Department of Physics, Hampton University, Hampton, VA 23668,
and Physics Division, Jefferson Lab, Newport News, VA 23606
E-mail: keppel@jlab.org

A newly-obtained data sample of inclusive electron-nucleon scattering from deuterium and hydrogen targets at Jefferson Lab has been analyzed for precision tests of quark-hadron duality. In all cases, duality appears to be a non-trivial dynamic property of the nucleon structure function. Assuming duality, the proton magnetic form factor is extracted from the inelastic data alone, and found to be in agreement with the world's data. Higher twist contributions are found to be small on average, even down to $Q^2 \approx 0.5$ GeV^2. The investigation yields a scaling curve from duality arguments which resembles deep-inelastic neutrino-nucleus scattering data, indicating a potential sensitivity to valence and valence-like structure.

1 Introduction

The interpretation of the resonance region in inclusive electron-proton scattering and its possible connection with deep inelastic scattering has been a subject of interest for nearly three decades since quark-hadron duality ideas, which successfully described hadron-hadron scattering, were first extended to electroproduction. Bloom and Gilman [1] showed that it was possible to equate the nucleon resonance region structure function $\nu W_2(\nu, Q^2)$ (at some typically low Q^2 value) to the structure function F_2 in the deep inelastic regime of electron-quark scattering (at some higher value of Q^2). The resonance structure function was demonstrated to be equivalent in average to the deep inelastic one, with these averages obtained over the same range in a scaling variable $\omega' = 1 + W^2/Q^2$.

This relationship between resonance electroproduction and the scaling behavior observed in deep inelastic scattering suggests a common origin for both phenomena. Inclusive deep inelastic scattering on nucleons is a firmly-established tool for the investigation of the quark-parton model. At large enough values of W and Q^2, QCD provides a rigorous description of the physics that generates the Q^2 behavior of the nucleon structure function $F_2 = \nu W_2$. An analysis of the resonance region in terms of QCD was initially presented [2,3] where Bloom and Gilman's approach was re-interpreted, and the integrals of the average scaling curves were equated to the non-singlet $n = 2$ QCD moments of the F_2 structure function. The moments can be expanded, according to the operator product expansion, in powers of $1/Q^2$, and the fall of the resonances along a smooth scaling curve with increasing Q^2 was explained in terms of this QCD twist expansion of the structure function. Changes in the lower moments of the F_2 structure function due to higher twist effects are small, so that averages of this function over a sufficient range in x at moderate and high Q^2 are approximately the same [2]. Duality is expected to hold so long as $O(1/Q^2)$ or higher inverse power scaling violations are small or cancelling.

Substantial progress has been made both theoretically in understanding QCD in the past twenty years and experimentally in determining the scaling behavior of the F_2 structure function. Combining the latter with the new precision resonance data here presented [4], it is now possible to revisit quark-hadron duality with a more quantitative approach, addressing the recent theoretical interest in the topic [5,6,7,8,9,10,11].

2 Verification of Quark-Hadron Duality

Sample νW_2 spectra extracted from the measured differential cross sections from hydrogen are plotted in Fig. 1 as a function of the Nachtmann scaling variable $\xi = 2x/(1 + 4M^2x^2/Q^2)$. It has been shown that ξ is the correct variable to use in studying QCD scaling violations in the nucleon [12,2]. The arrows indicate values of $x = 1$ (elastic scattering) for the four values of Q^2 shown. The solid and dashed curves are from a parameterization [13] of deep inelastic proton structure function data at $Q^2 = 10$ and 5 GeV^2, respectively. Notice that the resonance spectra at different Q^2 appear at different ξ on the deep inelastic scaling curve, but that the curve generally represents an average of the data at the disparate kinematics. This is a manifestation of the original Bloom and Gilman observation.

To quantify the observed duality, we show in Fig. 2 the Q^2 dependence of the integral ratio quantity I(Res/DIS), the ratio of the structure function F_2 obtained from the resonance data, integrated over the region from pion threshold to the onset of the deep inelastic regime ($W^2 = 4$ GeV^2), compared to the indicated deep inelastic structure functions integrated over the same region of ξ:

$$I(Res/DIS) = \frac{\int_{\xi(W^2=4.0)}^{\xi(W^2=1.1)} F_2^{data}(\xi, Q_o^2)d\xi}{\int_{\xi(a)}^{\xi(b)} F_2^{scaling}(\xi, Q^2 = 10)d\xi}. \quad (1)$$

Here, $\xi(a, b)$ correspond to the same value of ξ used in the numerator integral, which at the higher Q^2 of the deep inelastic data no longer correspond to $W^2 = (4.0, 1.1)$, but to some higher $W^2 = (a, b)$. The range $1.1 \leq W^2 \leq 4.0$ GeV^2 is the resonance region of the data. In all cases but the Resonance fit (discussed below), the scaling region structure functions are integrated as a function of ξ at fixed $Q^2 = 10$ (GeV/c)^2. The higher W^2 deep inelastic data are, for the same region in ξ, at higher Q^2. The integral ratio data are plotted as a function of the fixed $Q^2 = Q_o^2$ values associated with the measured resonance spectra.

The uncertainties shown represent the experimental uncertainty in the numerator integral only, obtained from the correlated systematic uncertainties in the resonance data. The high Q^2 (above 4 (GeV/c)^2) numerator values are generated from a global fit to inclusive SLAC resonance region data [14] and are assigned an uncertainty representing a combination of experimental uncertainty and normalization considerations involved in utilizing the older data sets [15]. Also shown are very low Q^2 (< 0.5 (GeV/c)^2) resonance data from SLAC [16].

When used to obtain the deep inelastic denominator of the ratios in Fig. 2, the well-known CTEQ4 LQ [17] and MRS(G) LQ [13] scaling curves display a marked deviation from unity which increases with Q^2. This is not necessarily due to higher twist effects, but rather to the difficulty in accurately modeling the large ξ behavior of the structure function. With increasing Q^2, the moments are determined by a smaller and smaller region near $\xi = 1$. Unfortunately, there is a limited amount of deep inelastic F_2 data currently available at large ξ and x and these curves fall below both the average resonance data and the NMC parameterization above $\xi > 0.7$.

The points on Fig. 2 labeled NMC represent ratios obtained using a parameterization of lower x deep inelastic F_2 data from CERN [18], which links smoothly to a global fit [15] to higher x deep inelastic data from SLAC, in the denominator integral. Because this is a fit to the data, it may implicitly contain higher twist effects. Using the NMC parameterization, the ratio (Res/DIS) is above unity, but constant at about 1.2. The ratio decreases to about 1.1 when the appropriate logarthmic Q^2 dependence is included to obtain the parameterization at $Q^2 = 5$ (GeV/c)^2. This effect is indicated by the dashed line in the figure.

To obtain the points labeled Resonance fit in Fig. 2, a fit to the *average* strength of *all* the hydrogen resonance spectra was obtained and utilized as a scaling curve. This approach assumes duality, and therefore that the average of the resonance data is equivalent to a proper scaling curve. For small bins in ξ, resonance νW_2 data were averaged, regardless of Q^2, W^2. These average points were then fit as a function of ξ only, and the resultant scaling curve was obtained using a form similar to the x-dependent part of the NMC parameterization [18]. The scaling curve denoted Resonance fit is, then:

$$F_2 = \xi^{0.870}(1-\xi)^{0.006}\,[0.005 - 0.058(1-\xi) -$$

$$0.017(1-\xi)^2 + 2.469(1-\xi)^3 - 0.240(1-\xi)^4\big] \,. \tag{2}$$

To constrain the fit in the kinematic regions beyond the scope of the data, data obtained from [16] were also used below $Q^2 = 0.5$ (GeV/c)^2 and $\xi = 0.2$. Resonance data in the range $5 < Q^2 < 8$ (GeV/c)^2 were generated from [14] and used to constrain the fit at large ξ. Note that the ratio I(Res/DIS) in Fig. 2 comparing the resonance strength to the Resonance fit scaling curve need not be unity here even though the curve was extracted from the resonance data. The average scaling curve strength at any given value of ξ was obtained from a kinematic range of data at variant values of W^2 and Q^2 oscillating around this curve. The individual spectra used to obtain the plotted ratios are, however, at the indicated fixed Q^2 values only.

Even at Q^2 values as low as ≈ 0.5, the integrated strengths of the spectra at *fixed* Q^2 are within 10% of the resonance average curve, obtained from data at numerous W^2 and Q^2 values, integrated over the same region in ξ. Similarly, the ratio I(Res/DIS) is a constant ≈ 1.2, within 10%, when the NMC deep inelastic parameterization is utilized. Additionally, it is indicated that this latter ratio would tend toward unity if it were possible to compare the resonance region data with lower Q^2 values of the deep inelastic. In the QCD-based explanation of duality [2], these results indicate that higher twist effects are surprisingly reduced if the data are integrated over the resonance region.

If the higher twist contributions are small or cancelling on average in the region of the data, then duality allows for the determination of the nucleon form factor from data obtained in the purely inelastic region. Fig. 3 depicts the proton magnetic form factor, G_M^p, extracted using the NMC and JLab scaling curves integrated over the *entire* range in ξ, i.e. from 1 to the deep inelastic. The integral obtained from the resonance data (which starts at pion threshold rather than $\xi = 1$) is then subtracted from the scaling integrals. The magnetic form factor is then extracted from the remaining integrated strength using:

$$G_M{}^2 = \frac{1 + 4M^2/\mu^2 Q^2}{1 + 4M^2/Q^2} \frac{2 - \xi_p}{\xi_p{}^2} \int_{\xi_{thr}}^{1} F_2 d\xi. \tag{3}$$

In both cases, the G_M^p extracted from the integrals are in remarkably good agreement with the Gari-Krümpelmann model [19] of the world's magnetic form factor data. For the JLab fit worst case, the proton magnetic form factor is reproduced to within 30% of the accepted value.

Additionally, the duality integral obtained more locally in restricted ξ ranges around the three prominent resonance enhancement regions observed in inclusive nucleon resonance electroproduction averages to the deep inelastic strength integrated over the same regions in ξ [4], manifesting duality behavior even in these limited ranges at low W^2, Q^2 where higher twist effects might be expected to be large.

3 Valence Sensitivity

It is to be noted that the duality-averaged scaling curve, given by Equation 2, has a valence-like shape. This is not unreasonable, given that the resonance data are at low Q^2 at low x, and high Q^2 at large x. To verify this, Fig. 4 shows a comparison of the similarly-averaged scaling curve from the deuterium resonance data (solid curve) with a selection of the world's data for the xF_3 structure function. The xF_3 structure function can be accessed by deep inelastic neutrino-iron scattering [20,21], and is associated with the parity-violating term in the hadronic current. Thus xF_3 measures in the quark-parton model the difference between quark and anti-quark distributions, and is to first order insensitive to sea quark distributions. To enable a direct comparison, the JLab average scaling curve is multiplied by a factor of 18/5 to account for the quark charges, and a straightforward nuclear correction to the xF_3 data is applied to obtain neutrino-deuterium data [22]. Although the agreement between the averaged F_2 scaling curve of the deuterium resonance region and the deep inelastic neutrino xF_3 data is not perfect, the similarity is striking. The observation of Bloom and Gilman that there may be a common origin between the electroproduction of resonances and deep inelastic scattering seems to be true for even the lowest values of Q^2 if one assumes sensitivity to a valence-like quark distribution only.

4 Conclusions

By utilizing new inclusive data in the resonance region at large x, it has been possible to revisit quark-hadron duality experimentally for the first time in nearly three decades. The original duality observations are verified, and the QCD moment explanation indicates that higher twist contributions to the F_2 structure function are small or cancelling, even in the low Q^2 regime. The magnetic form factor has been extracted from the inelastic data using duality techniques and is in good agreement with global measurements. Cross sections have been measured for both hydrogen and deuterium targets, and the F_2 structure function extracted from these oscillates around an average scaling curve, down to the lowest momentum transfers measured. This average curve resembles deep inelastic xF_3 structure function data, indicating a sensitivity to valence-like structure.

References

1. E.D. Bloom and F.J. Gilman, Phys. Rev. D **4** (1970) 2901
2. A. De Rujula, H. Georgi, and H.D. Politzer, Ann. Phys. **103** (1977) 315
3. H. Georgi and H.D. Politzer, Phys. Rev. D **14** (1976) 1829
4. I. Niculescu, Ph.D. Thesis, Hampton University, 1999
5. X. Ji and P. Unrau, Phys. Rev. D **52** (1995) 73
6. V.M. Belyaev and A. Radyushkin, Phys. Lett. **B359** (1995) 194
7. G. West, hep-ph/9612403 (1996)
8. G. Ricco *et al.*, Phys. Rev. C **57** (1998) 356
9. C. Coriano, H-N. Li, and C. Savkli, J. High En. Phys. 9807 (1998) 008
10. C. Carlson and N. Mukhopadhyay, Phys. Rev. Lett. **74** (1995) 1288
11. M. Anghinolfi *et al.*, Nucl. Phys. **A602** (1996) 405
12. O. Nachtmann, Nucl. Phys. **B63** (1975) 237
13. A.D. Martin, R.G. Roberts, and W.J. Stirling, Phys. Rev. D **50** (1994) 6734
14. C.E. Keppel, Proceedings of the Workshop on CEBAF at Higher Energies, Eds. N. Isgur and P. Stoler (1994) 237
15. L.W. Whitlow *et al.*, Phys. Lett. **B282** (1992) 475
16. S. Stein *et al.*, Phys. Rev. D **12** (1975) 1884
17. H.L. Lai *et al.*, Phys. Rev. D **55** (1997) 1280
18. M. Arneodo *et al.*, Phys. Lett. **B364** (1995) 107
19. M. Gari and W. Krümpelmann, Phys. Lett. **B141** (1984) 295
20. P. Berge *et al.*, Z. Phys. **C49** (1991) 187.
21. E. Oltman *et al.*, Z. Phys. **C53** (1992) 51; J.H. Kim *et al.*, Phys. Rev. Lett. **81** (1998) 3595.
22. J.J. Aubert *et al.*, Nucl. Phys. **B293** (1987) 740

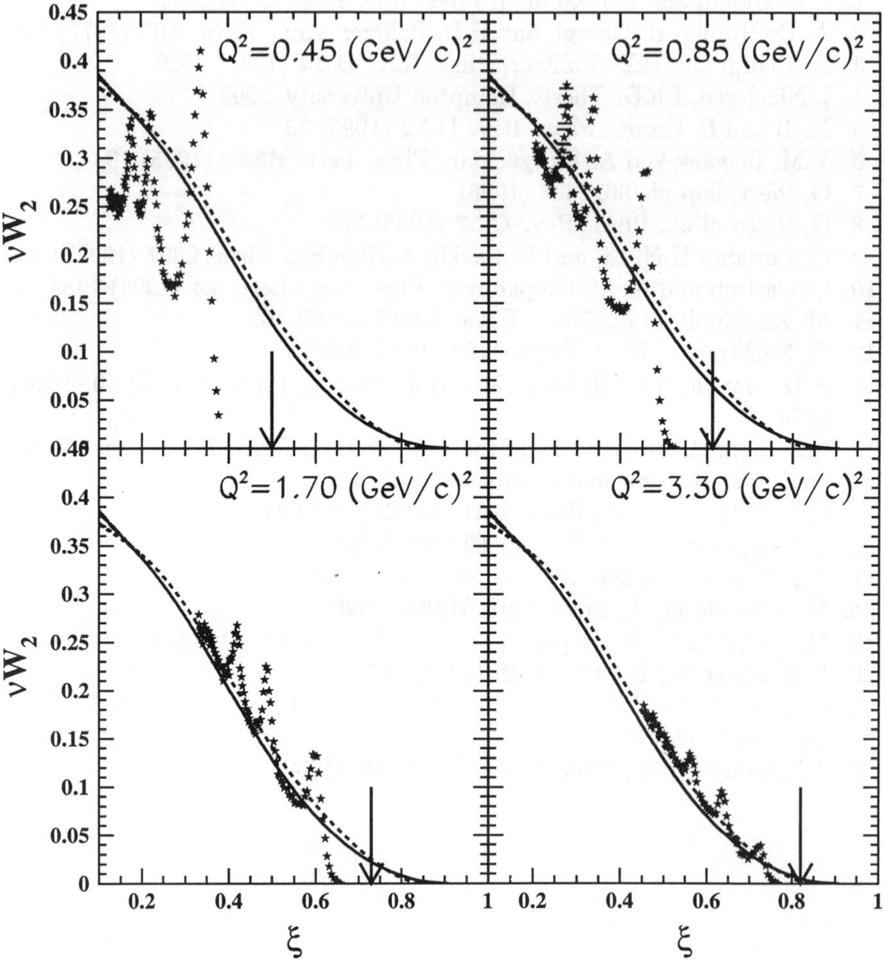

Figure 1: Sample hydrogen νW_2 structure function spectra obtained at $Q^2 = 0.45, 0.85, 1.70,$ and 3.30 GeV^2. Arrows indicate elastic kinematics. The solid (dashed) line represents the NMC fit (Ref. 9) of deep inelastic structure function data at $Q^2 = 10$ (GeV^2) ($Q^2 = 5$ GeV^2).

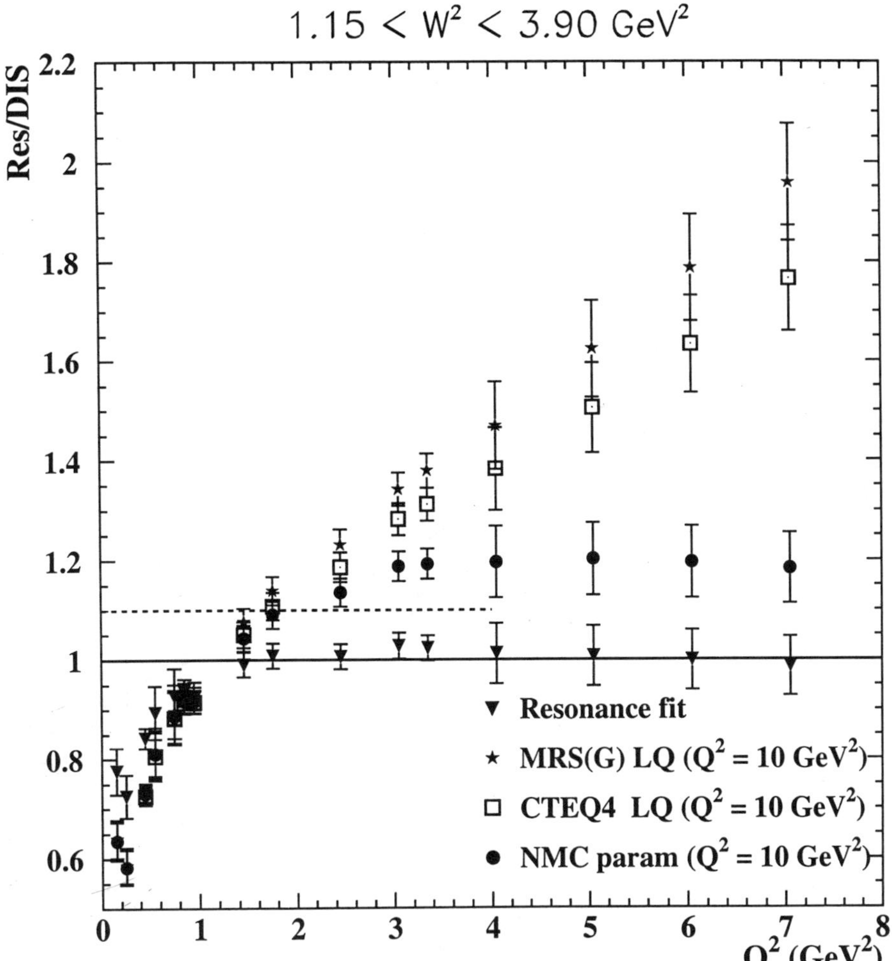

Figure 2: The ratios of integrals obtained over the hydrogen resonance structure function in the ξ range corresponding to invariant masses between $1.1 < W^2 < 3.9$ GeV^2 (Res) to integrals of the deep inelastic structure functions obtained from parameterizations (triangles are Resonance fit; circles are NMC at $Q^2 = 10$ (GeV/c)^2; squares are CTEQ4 low Q^2; stars are MRS(G) low Q^2) for the same range in ξ (DIS). The dashed line indicates what this ratio would be if the NMC curve were obtained at $Q^2 = 5$ (GeV/c)^2.

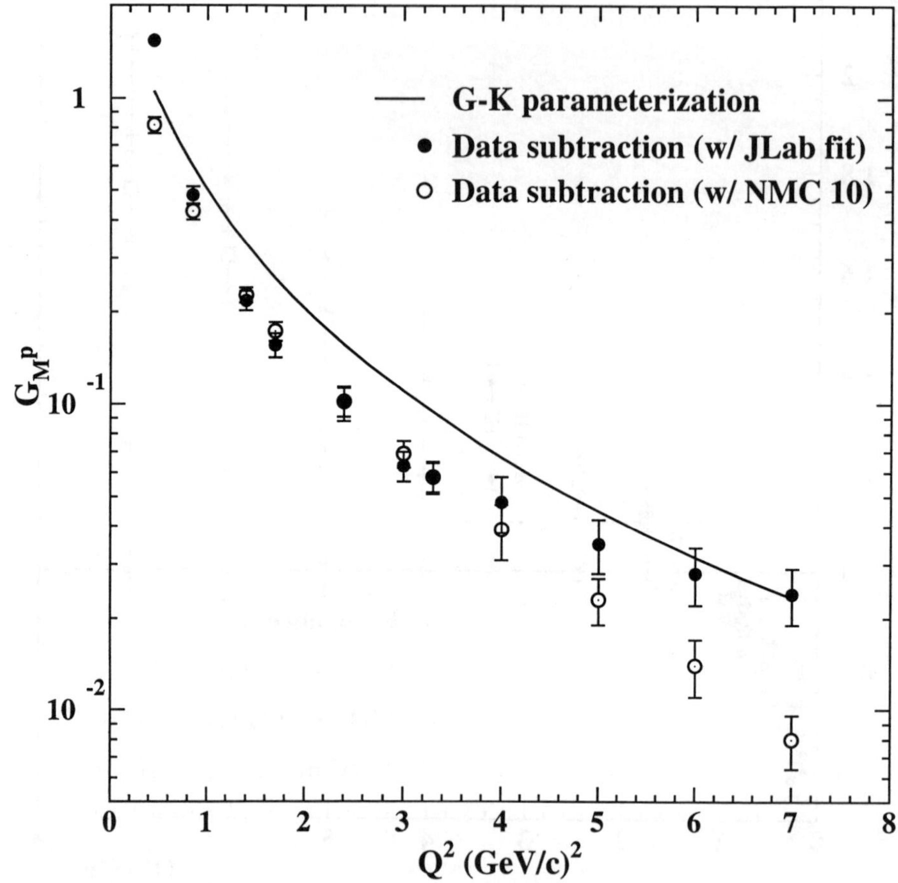

Figure 3: The proton magnetic form factor extracted from the inelastic data using duality assumptions as described in the text. The extracted data are compared to the model curve of Ref. 19

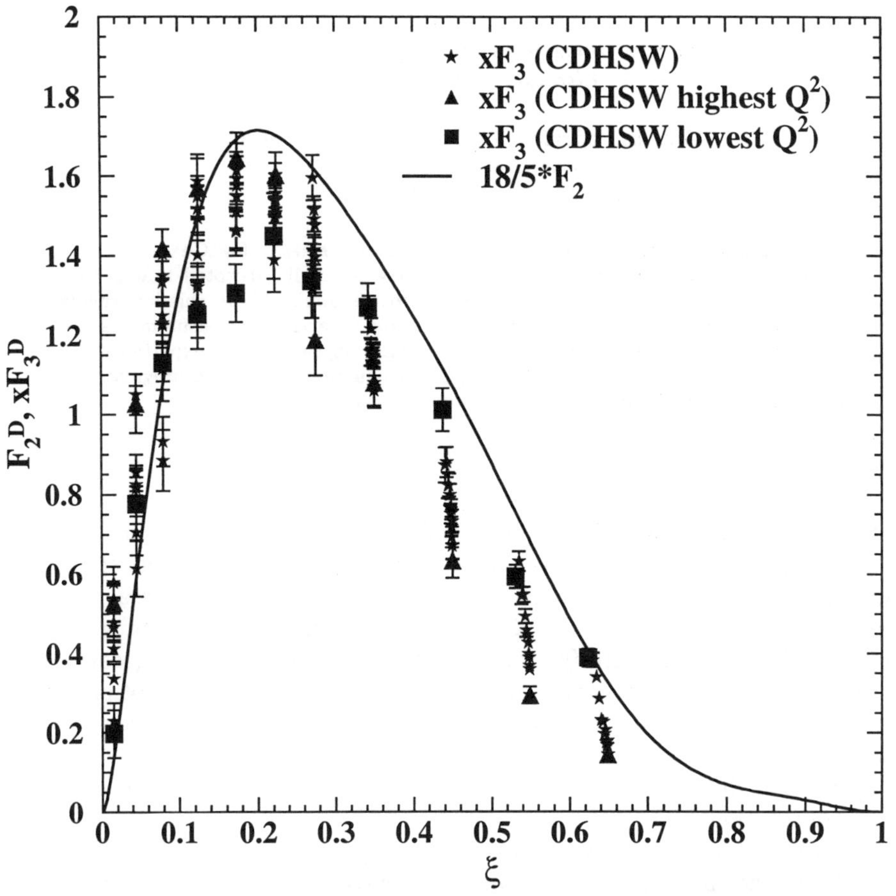

Figure 4: A comparison of the duality-averaged F_2 scaling curve, determined from the nucleon resonance region data with a deuterium target, with the CDHSW data on xF_3 from deep inelastic neutrino-nucleus scatttering data.

SPIN MEASUREMENTS IN HIGH $-t$ EXCLUSIVE PHOTOREACTIONS

R. GILMAN
Rutgers University Dept. Physics & Astronomy,
Piscataway, NJ 08855, USA
and
Thomas Jefferson National Accelerator Facility,
Newport News, VA 23606, USA
E-mail: gilman@jlab.org

The approximate validity of the constituent counting rules to few GeV photoreactions has been known for many years. It is now generally accepted that perturbative QCD, which leads to these rules, cannot be applicable in this low energy and momentum transfer regime. Cross section and polarization measurements now underway show promise to lead to the underlying physics. Initial experiments on deuteron photodisintegration and pion photoproduction are now in progress, and are expected to be followed by several other measurements.

1 Introduction

At beam energies above a few GeV and four momentum transfers $-t > 1$ $(\mathrm{GeV/c})^2$, cross sections [1] for several exclusive photoreactions demonstrate the approximate validity of the constituent counting rules (CCR). These rules can be derived from perturbative QCD [2], but this explanation is not generally accepted [3]. One can identify two shortcomings with the existing data set.[a] First, the relatively poor precision of the cross section data are not sufficient to distinguish between the CCR and more recent quark-based models [6,7] that predict modest scaling violations. Second, no polarization data exist at high $-t$, and thus the related prediction of hadron helicity conservation is untested in these reactions.

More precise cross-section measurements, now possible at Jefferson Lab, can determine the power law scaling of the cross sections systematically as a function of angle. Polarization measurements are possible with the focal plane proton polarimeter in Hall A. The induced polarization P and the polarization transfer C_x measure the imaginary and real parts of interferences between helicity violating and helicity conserving amplitudes. If hadron helicity violation measures orbital angular momentum [8], these data relate to the nucleon spin

[a]Two recent high-quality Jefferson Lab measurements of the proton electric [4] and deuteron [5] form factors do not lead to an unambiguous interpretation, as the deuteron form factor follows asymptotic predictions, while the proton form factor does not, at similar momentum transfer.

problem. Spin measurements are generally helpful to determine the reaction mechanism, as they are sensitive to small reaction amplitudes.

In the following sections, we will review the experiments underway and some of the expected future experimental efforts.

2 Deuteron Photodisintegration

Several experimental measurements of deuteron photodisintegration at SLAC and at Jefferson Lab have shown the applicability of the CCR at large scattering angles. These data [9,10] are compared in Fig. 1, taken from [9], to several theories. Multiplying the cross sections by s^{11} removes the asymptotic energy dependence expected from pQCD. The agreement at 90°_{cm} is better than one would expect, given that phase space factors should not be negligible in this kinematic regime. The reduced nuclear amplitude analysis [11,12] (RNA), given by Eq. (1), is shown as the long dashed line.

$$\frac{d\sigma}{d\Omega_{\mathrm{cm}}} \propto \frac{1}{s - m_{\mathrm{d}}^2} F_{\mathrm{p}}^2(\hat{t_{\mathrm{p}}}) F_{\mathrm{n}}^2(\hat{t_{\mathrm{n}}}) \frac{1}{p_T^2} f^2(\theta_{\mathrm{cm}}) \tag{1}$$

This approach can be viewed as an attempt to extend the validity of the CCR to lower energies and momentum transfers by including various expected kinematic and form factors.

The quark gluon string model [13], given as the dashed - dotted line, is in reasonable agreement with the forward angles. However, this calculation is based on Regge theory and is not expected to be valid at large angles.

Earlier meson-exchange calculations [14], shown as a solid line, do not fall as fast as the data at high energies. More recent asymptotic meson-exchange calculations [15], shown as the dotted line, do agree reasonably with the 90°_{cm} data, as well as at other angles (not shown).

More recently, Radyushkin [16] has proposed a quark model which naturally explains the energy and angle dependence of the existing data. He proposes that the reaction proceeds largely via quark exchange between the two nucleons, with the photon coupling to one of the exchanged quarks, and the photon's momentum shared between the two by some short-distance mechanism, such as gluon exchange. Coupling to a quark within the nucleon is suppressed because high-momentum components of the wave function are small. Preferentially the two quarks are different, e.g., one u and one d quark, since quarks of the same charge would lead to cancellations. In this picture, the energy and angle dependence of the amplitude arise largely from the transition form factors of $p \leftrightarrow n$, as the hard short-distance process should have a relatively slow angle and energy dependence. Assuming these transition form factors are like

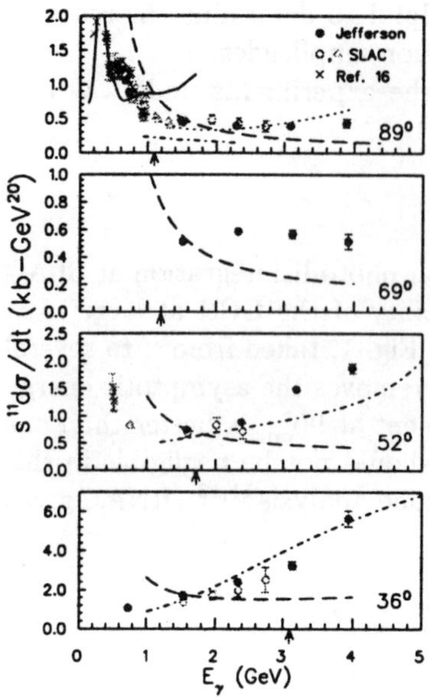

Figure 1: Deuteron photodisintegration data compared to several theories, as described in the text.

the nucleon dipole form factor, and including phase space factors, the result obtained is similar to the RNA result, Eq. (1).

Three measurements are being performed at Jefferson Lab this year to further stress the predictions. Cross sections at energies above 4 GeV and at forward angles are being measured in Hall C[17], while large angle cross sections at lower energies are being measured in Hall A[18].

Fig. 2 compares existing data [9] to calculations. Radyushkin [16] predicts that the angular distribution is largely symmetric about 90°_{cm}. The curve was generated from an expression similar to that of Eq. (1), in which the $1/p_T^2$ factor does not appear, the angle dependence was taken to be unity, and the absolute normalization was fit to the 1.6 GeV, 90°_{cm} data point. These slight differences reduce the energy and angle dependence, and allow the data to be well fitted, unlike in the RNA approach.

The asymptotic meson-exchange model [15] predicts an asymmetric angular distribution, as preferentially the photon couples to the proton, ejecting it in the forward direction. In the high-energy limit, assuming helicity conservation (hcon), coupling to the neutron is reduced, suppressing back-angle cross sections.

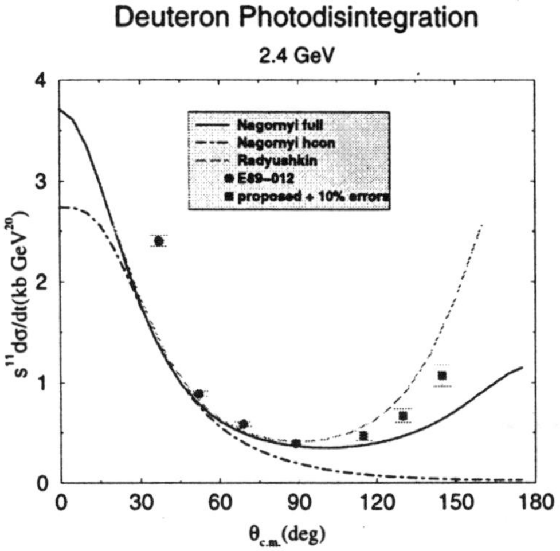

Figure 2: Comparison at one energy of cross-section predictions to existing data and planned measurements of E99-008; the curves are described in the text.

Further constraints on theories, including the first tests of helicity conservation, will be provided by experiment 89-019 [19]. This experiment measures recoil proton polarizations for energies up to $E_\gamma = 2.4$ GeV; the induced polarization P and the transverse polarization transfer C_x test helicity conservation, which requires both to vanish. In the case that one or both are nonzero, they provide a measure of whether the reaction amplitudes are largely real, as expected in most quark models, or have large imaginary parts, as one expects if there are resonance contributions. Lower-energy induced-polarization data [20] exist, and are shown in Fig. 3 compared to a Bonn calculation [21] and to the helicity conserving prediction, $P = 0$, generally assumed in quark models. In the Bonn meson-exchange calculations, the structures in the polarization result from the interference of the Born amplitude with the Δ resonance, and with the D_{13} and D_{15} resonances near 1 GeV.

3 Pion Photoproduction

While deuteron photodisintegration provides a large momentum transfer at modest energy [22], one naively expects that quark physics should be more evident in a simpler reaction involving fewer quarks, such as pion photoproduction. Cross sections in such reactions are known to approximately scale, but polarization data are sparse above 1 GeV and do not exist above the resonance

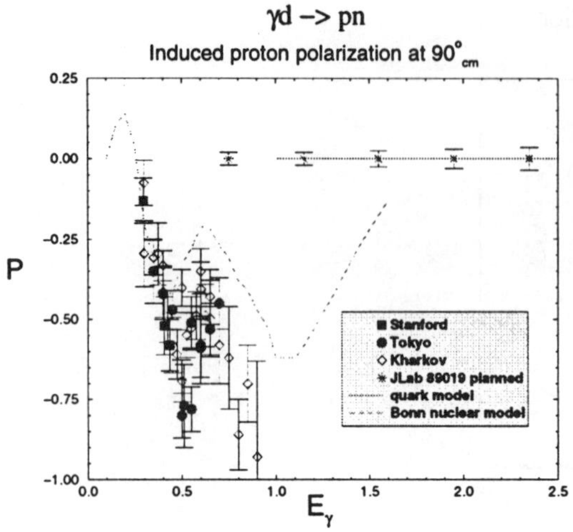

Figure 3: Induced polarization data for deuteron photodisintegration; the curves are described in the text.

region.

An example of this is shown in Fig. 4. The good agreement of the several data sets is evident, and the data are well fit up to about 1 GeV by the phase shift analysis of SAID [23]. Resonance structures are evident in the data. There is no indication in these data that the induced polarization vanishes at higher energies. While some high-energy induced polarization data exist, there are no polarization transfer data.

An experiment [24] will run this fall at Jefferson Lab to investigate the spin observables. The Hall A focal plane polarimeter will measure induced proton polarization and polarization transfer in neutral pion photoproduction from the proton. The experiment uses a circularly-polarized photon beam generated from the longitudinally-polarized electron beam. Measurements are planned for several energies, from near 1 GeV to over 4 GeV, and for a range of angles.

Various quark-model calculations, including pQCD [25] and non-forward parton distributions [26] (NFPD), have been compared to the cross section data. The pQCD calculations sum all leading-order diagrams, while the NFPD calculations assume the meson is produced by hard scattering from a quark. These cross section estimates tend to fall below the experimental values, indicating the need for additional soft physics. The experiment [24] will provide high quality cross sections, so that small scaling violations expected [26] may be detectable. Spin observables are also calculated within these models.

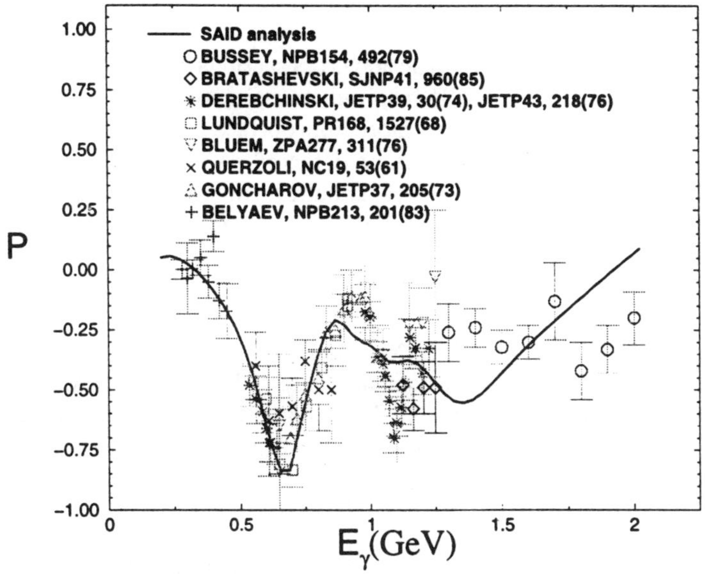

Figure 4: Induced polarization data for pion photoproduction from the proton, compared to the SAID phase shift analysis.

4 Future Experiments

The experiments described above become more difficult at higher energies, due to the decrease in cross section with energy, and to the decrease in polarimetry figure of merit with proton momentum. There is no simple solution to extending deuteron photodisintegration recoil-polarization measurements to higher energies. Pion photoproduction with untagged Bremsstrahlung suffers in addition from an increasing ep elastic scattering background. One can either measure a photon from the neutral pion decay in coincidence, as will be done to investigate real Compton scattering [27], or one can attempt an alternative pseudo-scalar meson production reaction.

Photoproduction of $K^+\Lambda^0$ final states [28] is particularly attractive. From a theoretical viewpoint, the Λ^0 wave function is particularly simple, as its spin is carried entirely by the strange quark, within the constituent quark model. From an experimental viewpoint, the self-analyzing decay of the Λ^0 allows accurate polarimetry with only a few thousand events. Thus, it is possible to extend precise measurements to energies well above 4 GeV, and to high momentum transfers.

5 Summary

It has for many years been a puzzle why reaction cross sections and form factors have tended to follow the simple, pQCD inspired, constituent counting rules. Recent theoretical efforts in quark models more appropriate for this kinematic regime indicate subtle scaling violations may be detected. A new generation of experiments, including those described above, should be able to detect scaling violations, and to provide spin observables that test theoretical assumptions.

Acknowledgments

This work has been supported by the U. S. National Science Foundation, grant PHY 9803860, and by the U. S. Department of Energy. The author acknowledges many interesting discussions with A. Afanasev, R. Holt, M. Liang, B. Liang-Gilman, Z.-E. Meziani, A. Radyushkin, and I. Strakovsky.

References

1. A. M. Boyarski *et al.*, *Phys. Rev. Lett.* **22**, 1131 (1969); G. Vogel *et al.*, *Phys. Lett.* B **40**, 513 (1972); and R. L. Anderson *et al.*, *Phys. Rev.* D **14**, 679 (1976).
2. S. J. Brodsky and G. R. Farrar, *Phys. Rev. Lett.* **31**, 1153 (1973); V. A. Mateev *et al.*, Lett. Nuovo Cimento **7**, 719 (1973); S. J. Brodsky and G. R. Farrar,*Phys. Rev.* D **11**, 1309 (1975). Also see S. J. Brodsky and G. P. Lepage in *Perturbative Quantum Chromodynamics*, ed. A. Mueller, page 93 (World Scientific, Singapore, 1989).
3. N. Isgur and C. Llewellyn Smith, *Nucl. Phys.* B **317**, 526 (1989); A. V. Radyushkin, *Nucl. Phys.* A **523**, 141c (1991).
4. M. Jones *et al.*, to be published.
5. L. C. Alexa *et al.*, *Phys. Rev. Lett.* **82**, 1374 (1999); *ibid.* nucl-ex/9812012.
6. A. V. Radyushkin,*Phys. Rev.* D **58**, 114008 (1998), hep-ph/9803316; *ibid.* hep-ph/9811223.
7. P. Kroll, M. Schurmann, K. Passek, and W. Schweiger, *Phys. Rev.* D **55**, 4315 (1997); *ibid.* hep-ph/960435.
8. Pankaj Jain and John P. Ralston, KU-HEP-93-27, hep-ph/9305250, in *Proceedings of Future directions in particle and nuclear physics at multi-GeV hadron beam facilities*, 217-222 (1993).
9. C. Bochna *et al. Phys. Rev. Lett.* **65**, 1401 (1998).
10. SLAC data are from J. E. Belz *et al.*, *Phys. Rev. Lett.* **74**, 646 (1995); S. J. Freedman *et al.*, *Phys. Rev.* C **48**, 1864 (1993); and J. Napolitano

et al., *Phys. Rev. Lett.* **61**, 2530 (1988). Earlier "Ref 16" data are from P. Dougan, *et al.*, Z. Physik **A276**, 55 (1976), R. Ching and C. Schaerf, Phys. Rev. **141**, 1320 (1966), H. Myers, *et al.*, Phys. Rev. **121**, 630 (1961), and J. Arends, *et al.*, Nucl. Phys. **A412**, 509 (1984).

11. S. J. Brodsky and J. R. Hiller, *Phys. Rev.* C **28**, 475 (1983).
12. S. J. Brodsky and B. T. Chertok, *Phys. Rev. Lett.* **37**, 269 (1976).
13. L. A. Kondratyuk *et al. Phys. Rev.* C **48**, 2491 (1993).
14. T.-S. H. Lee, Argonne National Laboratory Report No. PHY-5253-TH-88; T.-S.H. Lee, in *Proceedings of the International Conference on Medium and High Energy Nuclear Physics, Taipai, Taiwan, 1988* (World Scientific, Singapore, 1988), p.563.
15. S. I. Nagornyi, YU. A. Kasatkin, and I.K. Kirichenko, Sov. J. Nucl. Phys. **55**, 189 (1992); A. E. L. Dieperink and S. I. Nagorny, nucl-th/9904058.
16. A. V. Radyushkin, private communication. Note also a similar calculation, L. L. Frankfurt, G. A. Miller, M. M. Sargsian, and M. I. Strikman, hep-ph/9904222.
17. Jefferson Lab Exp. 96-003, R. Holt *et al.*
18. Jefferson Lab Exp. 99-008, R. Gilman, R. Holt, Z.-E. Meziani *et al.*
19. Jefferson Lab Exp. 89-019, R. Gilman, R. Holt, Z.-E. Meziani *et al.*
20. F. F. Liu *et al.*, *Phys. Rev.* **165**, 1478 (1968); T. Kamae *et al.*, *Phys. Rev. Lett.* **38**, 468 (1977); T. Kamae *et al.*, *Nucl. Phys.* B **139**, 394 (1978); H. Ikeda *et al.*, *Phys. Rev. Lett.* **42**, 1321 (1979); A. S. Bratashevskij *et al.*, *Nucl. Phys.* B **166**, 525 (1980); H. Ikeda *et al.*, *Nucl. Phys.* B **172**, 509 (1980); A. S. Bratashevskij *et al.*, Yad. Fiz. **32**, 418 (1980), [Sov. J. Nucl. Phys. **32**, 216 (1980)]; V. P. Barannik *et al.*, *Nucl. Phys.* A **451**, 751 (1986).
21. Y. Kang, P. Erbs, W. Pfeil, and H. Rollnik, *Abstracts of the Particle and Nuclear Intersections Conference,* (MIT, Cambridge, MA 1990); Y. Kang, Ph.D. dissertation, Bonn (1993); W. Pfeil, private communication.
22. R. J. Holt, *Phys. Rev.* C **41**, 2400 (1990).
23. The phase shift analysis was generated with the SM99K solution, valid to about 2 GeV; see http://www.phys.vt.edu/~igor/welcome.html.
24. Jefferson Lab Exp. 94-012, R. Gilman, R. Holt, *et al.*
25. G. R. Farrar, K. Huleihel, and H. Zhang, *Nucl. Phys.* B **349**, 655 (1991).
26. Andrei Afanasev, private communication.
27. Jefferson Lab Exp. 99-114, B. Wojtsekhowski, A. M. Nathan, C. Hyde-Wright, *et al.*.
28. Jefferson Lab Exp. 99-116, M. Liang, R. Gilman, A. Afanasev, *et al.*

MEASUREMENT OF THE SPIN ASYMMETRY IN THE PHOTOPRODUCTION OF PAIRS OF HIGH-P_T HADRONS AT HERMES

RICHARD G. MILNER
MIT-BATES Linear Accelerator Center,
Laboratory for Nuclear Science,
21 Manning Avenue, Middleton,
MA 01949, USA
E-mail: milner@mitlns.mit.edu

1 Introduction

From polarized deep inelastic lepton-nucleon scattering (DIS) experiments, it has been inferred that the quark spins account for only a fraction of the nucleon spin. One possible explanation is a significant gluon polarization in the nucleon. In principle, the polarized gluon distribution $\Delta G(x_G)$ (x_G is the fraction of the nucleon momentum carried by the struck gluon) can be probed by a measurement of the scaling violation of the polarized structure functions. However, the presently available data on polarized inclusive deep inelastic scattering only poorly constrain $\Delta G(x_G)$, although there is some indication for the integral to be positive [1,2,3]. On the other hand, two theoretical calculations in the bag model obtain different predictions for the sign of the integral of $\Delta G(x_G)$ [4,5]. Several recent experimental proposals have concentrated on ways to measure $\Delta G(x_G)$ directly[6,7,8].

One way to measure $\Delta G(x_G)$ directly is via the photon gluon fusion process. Two useful experimental signatures of this process are charm production and production of jets with high transverse momentum p_T. In the former case, the large mass of the charm quark suppresses its production in the fragmentation process. A similar argument applies to the production of jets: the transverse momentum produced in the fragmentation process is small and two back-to-back jets with sufficiently high p_T thus reflect the high p_T of the quark and anti-quark produced in the photon gluon fusion process. Both charm production and high-p_T jet production in DIS have resulted in direct measurements of the unpolarized gluon structure function $G(x_G)$ [9,10,11].

At lower energy fixed target experiments, high-p_T hadrons must serve in place of jets [12]. Several phenomenological studies of the potential of high-p_T meson photoproduction as a probe of $\Delta G(x_G)$ have been performed[13,14].

2 The HERMES Measurement

The HERMES collaboration has carried out the first measurement of a spin asymmetry in photoproduction of pairs of high-p_T hadrons. The data were collected in 1996 and 1997 at the HERA storage ring of the DESY laboratory. Polarized positrons of energy 27.5 GeV were scattered off a polarized internal hydrogen gas target. The beam polarization was continuously measured by Compton back scattering and had an average value of 0.55 ± 0.02 [15,16]. The average target polarization was 0.86 ± 0.04 [15,17]. In both cases the quoted uncertainty is predominantly systematic. The HERMES detector [18] is a forward spectrometer that identifies charged particles in the scattering angle range of $0.04 < \theta < 0.22$ rad. Particle identification (PID) is accomplished using an electromagnetic calorimeter, a scintillator hodoscope preceded by two radiation lengths of lead, a transition radiation detector, and a C_4F_{10}/N_2(70:30) gas threshold Čerenkov counter. A likelihood method, based on the empirical responses of each of the four PID detectors, is used to discriminate between positrons and hadrons. The luminosity is measured in a pair of $NaBi(WO_4)_2$ electromagnetic calorimeters that detect Bhabha-scattering from target electrons.

The longitudinal cross section asymmetry $A_{||}$ was determined using the formula:

$$A_{||} = \frac{N^{\uparrow\downarrow}L^{\uparrow\uparrow} - N^{\uparrow\uparrow}L^{\uparrow\downarrow}}{N^{\uparrow\downarrow}L_P^{\uparrow\uparrow} + N^{\uparrow\uparrow}L_P^{\uparrow\downarrow}} \,. \tag{1}$$

Here $N^{\uparrow\uparrow}(N^{\uparrow\downarrow})$ is the number of oppositely charged hadron pairs observed for target spin parallel (anti-parallel) to the beam spin orientation. The luminosities for each target spin state are $L^{\uparrow\uparrow(\uparrow\downarrow)}$ and $L_P^{\uparrow\uparrow(\uparrow\downarrow)}$, the latter being weighted by the product of the beam and target polarization values for each spin state.

Events were selected that contained at least one positively charged hadron h^+ and at least one negatively charged hadron h^-. The observation of the scattered positron was not required, in order to include the very low Q^2 region which dominates the cross section. Here Q^2 is the negative square of the 4-momentum of the virtual photon. The highest momentum hadrons of each charge were required to have a momentum above 4.5 GeV/c and a transverse momentum p_T above 0.5 GeV/c. Here p_T is defined as the momentum transverse to the positron beam direction and is approximately equal to the momentum transverse to the photon direction when $Q^2 \approx 0$. To suppress contributions from vector meson resonances from the data sample, a minimum value of the invariant mass of the two hadrons (assuming both hadrons to be pions) $M(2\pi) > 1.0$ GeV/c^2 was imposed. Additionally, both hadrons were required to originate from the target region and to have a common vertex. A detailed account of the analysis may be found in [19].

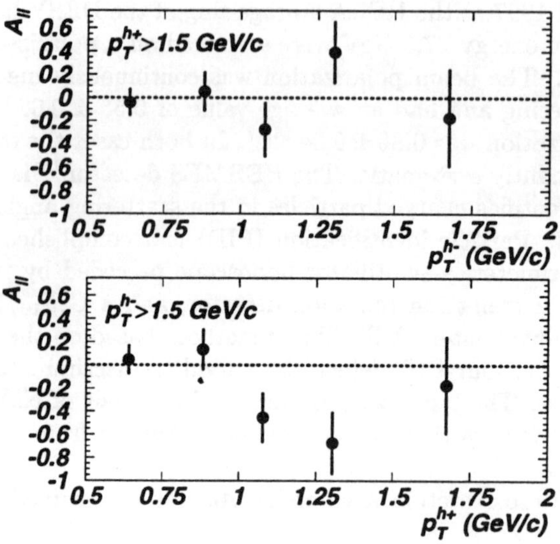

Figure 1: $A_{||}(p_T^{h^+}, p_T^{h^-})$ for $p_T^{h^+} > 1.5$ GeV/c (top) and for $p_T^{h^-} > 1.5$ GeV/c (bottom). Note that the rightmost data point is identical in both plots.

Fig. 1 presents the measured $A_{||}$ for the highest values of transverse momenta accessible at HERMES; in the top (bottom) panel the positive (negative) hadron was required to have a p_T greater than 1.5 GeV/c and the asymmetry $A_{||}$ is then plotted as a function of the p_T of the hadron of opposite charge. The data suggest a more negative asymmetry when the transverse momentum of the negative hadron is higher than that of the positive hadron. Ignoring this charge asymmetry and averaging over the five bins satisfying the requirement $p_T^{h_1} > 1.5$ GeV/c and $p_T^{h_2} > 1.0$ GeV/c, a negative asymmetry $A_{||} = -0.28 \pm 0.12$ (stat.) $\pm\, 0.02$ (syst.) is observed. (The symbol h_1 signifies the hadron with the higher p_T.) When the requirement $p_T^{h_1} > 1.5$ GeV/c is not enforced, the asymmetry is consistent with zero. The observed negative asymmetry is in contrast to the positive asymmetries typically measured in deep inelastic scattering from protons.

A possible background to the observed asymmetry arises from coincident detection of a negative hadron and the scattered positron, the latter being misidentified as a positive hadron. From studies of other processes, the probability for positron/hadron misidentification has been determined to be less than 0.2%. By comparing yields of h^+h^- pairs to those of e^+h^- pairs detected in the final state, the background arising from this misidentification has been estimated to be less than 0.1%, for the kinematics selected by this analysis. Other sources of background include high-p_T particles from charm decays. Contributions from both open charm and J/ψ decays have been found to be negligible using the AROMA[20] Monte Carlo generator.

The systematic uncertainty arising from the measurement of the beam and target polarizations is about 6% of $A_{||}$, much smaller than the statistical error and independent of p_T. Resolution effects and alignment uncertainties were found to be negligible. Electroweak radiative corrections are expected to be very small compared to the statistical uncertainty.

3 Interpretation

The measured asymmetry was interpreted assuming that several different processes could contribute to the two-hadron cross section: lowest order deep inelastic scattering (containing no hard QCD vertex), interaction via the hadronic structure of the photon – described by the vector meson dominance model (VMD) and by non-resonant hadronic "anomalous" photon structure, and the two first order QCD processes (termed "direct") which describe the interaction of a pointlike photon. These are photon gluon fusion (PGF) and the QCD Compton effect (QCDC).

The contribution from lowest order DIS is suppressed by the requirement of high p_T, and was confirmed to be negligible by a simulation based on the LEPTO Monte Carlo generator [21]. Contributions from VMD were assumed to have a negligible spin asymmetry, and were thus treated as a dilution of the other asymmetries. Finally, we neglect possible contributions from anomalous photon structure, where the photon fluctuates into a non-resonant $q\bar{q}$ pair which interacts via hard processes with the partons inside the nucleon. This is supported by a model[22] that explains the excess of forward hadrons with high p_T observed in γp reactions at $70-90$ GeV, relative to those from πp and Kp scattering[23]. At this energy, the model prediction at high p_T is dominated by direct processes involving hard coupling of the photon to the partons in the proton. At the lower energy of the present experiment, a negligible contribution from anomalous photon structure is predicted by the model.

Under the assumptions described above, only two of the five possible spin asymmetries A_i contribute significantly to the measured asymmetry:

$$A_{||} \approx (A_{\text{PGF}} f_{\text{PGF}} + A_{\text{QCDC}} f_{\text{QCDC}}) D \tag{2}$$

where f_i is the unpolarized fraction of events from subprocess i ($f_{\rm PGF}+f_{\rm QCDC}+f_{\rm VMD}=1$), and D is the virtual photon depolarization factor. In the small region of phase space selected by the present analysis, the A_i's may be approximated by the products of the hard subprocess asymmetries and the quark and gluon polarizations. The subprocess asymmetries $\hat{a}_{\rm PGF}=\hat{a}(\gamma g\to q\bar{q})$ and $\hat{a}_{\rm QCDC}=\hat{a}(\gamma q\to qg)$ are directly calculable in leading order (LO) QCD [13]. For real photons and massless quarks, $\hat{a}_{\rm PGF}=-1$, while $\langle\hat{a}_{\rm QCDC}\rangle$ is about +0.5 (averaged over the kinematics selected by this analysis) and is independent of the quark flavor. The effective quark polarization $\Delta q/q$ is computed as a suitably weighted combination of $\Delta u/u$ and $\Delta d/d$, which are known from inclusive and semi-inclusive polarized DIS measurements [24,25]. The measured asymmetry can therefore be expressed as follows:

$$A_{||}\approx\left(\hat{a}_{\rm PGF}\frac{\Delta G}{G}f_{\rm PGF}+\hat{a}_{\rm QCDC}\frac{\Delta q}{q}f_{\rm QCDC}\right)D, \tag{3}$$

where the kinematic dependences have been suppressed for brevity. This equation can be solved for $\Delta G/G$ after appropriate averaging over the selected kinematics.

The PYTHIA Monte Carlo generator [26] was used to provide a model for the data. An important parameter in the simulation of the direct processes, the minimum transverse momentum of the outgoing partons ($\hat{p}_T^{\rm min}$), was chosen following Ref. [22] to be 0.5 GeV/c. The kinematic region used in the interpretation of the measurement ($p_T^{h_1}>1.5$ GeV/c and $p_T^{h_2}>0.8$ GeV/c) was chosen so that the final results depend only weakly on the choice of $\hat{p}_T^{\rm min}$. The Lund fragmentation parameters used in the simulation have been adjusted to fit the HERMES semi-inclusive hadron multiplicity data [27].

The normalized yield for the production of two high-p_T hadrons is compared to the Monte Carlo simulation in Fig. 2. Here, the Weizsäcker-Williams approximation has been used to relate the photoproduction cross section simulated by PYTHIA to the measured electroproduction cross section. Also shown in Fig. 2 are the contributions from the three subprocesses included in the simulation. The simulated yield has a p_T-dependence similar to that of the data, but is significantly smaller in magnitude. Good agreement is found for the distributions in other kinematic variables, such as the azimuthal angle between the two hadrons and $\Delta p_T=|\vec{p}_T^{\,h^-}|-|\vec{p}_T^{\,h^+}|$. As the simulation of the direct QCD processes is restricted to leading order, the observed difference in normalization might be due to contributions from higher-order QCD processes and/or contributions from hard interactions of the hadronic structure of the photon. It should be noted that the agreement becomes much better if the default Lund fragmentation parameters are used. However, the final re-

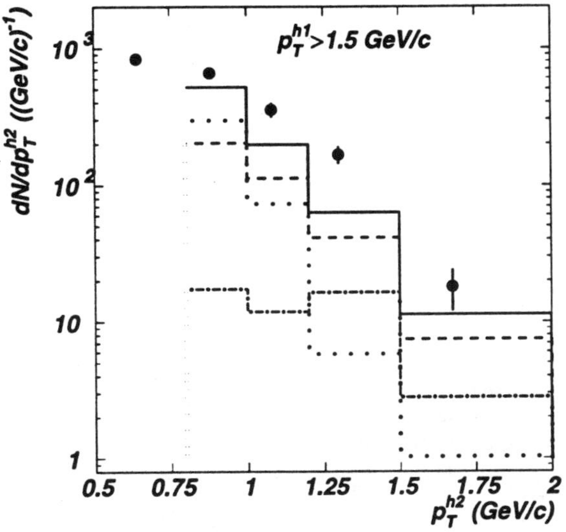

Figure 2: Comparison of data (circles) and Monte Carlo simulation (full histogram) for $dN/dp_T^{h_2}$ for $p_T^{h_1} > 1.5$ GeV/c. The dashed, dashed-dotted and dotted lines represent the contributions from the PGF, VMD and QCDC processes, respectively; the solid line represents their sum.

sult for $\Delta G/G$ is found to depend only weakly on the choice of fragmentation parameters.

In the same region of phase space where a negative asymmetry is observed ($p_T^{h_1} > 1.5$ GeV/c and $p_T^{h_2} > 1.0$ GeV/c), the simulated cross section is dominated by photon gluon fusion. The consequent sensitivity of the measured asymmetry to the polarized gluon distribution is demonstrated in Fig. 3, where $A_{||}$ at high transverse momenta (i.e. the average of the two panels of Fig. 1) is compared with Monte Carlo predictions for different distributions of $\Delta G/G$.

The Monte Carlo simulation was used to determine the quantities necessary to relate the data to $\langle \Delta G/G \rangle$, where the angle brackets indicate averaging over the kinematics of the measurement. These quantities have been determined from PYTHIA to be $\langle D\hat{a}_{\rm QCDC}\frac{\Delta q}{q}\rangle = 0.15$, $\langle D \rangle = 0.93$, $\langle x_G \rangle = 0.17$, $\langle Q^2 \rangle = 0.06$ $(\mathrm{GeV/c})^2$, and $\langle \hat{p}_T^2 \rangle = 2.1$ $(\mathrm{GeV/c})^2$. The distribution $\Delta G(x_G)$ is probed principally in the range $0.06 < x_G < 0.28$. Note that the hard scale of this process is not given by Q^2, but rather by $\hat{p}_T^2$, the square of the transverse momentum carried by each of the outgoing quarks.

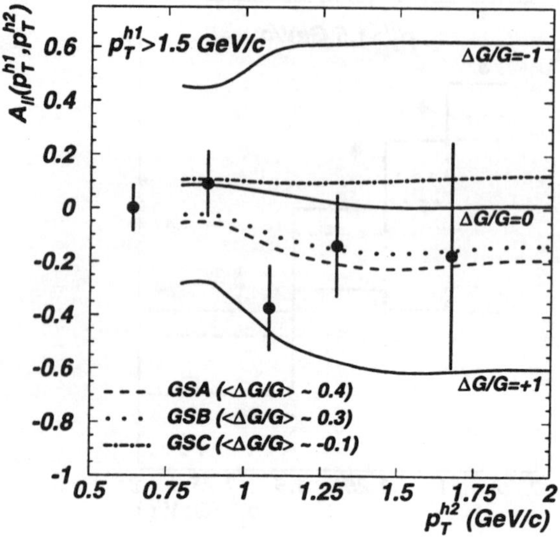

Figure 3: $A_{||}$ for high-p_T hadron production measured at HERMES compared with Monte Carlo predictions for $\Delta G/G = \pm 1$ (lower/upper solid curves), $\Delta G/G = 0$ (middle solid curve), and the phenomenological LO QCD fits of Ref. [28] (dashed, dotted, and dot-dashed curves).

For the four values of $A_{||}$ at $p_T^{h_2} > 0.8$ GeV/c presented in Fig. 3, $\langle \Delta G/G \rangle$ was extracted according to equation (3). Since these four measurements probed essentially the same range of x_G, the results for $\langle \Delta G/G \rangle$ were averaged. Using the assumptions and model parameters described above, a value for $\langle \Delta G/G \rangle$ was determined in LO QCD to be 0.41 ± 0.18 (stat.) $\pm\, 0.03$ (syst.), where the systematic uncertainty represents the experimental contribution only.

The extracted value of $\langle \Delta G/G \rangle$ is compared in Fig. 4 with several phenomenological LO QCD fits of a subset of the world's data on $g_1(x, Q^2)$ [28,29]. The horizontal error bar represents the standard deviation of the x_G distribution for the cited kinematical constraints on the produced hadrons, as given by the Monte Carlo.

4 Conclusions

In summary, a positive value for the gluon polarization has been extracted from a measurement of the spin asymmetry in the photoproduction of pairs of

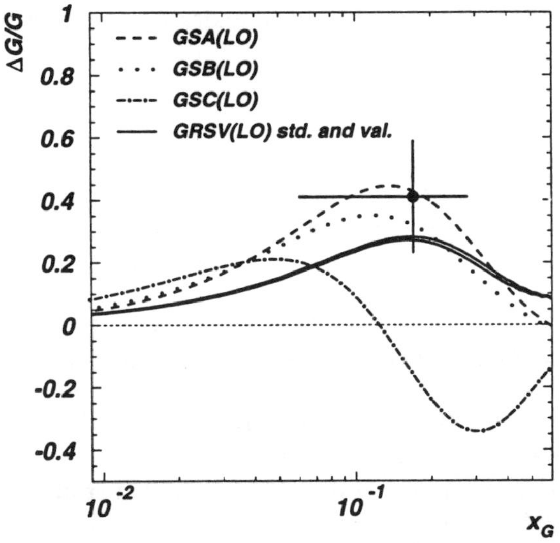

Figure 4: The extracted value of $\Delta G/G$ compared with phenomenological QCD fits to a subset of the world's data on $g_1^{\mathrm{p,n}}(x,Q^2)$. The curves are from Refs. [28,29], evaluated at a scale of 2 $(\mathrm{GeV/c})^2$. The indicated error on $\Delta G/G$ represents statistical and experimental systematic uncertainties only; no theoretical uncertainty is included.

hadrons at high p_T. This interpretation of the observed negative asymmetry is based on a model which takes into account leading order QCD processes and VMD contributions to the cross section. At the kinematics of this measurement, no spin-dependent analyses of higher order QCD processes or contributions from anomalous photon structure are presently available; these processes have therefore been neglected in the model presented here. If such processes would be important but have no significant spin asymmetry, the extracted value of $\langle \Delta G/G \rangle$ would increase, but still differ from zero by 2.3σ. To alter the principal conclusion of this analysis, i.e., that $\langle \Delta G/G \rangle$ at $\langle x_G \rangle = 0.17$ is positive, a significant contribution from a neglected process with a large negative spin asymmetry would be needed.

Acknowledgements

I gratefully acknowledge the contributions of my HERMES colleagues, particularly M. Amarian, A. Brüll, and J. Martin. The author's research is supported by the United States Department of Energy under Cooperative Agreement DE-FC02-94-ER40818.

References

1. R. D. Ball, S. Forte, G. Ridolfi, Phys. Lett. **B378** (1996) 255.
2. E154, K. Abe et al., Phys. Lett. **B405** (1997) 180.
3. SMC, B. Adeva et al., Phys. Rev. **D58** (1998) 112002.
4. R. L. Jaffe, Phys. Lett. **B365** (1996) 359.
5. V. Barone, T. Calarco, A. Drago, Phys. Lett. **B431** (1998) 405.
6. PHENIX, BNL-PROPOSAL-R5, Aug. 1992; STAR, S.E. Vigdor, hep-ex/9905034.
7. COMPASS, CERN/SPSLC 96-14, Mar. 1996.
8. HERMES, DESY-PRC 97-03, Jan. 1997.
9. H1, S. Aid et al., Nucl. Phys. **B472** (1996) 3; S. Aid et al., Nucl. Phys. **B472** (1996) 32; S. Aid et al., Nucl. Phys. **B449** (1995) 3.
10. ZEUS, J. Breitweg et al., Z. Phys. **C76** (1997) 599; J. Breitweg et al., Phys. Lett. **B407** (1997) 402.
11. NMC, D. Allasia et al., Phys. Lett. **B258** (1991) 493.
12. A. Bravar, D. von Harrach, A. Kotzinian, Phys. Lett. **B421** (1998) 349.
13. M. Fontannaz, D. Schiff, B. Pire, Z. Phys. **C8** (1981) 349.
14. A. Afanasev, C. E. Carlson, C. Wahlquist, Phys. Rev. **D58** (1998) 054007.
15. HERMES, A. Airapetian et al., Phys. Lett. **B442** (1998) 484.
16. W. Lorenzon, Proc. of the Workshop "Polarized gas targets and polarized beams," edited by R. J. Holt and M. A. Miller, Urbana-Champaign, USA, AIP Conf. Proc. 421 (1997) 181.
17. C. Baumgarten, Proc. of "13th International Symposium on High-Energy Spin Physics (SPIN98)," Protvino, Russia, (1998).
18. HERMES, K. Ackerstaff et al., Nucl. Instrum. Meth. **A417** (1998) 230.
19. J.W. Martin, Ph.D. thesis, Massachusetts Institute of Technology, 1999 (unpublished).
20. G. Ingelman, J. Rathsman, G. Schuler, Comp. Phys. Comm. **101** (1997) 135.
21. G. Ingelman, A. Edin, J. Rathsman, Comp. Phys. Comm. **101** (1997) 108.
22. G. A. Schuler, T. Sjöstrand, hep-ph/9403393 and CERN-TH-7193-94; G. A. Schuler, T. Sjöstrand, Nucl. Phys. **B407** (1993) 539.
23. OMEGA, R. J. Apsimon et al., Z. Phys. **C43** (1989) 63.
24. HERMES, K. Ackerstaff et al., hep-ex/9906035 and DESY 99-048.

25. SMC, B. Adeva et al., Phys. Lett. **B420** (1998) 180.
26. T. Sjöstrand, Comp. Phys. Comm. **82** (1994) 74. The Monte Carlo results for this paper were obtained using PYTHIA version 5.724 and JETSET version 7.410.
27. H.A.M. Tallini, Ph.D. thesis, Univ. Liverpool, (1998).
28. T. Gehrmann, W. J. Stirling, Phys. Rev. **D53** (1996) 6100.
29. M. Glück, E. Reya, M. Stratmann, W. Vogelsang, Phys. Rev. **D53** (1996) 4775.

EXTRACTION OF POLARIZED PARTON DISTRIBUTIONS AT HERMES

B. Tipton
26-547 MIT,
77 Massachusetts Avenue,
Cambridge, MA 02139
E-mail: tipton@mit.edu

On behalf of the HERMES collaboration

HERMES has measured inclusive and semi-inclusive hadron deep-inelastic scattering asymmetries with a polarized positron beam and polarized ^{3}He and ^{1}H targets. Results for these asymmetries in the kinematic range $0.023 < x_{Bj} < 0.6$ and $1\text{GeV}^2 < Q^2 < 10\text{GeV}^2$ from the 1995, 1996, and 1997 data are presented here. Polarized up, down, and sea quark distributions are extracted from the asymmetries as a function of x. The extraction uses models of the unpolarized parton distributions, the fragmentation process, the HERMES acceptance, and the average sea polarization. First moments of the polarized quark distributions are determined.

1 Introduction

The HERMES experiment studies the spin structure of the nucleon using semi-inclusive, polarized deep inelastic scattering (DIS):

$$\vec{l} + \vec{N} \rightarrow l' + X. \tag{1}$$

In the quark parton model, the spin of the nucleon may be schematically decomposed as:

$$< s_z^N > \; = \; \frac{1}{2} = \; \frac{1}{2}\Delta\Sigma + L_z^q + \Delta G + L_z^G, \tag{2}$$

where $\Delta\Sigma$, ΔG, L_z^q and L_z^G are contributions from quark and gluon spin and quark and gluon orbital angular momentum. $\Delta\Sigma$ can be further divided among the quark flavors, and between valence and sea contributions:

$$\Delta\Sigma = \Delta u_v + \Delta d_v + \Delta u_s + \Delta\bar{u} + \Delta d_s + \Delta\bar{d} + \Delta s + \Delta\bar{s}. \tag{3}$$

Previous inclusive experiments precisely measured the polarized structure function $g_1(x)$, which can be interpreted in the Quark-Parton Model(QPM) as the sum over all flavors of polarized parton distribution functions(PDFs), Δq:

$$g_1^p(x) = \frac{1}{2}\sum_q e_q^2(q^+(x) - q^-(x)), = \frac{1}{2}\sum_q e_q^2 \Delta q(x). \tag{4}$$

The function, $q^{+(-)}$, represents the distribution of quarks of flavor q with spin parallel (anti-parallel) to the nucleon spin, and e_q is the quark charge. Here, x is the Bjorken scaling variable $x = Q^2/2M_p\nu$, measured by scattered lepton kinematics; Q^2 and ν are the negative squared four-momentum and energy of the exchanged virtual photon. By measuring $g_1(x)$ from various targets and integrating over x, it has been found that $\Delta\Sigma$ accounts for only about a third of the nucleon spin, and that the strange sea seems negatively polarized.

Measuring asymmetries in semi-inclusive reactions yields further insights into the individual contributions of the quark flavors to $\Delta\Sigma$. In these reactions, one also detects the properties of the fragments of the nucleon breakup after the initial scattering, which are correlated to the flavor of the struck quark in the reaction. The HERMES experiment was designed to optimize semi-inclusive DIS studies.[1] The experiment is located at the DESY laboratory in Hamburg, Germany, where the HERA ring circulates a 27.5 GeV positron(or electron) beam. The positron beam is naturally polarized transverse to the beam direction due to an asymmetry in the emission of synchrotron radiation, and longitudinal polarization at the HERMES interaction point is provided by a pair of matched spin rotators. Beam polarizations averaged 55% during the run, with a relative systematic uncertainty in the measurement of 4.0% (3.4%) for the ^{3}He (H) data.

The HERA beam passes through the center of a 40 cm long, cylindrical storage cell containing polarized ^{3}He or ^{1}H. In 1995, the ^{3}He target atoms become polarized through metastability-exchange optical pumping.[2] Polarizations averaged 46% with a fractional uncertainty of 5%. In 1996 and 1997, an Atomic Beam Source(ABS) used permanent sextupole magnets and radio-frequency units to select desired hyperfine states of ^{1}H. A Breit-Rabi polarimeter(BRP) monitors the proton polarization, which averaged 86% during the run.

Extending 8.5 meters from the target, the HERMES spectrometer detects particles scattered in the forward direction. A series of tracking chambers placed before and after a dipole magnet determine momenta and directions of those particles. A threshold Čerenkov detector, a Transition Radiation Detector, a Preshower Counter, and an Electromagnetic Calorimeter allow separation of electrons from hadrons. The Čerenkov further provides identification of pions above particle momenta of 5.6 and 3.8 GeV/c in 1995 and 1996-7.

2 Semi-Inclusive Deep-inelastic Scattering Asymmetries

In polarized deep-inelastic scattering, one can construct a cross-section asymmetry of parallel and anti-parallel orientations of the nucleon spin to the lepton

beam spin, $A_{\parallel}$:

$$A_{\parallel} = \frac{\sigma^{\uparrow\Downarrow} - \sigma^{\uparrow\Uparrow}}{\sigma^{\uparrow\Downarrow} + \sigma^{\uparrow\Uparrow}}. \quad (5)$$

Under the assumption that $g_2(x)$ is zero, the measured double-spin asymmetry directly relates to the asymmetry in nucleon virtual photoabsorption, $A_1(x)$, and to the structure function ratio, $\frac{g_1}{F_1}(x)$:

$$A_1^h = \frac{g_1^h}{F_1^h} = \frac{A_{\parallel}^h}{D(1+\gamma\eta)}. \quad (6)$$

For this discussion, the asymmetry and structure function index h includes both inclusive positron data and coincident positive and negative unidentified hadrons and identified pions (h^+, h^-, π^+, and π^-), and D, γ, and η are kinematic factors.

HERMES has measured asymmetries for inclusive and semi-inclusive h^+ and h^- reactions from 1995, 1996, and 1997 running. These asymmetries are shown in Figure 1. Inclusive DIS processes are selected from events with cuts on $Q^2 > 1$ GeV^2, $y = \frac{\nu}{E} < 0.85$, and the final hadronic state mass, $W^2 > 4$ GeV^2. The sizes of the inclusive event samples satisfying these requirements were $2.2 \cdot 10^6$ and $2.3 \cdot 10^6$ events for ^{3}He and ^{1}H respectively. The analyzed semi-inclusive events form a subsample having a detected charged hadron and $W^2 > 10$ GeV^2. Cuts on the hadron energy fraction, $z > 0.2$, and the Feynman variable, $x_f > 0.1$, also select nucleon fragments from the current region. Beam and target polarization measurement uncertainties, uncertainties in $R(x, Q^2)$, and experimental yield fluctuation uncertainties in 1995 are the dominant contributions to the systematic uncertainty of the measured asymmetries.

3 Interpretation of the Asymmetries

In the QPM and under the assumption of factorization, one sees that the photoabsorption asymmetry probes the helicity difference parton distributions $\Delta q(x)$:

$$A_1^h(x,z) = \frac{\sum_q e_q^2\, \Delta q(x,Q^2) D_q^h(z,Q^2)}{\sum_q e_q^2\, q(x,Q^2) D_q^h(z,Q^2)} \frac{(1+R(x,Q^2))}{1+\gamma^2}. \quad (7)$$

The asymmetries are assumed Q^2 independent as the experimental values show no significant Q^2 dependence. The fragmentation functions $D_q^h(z)$ are spin-independent and represent the probability that a struck quark q fragments into a hadron type h.

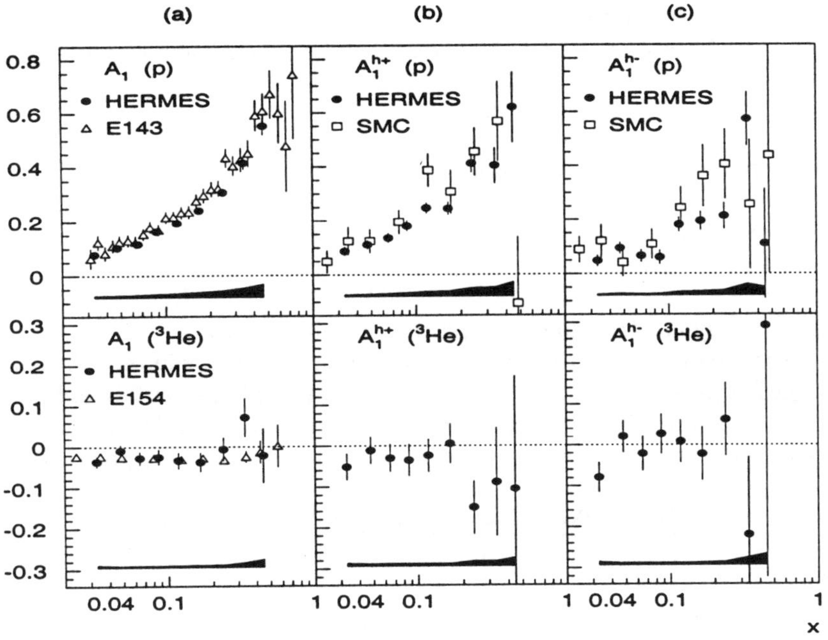

Figure 1: Results for the virtual photon asymmetries from HERMES 1995-7 data. The top row shows the asymmetries on the ^{1}H target; the bottom row, the ^{3}He target. The columns are, from left to right, the inclusive, semi-inclusive h^+, and semi-inclusive h^- asymmetries. The inclusive asymmetries are compared with the total uncertainties of the SLAC E143 [4] and E154 [5] g_1/F_1 measurements. For the proton semi-inclusive asymmetries, a comparison with the SMC [6] results are shown. All points are plotted at measured x and Q^2.

This expression may be rearranged by defining a quark purity for each event type h as

$$P_q^h(x,z) = \frac{e_q^2\, q(x)\, \int_{0.2}^{1.0} dz D_q^h(z)}{\sum_{q'} e_{q'}^2\, q'(x)\, \int_{0.2}^{1.0} dz D_{q'}^h(z)}. \tag{8}$$

P_q^h represents the probability that a detected event originated from a struck quark of flavor q in the nucleon.

Using these purities, the asymmetries in Equation 7 may be rewritten as

$$A_1^h(x) = \sum_q P_q^h(x) \cdot \left(\frac{\Delta q(x)}{q(x)}\right) \frac{1+R}{1+\gamma^2}. \tag{9}$$

The asymmetries, purities, and quark polarizations may be grouped into matrices, and the system of equations may then be solved using matrix algebra,

$$\vec{\mathbf{A}}(x) = \mathbf{P} \cdot \vec{\mathbf{Q}}(x). \tag{10}$$

The quark polarizations are obtained by least squares minimization techniques.

The quark purities relevant for the HERMES experiment have been estimated with a Monte Carlo simulation of polarized DIS. The simulation model is based on LUND string fragmentation, [7] CTEQ low Q^2 unpolarized parton distributions, [8] and a model of the HERMES detector. Nuclear corrections relate ^{3}He asymmetries to quark distributions in the proton. [9]

In solving Equation 10, one would like to separate the spin contributions of six quark flavors: u, d, s, $\bar{u}$, $\bar{d}$, and $\bar{s}$. In practice, limited statistics and limited sensitivity to the sea quark flavors require selection of a model of average sea polarization. A simple assumption of flavor independent sea polarization is used here:

$$\frac{\Delta q_s}{q_s} \equiv \frac{\Delta d_s}{d_s} = \frac{\Delta u_s}{u_s} = \frac{\Delta \bar{d}}{\bar{d}} = \frac{\Delta \bar{u}}{\bar{u}} = \frac{\Delta \bar{s}}{\bar{s}} = \frac{\Delta s}{s}. \tag{11}$$

4 Extracted Polarized Quark Distributions

Using the six experimental asymmetries shown in Figure 1, three quark polarizations have been extracted: $(\Delta u + \Delta\bar{u})/(u + \bar{u})(x)$, the total up flavor polarization; $(\Delta d + \Delta\bar{d})/(d + \bar{d})(x)$, the total down flavor polarization; and $(\Delta q_s/q_s)(x)$, the average polarization of the quark sea. After multiplying by x and the unpolarized PDF parameterization, Figure 2 shows the results from this extraction in nine x_{Bj} bins. The extraction has been repeated with GRV [10] parton distributions, with the independent fragmentation model, and with different parameterizations for the fit of the LUND string model to HERMES multiplicities; a systematic uncertainty is assigned for the variation in the resulting quark polarizations. For $x > 0.3$, it is assumed that sea polarization does not contribute significantly to the measured asymmetries; the sea polarization, $\frac{\Delta q_s}{q_s}(x)$, is set to zero with a small systematic uncertainty.

In Figure 2, the parameterizations of De Florian *et. al.* ($0.1 < \Delta G < 0.8$, LO) [11] Gehrmann and Stirling (Gluon A, LO),[13] and Glück *et. al.* (Standard, LO) [12] are provided for comparison. A correction of 1+R to the De Florian and Glück parameterizations is necessary to achieve consistency in model assumptions. The HERMES data agrees with these parameterizations within uncertainties. Figure 3 compares the extracted valence quark polarization to a similar extraction by the SMC [6] experiment. Note that the HERMES definition of the flavor independence of sea polarization differs slightly from the

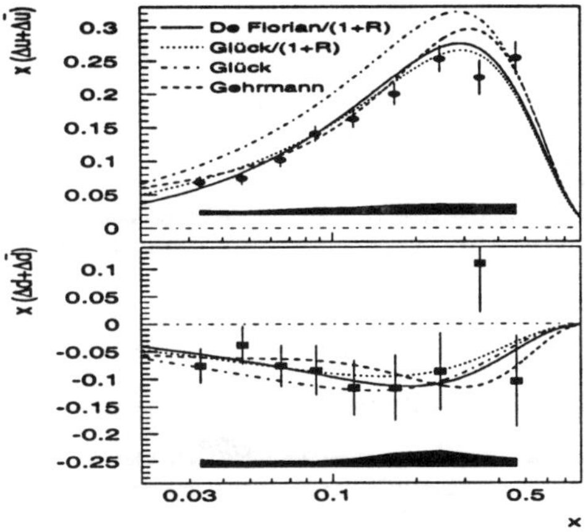

Figure 2: Polarized parton distributions for up and down quarks, $x(\Delta u+\Delta\bar{u})$ and $x(\Delta d+\Delta\bar{d})$, compared to parameterizations of world data. The data points have been evolved to Q^2=2.5 GeV^2 for the comparison. The parameterizations are explained in the text.

sea model chosen by SMC; it has been verified that the results are insensitive to the slight differences between the models. Within statistical and systematic uncertainties, the results of both experiments are equivalent, though high precision HERMES proton data has yielded more sensitivity to Δu_v and $\Delta\bar{u}$.

Moments of the quark distributions in the measured x range are determined from the area under the measured points. For comparison to previous measurements, the distributions are extrapolated to low x by constant fits to the data; the low x extrapolation is quoted but no value is quoted for the uncertainty due to theoretical ambiguities. The resulting measured region integrals are listed in Table 1. The integrals are compared to predictions from Reference 14, which are corrected to fourth order in QCD; these predictions have been extracted from inclusive data assuming $SU(3)_{\text{flavor}}$ symmetry. The HERMES values are slightly smaller in magnitude than these predictions.

5 Conclusions and Outlook

The first three years of HERMES data taking have yielded high statistics polarized deep-inelastic scattering data on ^{3}He and ^{1}H targets. The polarized measurements yield inclusive and semi-inclusive charged hadron asymmetries, which have been presented as a function of x. In the framework of the Quark-

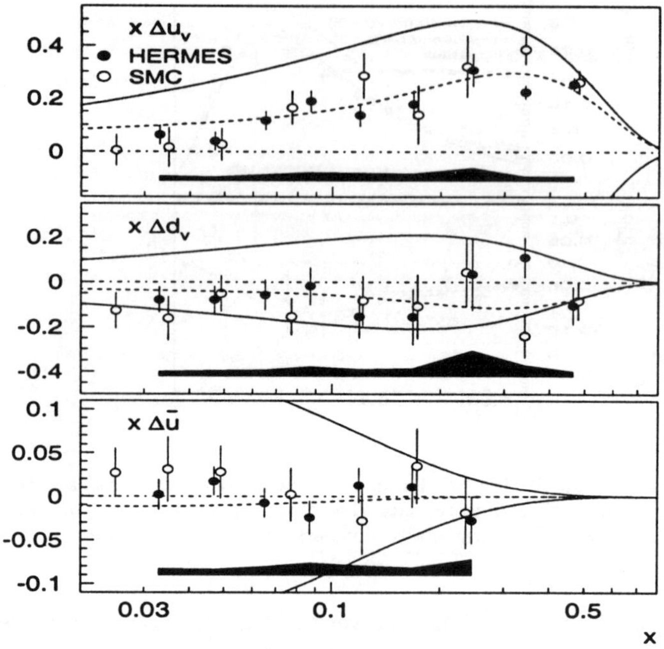

Figure 3: Valence separation of parton distributions at Q^2=2.5 GeV^2 for up and down and sea quarks, $x(\Delta u_v)$ and $x(\Delta d_v)$ and $x(\Delta q_s)$, compared to results from the SMC experiment. The SMC data points have been extrapolated to Q^2=2.5 GeV^2 for the comparison.

Parton Model, these asymmetries are simultaneously fit to yield helicity distributions for up, down and sea quarks.

Further sensitivity to sea polarizations is gained by identification of different coincident hadron species. Identification of pions from hadrons, e.g. by the HERMES threshold Čerenkov, may provide better sensitivity to Δs in the quark polarization extraction.[15] A concern in this analysis involves the knowledge of light quark fragmentation properties in separating strange quark influences on the asymmetry. A small breaking in isospin or charge conjugation symmetry in light quark fragmentation influences the semi-inclusive asymmetries in a manner similar to a large strange quark polarization.

HERMES will soon have a new data set with high statistics on a deuterium target, which will significantly improve the precision on the d-quark helicity distribution. Before the deuterium run began, the Threshold Čerenkov was replaced with a Ring Imaging Čerenkov(RICH). This new detector will identify

Table 1: The first moments of the extracted polarized parton distributions. The results are given at Q^2=2.5 GeV^2 for the measured region $0.023 < x < 0.6$, for the low-x extrapolation and for the total integral. Note that the values for $\Delta s + \Delta\bar{s}$ relies on the assumption in Eq. (11).

	measured region	low-x	total integral	prediction
$\Delta u + \Delta\bar{u}$	$0.51 \pm 0.02 \pm 0.03$	0.04	$0.57 \pm 0.02 \pm 0.03$	0.66 ± 0.03
$\Delta d + \Delta\bar{d}$	$-0.22 \pm 0.06 \pm 0.05$	-0.03	$-0.25 \pm 0.06 \pm 0.05$	-0.35 ± 0.03
$\Delta s + \Delta\bar{s}$	$-0.01 \pm 0.03 \pm 0.04$	0.00	$-0.01 \pm 0.03 \pm 0.04$	-0.08 ± 0.02

π, K and p particles over nearly all momenta accepted by the spectrometer. As kaon asymmetries provide direct sensitivity to strange quark polarization in the nucleon, a much higher precision determination of sea quark polarization is expected after the new run.

Acknowledgments

The author would like to acknowledge the support of the U.S. Department of Energy.

References

1. K. Ackerstaff *et al*, *Nucl. Instrum. Methods* A **417**, 230 (1998).
2. D. De Schepper *et al*, *Nucl. Instrum. Methods* A **419**, 16 (1998).
3. B. Braun in Proc. of the Workshop "Polarized gas targets and polarized beams", ed. R.J. Holt and M.A. Miller, AIP Conf. Proc. **421** (1997) 69.
4. K. Abe *et al*, *Phys. Rev.* D **58**, 112003 (1998).
5. K. Abe *et al*, *Phys. Rev. Lett.* **79**, 26 (1997).
6. B. Adeva *et al*, *Phys. Rev.* D **58**, 112001 (1998).
7. B. Andersson *et al*, *Phys. Rept.* **97**, 33 (1983).
8. H.L. Lai *et al*, *Phys. Rev.* D **126**, 1280 (1997).
9. J.L. Friar *et al*, *Phys. Rev.* C **42**, 2310 (1990).
10. M. Glück *et al*, *Z. Phys.* C **67**, 433 (1995).
11. D. de Florian *et al*, *Phys. Rev.* D **57**, 5803 (1998).
12. M. Glück *et al*, *Phys. Rev.* D **53**, 4775 (1996).
13. T. Gehrmann and W. J. Stirling, *Phys. Rev.* D **53**, 6100 (1996).
14. J. Ellis and M. Karliner, Invited Lectures at the International School of Nucleon Spin Structure, Erice, August 1995, hep-ph/9601280.
15. L.L. Frankfurt *et al*, *Phys. Lett.* B **230**, 141 (1989).

HARD EXCLUSIVE AND DIFFRACTIVE PROCESSES IN QCD

S. J. BRODSKY
Stanford Linear Accelerator Center
Stanford University, Stanford, California 94309

Exclusive and semi-exclusive processes, the diffractive dissociation of hadrons into jets, and hard diffractive processes such as vector meson leptoproduction provide new testing grounds for QCD and essential information on the structure of light-cone wavefunctions of hadrons, particularly the pion distribution amplitude. I review the basic features of the leading-twist QCD predictions and the problems and challenges of studying QCD at the amplitude level. The application of the light-cone formalism to the exclusive semi-leptonic decay of heavy hadrons is also discussed.

1 Introduction

Exclusive hard-scattering reactions and hard diffractive reactions are now providing an invaluable window into the structure and dynamics of hadronic amplitudes. Recent measurements of the photon-to-pion transition form factor at CLEO,[1] the diffractive dissociation of pions into jets at Fermilab,[2] diffractive vector meson leptoproduction at Fermilab and HERA, and the new program of experiments on exclusive proton and deuteron processes at Jefferson Laboratory are now yielding fundamental information on hadronic wavefunctions, particularly the distribution amplitude of mesons. Such information is also critical for interpreting exclusive heavy hadron decays and the matrix elements and amplitudes entering CP-violating processes at the B factories.

The natural formalism for describing the hadronic wavefunctions which enter exclusive and diffractive amplitudes is the light-cone Fock representation obtained by quantizing the theory at fixed at fixed "light-cone" time $\tau = t + z/c$.[3] This representation is the extension of Schrödinger many-body theory to the relativistic domain. In a quantum theory a bound state cannot have a fixed number of constituents. For example, the proton state has the Fock expansion

$$\begin{aligned} |p\rangle &= \sum_n \langle n\,|\,p\rangle\,|\,n\rangle \\ &= \psi^{(\Lambda)}_{3q/p}(x_i, \vec{k}_{\perp i}, \lambda_i)\,|\,uud\rangle \\ &\quad + \psi^{(\Lambda)}_{3qg/p}(x_i, \vec{k}_{\perp i}, \lambda_i)\,|\,uudg\rangle + \cdots \end{aligned} \tag{1}$$

representing the expansion of the exact QCD eigenstate on a non-interacting quark and gluon basis. The probability amplitude for each such n-particle state of on-mass shell quarks and gluons in a hadron is given by a light-cone Fock state wavefunction $\psi_{n/H}(x_i, \vec{k}_{\perp i}, \lambda_i)$, where the constituents have longitudinal light-cone momentum fractions $x_i = k_i^+/p^+ = (k_i^0+k_i^z)/(p^0+p^z)$, $\sum_{i=1}^n x_i = 1$, relative transverse momentum $\vec{k}_{\perp i}$, $\sum_{i=1}^n \vec{k}_{\perp i} = \vec{0}_\perp$, and helicities λ_i. The effective lifetime of each configuration in the laboratory frame is $2P_{lab}/(\mathcal{M}_n^2 - M_p^2)$ where $\mathcal{M}_n^2 = \sum_{i=1}^n (k_{\perp i}^2 + m_i^2)/x_i < \Lambda^2$ is the off-shell invariant mass and Λ is a global ultraviolet regulator. A crucial feature of the light-cone formalism is the fact that the form of the $\psi_{n/H}^{(\Lambda)}(x_i, \vec{k}_{\perp i}, \lambda_i)$ is invariant under longitudinal boosts; *i.e.*, the light-cone wavefunctions expressed in the relative coordinates x_i and $k_{\perp i}$ are independent of the total momentum P^+, $\vec{P}_\perp$ of the hadron. The ensemble $\{\psi_{n/H}\}$ of such light-cone Fock wavefunctions is a key concept for hadronic physics, providing a conceptual basis for representing physical hadrons (and also nuclei) in terms of their fundamental quark and gluon degrees of freedom. Given the $\psi_{n/H}^{(\Lambda)}$, we can construct any spacelike electromagnetic or electroweak form factor from the diagonal overlap of the LC wavefunctions.[4] Similarly, the matrix elements of the currents that define quark and gluon structure functions can be computed from the integrated squares of the LC wavefunctions.[5]

There has been much progress analyzing exclusive and diffractive reactions at large momentum transfer from first principles in QCD. Rigorous statements can be made on the basis of asymptotic freedom and factorization theorems which separate the underlying hard quark and gluon subprocess amplitude from the nonperturbative physics incorporated into the process-independent hadron distribution amplitudes $\phi_H(x_i, Q)$,[5] the valence light-cone wavefunctions integrated over $k_\perp^2 < Q^2$. An important new application is the recent analysis of hard exclusive B decays by Beneke, *et al.*[6] Key features of such analyses are: (a) evolution equations for distribution amplitudes which incorporate the operator product expansion, renormalization group invariance, and conformal symmetry;[5,7,8,9,10,11] (b) hadron helicity conservation which follows from the underlying chiral structure of QCD;[12] (c) color transparency, which eliminates corrections to hard exclusive amplitudes from initial and final state interactions at leading power and reflects the underlying gauge theoretic basis for the strong interactions;[13,14] and (d) hidden color degrees of freedom in nuclear wavefunctions, which reflects the color structure of hadron and nuclear wavefunctions.[15] There have also been recent advances eliminating renormalization scale ambiguities in hard-scattering amplitudes via commensurate scale relations[16,17,18] which connect the couplings entering exclusive amplitudes to the α_V coupling which controls the QCD heavy quark potential.[19] The postulate that the QCD coupling has an infrared fixed-point provides an understanding of the applicability of conformal scaling and dimensional counting rules to physical QCD processes.[20,21,19] The field of analyzable exclusive processes has recently been expanded to a new range of QCD processes, such as electroweak decay amplitudes, highly virtual diffractive processes such as $\gamma^* p \to \rho p$,[22,23] and semi-exclusive processes such as $\gamma^* p \to \pi^+ X$ [24,25,26] where the π^+ is produced in isolation at large p_T.

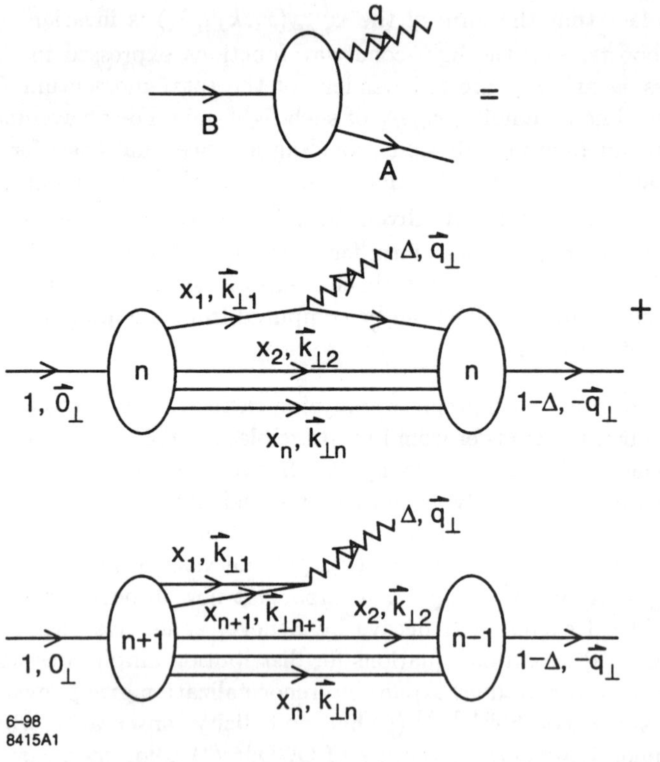

Figure 1: Exact representation of electroweak decays and time-like form factors in the light-cone Fock representation.

2 Electoweak Decays and the Light-Cone Fock Expansion

Dae Sung Hwang and I have recently shown how exclusive semi-leptonic B-decay amplitudes, such as $B \to A\ell\bar{\nu}$ can be evaluated exactly in the light-cone formalism.[27] These timelike decay matrix elements require the computation of the diagonal matrix element $n \to n$ where parton number is conserved, and the off-diagonal $n+1 \to n-1$ convolution where the current operator annihilates a $q\overline{q'}$ pair in the initial B wavefunction. See Fig. 1. This term is a consequence of the fact that the time-like decay $q^2 = (p_\ell + p_{\overline{\nu}})^2 > 0$ requires a positive light-cone momentum fraction $q^+ > 0$. Conversely for space-like currents, one can choose $q^+ = 0$, as in the Drell-Yan-West representation of the space-like electromagnetic form factors.[28,4,29] However, the off-diagonal convolution can yield a nonzero q^+/q^+ limiting form as $q^+ \to 0$. This extra term appears specifically in the case of "bad" currents such as J^- in which the coupling to $q\overline{q}$ fluctuations in the light-cone wavefunctions are favored. In effect, the $q^+ \to 0$ limit generates $\delta(x)$ contributions as residues of the $n+1 \to n-1$ contributions. The necessity for this zero mode $\delta(x)$ terms has been noted by Chang, Root and Yan,[30] and Burkardt.[31]

The off-diagonal $n+1 \to n-1$ contributions provide a new perspective for the physics of B-decays. A semi-leptonic decay involves not only matrix elements where a quark changes flavor, but also a contribution where the leptonic pair is created from the annihilation of a $q\overline{q'}$ pair within the Fock states of the initial B wavefunction. The semi-leptonic decay thus can occur from the annihilation of a nonvalence quark-antiquark pair in the initial hadron. This feature will carry over to exclusive hadronic B-decays, such as $B^0 \to \pi^- D^+$. In this case the pion can be produced from the coalescence of a $d\overline{u}$ pair emerging from the initial higher particle number Fock wavefunction of the B. The D meson is then formed from the remaining quarks after the internal exchange of a W boson.

In principle, a precise evaluation of the hadronic matrix elements needed for B-decays and other exclusive electroweak decay amplitudes requires knowledge of all of the light-cone Fock wavefunctions of the initial and final state hadrons. In the case of model gauge theories such as QCD(1+1) [32] or collinear QCD [33] in one-space and one-time dimensions, the complete evaluation of the light-cone wavefunction is possible for each baryon or meson bound-state using the DLCQ method.[34,33] It would be interesting to use such solutions as a model for physical B-decays.

3 The Transition from Soft to Hard Physics

The existence of an exact formalism provides a basis for systematic approximations and a control over neglected terms. For example, one can analyze exclusive semi-leptonic B-decays which involve hard internal momentum transfer using a perturbative QCD formalism[35,6] patterned after the analysis of form factors at large momentum transfer.[5] The hard-scattering analysis proceeds by writing each hadronic wavefunction as a sum of soft and hard contributions

$$\psi_n = \psi_n^{\rm soft}(\mathcal{M}_n^2 < \Lambda^2) + \psi_n^{\rm hard}(\mathcal{M}_n^2 > \Lambda^2), \tag{2}$$

where $\mathcal{M}_n^2$ is the invariant mass of the partons in the n-particle Fock state and Λ is the separation scale. The high internal momentum contributions to the wavefunction $\psi_n^{\rm hard}$ can be calculated systematically from QCD perturbation theory by iterating the gluon exchange kernel. The contributions from high momentum transfer exchange to the B-decay amplitude can then be written as a convolution of a hard-scattering quark-gluon scattering amplitude T_H with the distribution amplitudes $\phi(x_i, \Lambda)$, the valence wavefunctions obtained by integrating the constituent momenta up to the separation scale $\mathcal{M}_n < \Lambda < Q$. This is the basis for the perturbative hard-scattering analyses.[35,36,37,6] In the exact analysis, one can identify the hard PQCD contribution as well as the soft contribution from the convolution of the light-cone wavefunctions. Furthermore, the hard-scattering contribution can be systematically improved.

4 Hard Exclusive Processes

In general, hard exclusive hadronic amplitudes such as quarkonium decay, heavy hadron decay, and scattering amplitudes where hadrons are scattered with large momentum transfer can be factorized at leading power as a convolution of distribution amplitudes and hard-scattering quark/gluon matrix elements[5]

$$\begin{aligned} \mathcal{M}_{\rm Hadron} &= \prod_H \sum_n \int \prod_{i=1}^{n} d^2k_\perp \prod_{i=1}^{n} dx\, \delta\left(1 - \sum_{i=1}^{n} x_i\right) \delta\left(\sum_{i=1}^{n} \vec{k}_{\perp i}\right) \\ &\quad \times \psi_{n/H}^{(\Lambda)}(x_i, \vec{k}_{\perp i}, \Lambda_i)\, T_H^{(\Lambda)} \,. \end{aligned} \tag{3}$$

Here $T_H^{(\Lambda)}$ is the underlying quark-gluon subprocess scattering amplitude in which the (incident and final) hadrons are replaced by their respective quarks and gluons with momenta $x_i p^+$, $x_i \vec{p}_\perp + \vec{k}_{\perp i}$ and invariant mass above the separation scale $\mathcal{M}_n^2 > \Lambda^2$. The essential part of the wavefunction is the hadronic distribution amplitudes, [5] defined as the integral over transverse momenta of the valence (lowest particle number) Fock wavefunction; *e.g.* for the pion

$$\phi_\pi(x_i, Q) \equiv \int d^2k_\perp\, \psi_{q\bar{q}/\pi}^{(Q)}(x_i, \vec{k}_{\perp i}, \lambda) \tag{4}$$

where the global cutoff Λ is identified with the resolution Q. The distribution amplitude controls leading-twist exclusive amplitudes at high momentum transfer, and it can be related to the gauge-invariant Bethe-Salpeter wavefunction at equal light-cone time $\tau = x^+$. The $\log Q$ evolution of the hadron distribution amplitudes $\phi_H(x_i, Q)$ can be derived from the perturbatively-computable tail of the valence light-cone wavefunction in the high transverse momentum regime. The LC ultraviolet regulators provide a factorization scheme for elastic and inelastic scattering, separating the hard dynamical contributions with invariant mass squared $\mathcal{M}^2 > \Lambda^2_{\rm global}$ from the soft physics with $\mathcal{M}^2 \leq \Lambda^2_{\rm global}$ which is incorporated in the nonperturbative LC wavefunctions. The DGLAP evolution of quark and gluon distributions can also be derived in an analogous way by computing the variation of the Fock expansion with respect to Λ^2. The natural renormalization scheme for the QCD coupling in hard exclusive processes is $\alpha_V(Q)$, the effective charge defined from the scattering of two infinitely-heavy quark test charges. The renormalization scale can then be determined from the virtuality of the exchanged momentum of the gluons, as in the BLM and commensurate scale methods.[38,16,17,18]

The features of exclusive processes to leading power in the transferred momenta are well known:

(1) The leading power fall-off is given by dimensional counting rules for the hard-scattering amplitude: $T_H \sim 1/Q^{n-1}$, where n is the total number of fields (quarks, leptons, or gauge fields) participating in the hard scattering.[20,21] Thus the reaction is dominated by subprocesses and Fock states involving the minimum number of interacting fields. The hadronic amplitude follows this fall-off modulo logarithmic corrections from the running of the QCD coupling, and the evolution of the hadron distribution amplitudes. In some cases, such as large angle $pp \to pp$ scattering, pinch contributions from multiple hard-scattering processes must also be included.[39] The general success of dimensional counting rules implies that the effective coupling $\alpha_V(Q^*)$ controlling the gluon exchange propagators in T_H are frozen in the infrared, *i.e.*, have an infrared fixed point, since the effective momentum transfers Q^* exchanged by the gluons are often a small fraction of the overall momentum transfer.[19] The pinch contributions are suppressed by a factor decreasing faster than a fixed power.[20]

(2) The leading power dependence is given by hard-scattering amplitudes T_H which conserve quark helicity.[12,40] Since the convolution of T_H with the light-cone wavefunctions projects out states with $L_z = 0$, the leading hadron amplitudes conserve hadron helicity; *i.e.*, the sum of initial and final hadron helicities are conserved.

(3) Since the convolution of the hard scattering amplitude T_H with the light-cone wavefunctions projects out the valence states with small impact pa-

rameter, the essential part of the hadron wavefunction entering a hard exclusive amplitude has a small color dipole moment. This leads to the absence of initial or final state interactions among the scattering hadrons as well as the color transparency. of quasi-elastic interactions in a nuclear target.[13,14] For example, the amplitude for diffractive vector meson photoproduction $\gamma^*(Q^2)p \to \rho p$, can be written as convolution of the virtual photon and the vector meson Fock state light-cone wavefunctions the $gp \to gp$ near-forward matrix element.[22] One can easily show that only small transverse size $b_\perp \sim 1/Q$ of the vector meson distribution amplitude is involved. The sum over the interactions of the exchanged gluons tend to cancel reflecting its small color dipole moment. Since the hadronic interactions are minimal, the $\gamma^*(Q^2)N \to \rho N$ reaction at large Q^2 can occur coherently throughout a nuclear target in reactions without absorption or final state interactions. The $\gamma^* A \to VA$ process thus provides a natural framework for testing QCD color transparency. Evidence for color transparency in such reactions has been found by Fermilab experiment E665.[41]

5 Measurement of Light-cone Wavefunctions and Tests of Color Transparency via Diffractive Dissociation.

Diffractive multi-jet production in heavy nuclei provides a novel way to measure the shape of the LC Fock state wavefunctions and test color transparency. For example, consider the reaction [42,43,44] $\pi A \to \mathrm{Jet}_1 + \mathrm{Jet}_2 + A'$ at high energy where the nucleus A' is left intact in its ground state. The transverse momenta of the jets have to balance so that $\vec{k}_{\perp i} + \vec{k}_{\perp 2} = \vec{q}_\perp < R^{-1}{}_A$, and the light-cone longitudinal momentum fractions have to add to $x_1 + x_2 \sim 1$ so that $\Delta p_L < R_A^{-1}$. The process can then occur coherently in the nucleus. Because of color transparency, *i.e.*, the cancelation of color interactions in a small-size color-singlet hadron, the valence wavefunction of the pion with small impact separation will penetrate the nucleus with minimal interactions, diffracting into jet pairs.[42] The $x_1 = x$, $x_2 = 1 - x$ dependence of the di-jet distributions will thus reflect the shape of the pion distribution amplitude; the $\vec{k}_{\perp 1} - \vec{k}_{\perp 2}$ relative transverse momenta of the jets also gives key information on the underlying shape of the valence pion wavefunction.[43,44] The QCD analysis can be confirmed by the observation that the diffractive nuclear amplitude extrapolated to $t = 0$ is linear in nuclear number A, as predicted by QCD color transparency. The integrated diffractive rate should scale as $A^2/R_A^2 \sim A^{4/3}$. A diffractive dissociation experiment of this type, E791, is now in progress at Fermilab using 500 GeV incident pions on nuclear targets.[2] The preliminary results from E791 appear to be consistent with color transparency. The momentum fraction distribution of the jets is consistent with a valence light-cone wavefunction of the pion consistent with the shape of the asymptotic distribution amplitude, $\phi_\pi^{\rm asympt}(x) = \sqrt{3} f_\pi x(1-x)$. As discussed below, data from CLEO[1] for the $\gamma\gamma^* \to \pi^0$ transition form factor also favor a form for the pion distribution amplitude close to the asymptotic solution[5] to the perturbative QCD evolution equation.[45,46,19,47,48] It will also be interesting to study diffractive tri-jet production using proton beams $pA \to {\rm Jet}_1 + {\rm Jet}_2 + {\rm Jet}_3 + A'$ to determine the fundamental shape of the 3-quark structure of the valence light-cone wavefunction of the nucleon at small transverse separation.[43] One interesting possibility is that the distribution amplitude of the $\Delta(1232)$ for $J_z = 1/2, 3/2$ is close to the asymptotic form $x_1 x_2 x_3$, but that the proton distribution amplitude is more complex. This would explain why the $p \to \Delta$ transition form factor appears to fall faster at large Q^2 than the elastic $p \to p$ and the other $p \to N^*$ transition form factors.[49] Conversely, one can use incident real and virtual photons: $\gamma^* A \to {\rm Jet}_1 + {\rm Jet}_2 + A'$ to confirm the shape of the calculable light-cone wavefunction for transversely-polarized and longitudinally-polarized virtual photons. Such experiments will open up a direct window on the amplitude structure of hadrons at short distances.

6 Leading Power Dominance in Exclusive QCD Processes

There are a large number of measured exclusive reactions in which the empirical power law fall-off predicted by dimensional counting and PQCD appears to be accurate over a large range of momentum transfer. These include processes such as the proton form factor, time-like meson pair production in e^+e^- and $\gamma\gamma$ annihilation, large-angle scattering processes such as pion photoproduction $\gamma p \to \pi^+ p$, and nuclear processes such as the deuteron form factor at large momentum transfer and deuteron photodisintegration.[50] A spectacular example is the recent measurements at CESR of the photon to pion transition form factor in the reaction $e\gamma \to e\pi^0$.[1] As predicted by leading twist QCD[5] $Q^2 F_{\gamma\pi^0}(Q^2)$ is essentially constant for 1 GeV$^2 < Q^2 <$ 10 GeV^2. Further, the normalization is consistent with QCD at NLO if one assumes that the pion distribution amplitude takes on the form $\phi_\pi^{\rm asympt}(x) = \sqrt{3} f_\pi x(1-x)$ which is the asymptotic solution[5] to the evolution equation for the pion distribution amplitude.[45,46,19,48]

The measured deuteron form factor and the deuteron photodisintegration cross section appear to follow the leading-twist QCD predictions at large momentum transfers in the few GeV region.[51,52] The normalization of the measured deuteron form factor is large compared to model calculations [53] assuming that the deuteron's six-quark wavefunction can be represented at short distances with the color structure of two color singlet baryons. This provides indirect evidence for the presence of hidden color components as required by PQCD.[15]

There are, however, experimental exceptions to the general success of the leading twist PQCD approach, such as (a) the dominance of the $J/\psi \to \rho\pi$ decay which is forbidden by hadron helicity conservation and (b) the strong normal-normal spin asymmetry A_{NN} observed in polarized elastic $pp \to pp$ scattering and an apparent breakdown of color transparency at large CM angles and $E_{CM} \sim 5$ GeV. These conflicts with leading-twist PQCD predictions can be used to identify the presence of new physical effects. For example, It is usually assumed that a heavy quarkonium state such as the J/ψ always decays to light hadrons via the annihilation of its heavy quark constituents to gluons. However, as Karliner and I [54] have recently shown, the transition $J/\psi \to \rho\pi$ can also occur by the rearrangement of the $c\bar{c}$ from the J/ψ into the $|\, q\bar{q}c\bar{c}\rangle$ intrinsic charm Fock state of the ρ or π. On the other hand, the overlap rearrangement integral in the decay $\psi' \to \rho\pi$ will be suppressed since the intrinsic charm Fock state radial wavefunction of the light hadrons will evidently not have nodes in its radial wavefunction. This observation provides a natural explanation of the long-standing puzzle why the J/ψ decays prominently to two-body pseudoscalar-vector final states, whereas the ψ' does not. The unusual effects seen in elastic proton-proton scattering at $E_{CM} \sim 5$ GeV and large angles could be related to the charm threshold and the effect of a $|\, uuduudc\bar{c}\rangle$ resonance which would appear as in the $J = L = S = 1$ pp partial wave.[55]

If the pion distribution amplitude is close to its asymptotic form, then one can predict the normalization of exclusive amplitudes such as the space-like pion form factor $Q^2 F_\pi(Q^2)$. Next-to-leading order predictions are now becoming available which incorporate higher order corrections to the pion distribution amplitude as well as the hard scattering amplitude.[9,56,57] However, the normalization of the PQCD prediction for the pion form factor depends directly on the value of the effective coupling $\alpha_V(Q^*)$ at momenta $Q^{*2} \simeq Q^2/20$. Assuming $\alpha_V(Q^*) \simeq 0.4$, the QCD LO prediction appears to be smaller by approximately a factor of 2 compared to the presently available data extracted from the original pion electroproduction experiments from CEA.[58] A definitive comparison will require a careful extrapolation to the pion pole and extraction of the longitudinally polarized photon contribution of the $ep \to \pi^+ n$ data.

A debate has continued on whether processes such as the pion and proton form factors and elastic Compton scattering $\gamma p \to \gamma p$ might be dominated by higher twist mechanisms until very large momentum transfers.[59,60,61] For example, if one assumes that the light-cone wavefunction of the pion has the form $\psi_{\rm soft}(x, k_\perp) = A \exp(-b \frac{k_\perp^2}{x(1-x)})$, then the Feynman endpoint contribution to the overlap integral at small $k_\perp$ and $x \simeq 1$ will dominate the form factor compared to the hard-scattering contribution until very large Q^2. However, the above form of $\psi_{\rm soft}(x, k_\perp)$ has no suppression at $k_\perp = 0$ for any x; *i.e.*, the wavefunction in the hadron rest frame does not fall-off at all for $k_\perp = 0$ and $k_z \to -\infty$. Thus such wavefunctions do not represent well soft QCD contributions. Furthermore, such endpoint contributions will be suppressed by the QCD Sudakov form factor, reflecting the fact that a near-on-shell quark must radiate if it absorbs large momentum. If the endpoint contribution dominates proton Compton scattering, then both photons will interact on the same quark line in a local fashion and the amplitude is real, in strong contrast to the QCD predictions which have a complex phase structure. The perturbative QCD predictions[62] for the Compton amplitude phase can be tested in virtual Compton scattering by interference with Bethe-Heitler processes.[63] It should be noted that there is no apparent endpoint contribution which could explain the success of dimensional counting in large angle pion photoproduction.

It is interesting to compare the corresponding calculations of form factors of bound states in QED. The soft wavefunction is the Schrödinger-Coulomb solution $\psi_{1s}(\vec{k}) \propto (1 + \vec{p}^2/(\alpha m_{\rm red})^2)^{-2}$, and the full wavefunction, which incorporates transversely polarized photon exchange, only differs by a factor $(1 + \vec{p}^2/m_{\rm red}^2)$. Thus the leading twist dominance of form factors in QED occurs at relativistic scales $Q^2 > m_{\rm red}^2$.[64] Furthermore, there are no extra relative factors of α in the hard-scattering contribution. If the QCD coupling α_V has an infrared fixed point, then the fall-off of the valence wavefunctions of hadrons will have analogous power-law forms, consistent with the Abelian correspondence principle.[65] If such power-law wavefunctions are indeed applicable to the soft domain of QCD then, the transition to leading-twist power law behavior will occur in the nominal hard perturbative QCD domain where $Q^2 \gg \langle k_\perp^2 \rangle, m_q^2$.

Acknowledgments

I thank Carl Carlson, Anatoly Radyushkin, and their colleagues at Jefferson Laboratory for their kind hospitality at this conference.

References

1. J. Gronberg *et al.* [CLEO Collaboration], *Phys. Rev.* **D57**, 33 (1998) hep-ex/9707031.
2. D. Ashery, *et al.*, Fermilab E791 Collaboration, to be published.

3. S. J. Brodsky, H. Pauli and S. S. Pinsky, *Phys. Rept.* **301**, 299 (1998) hep-ph/9705477.
4. S. J. Brodsky and S. D. Drell, *Phys. Rev.* **D22**, 2236 (1980).
5. G. P. Lepage and S. J. Brodsky, *Phys. Rev.* **D22**, 2157 (1980); *Phys. Lett.* **B87**, 359 (1979); *Phys. Rev. Lett.* **43**, 545, 1625(E) (1979).
6. M. Beneke, G. Buchalla, M. Neubert and C.T. Sachrajda, hep-ph/9905312.
7. S. J. Brodsky, Y. Frishman, G. P. Lepage and C. Sachrajda, *Phys. Lett.* **91B**, 239 (1980).
8. S. J. Brodsky, Y. Frishman and G. P. Lepage, *Phys. Lett.* **167B**, 347 (1986).
9. D. Müller, *Phys. Rev.* **D49**, 2525 (1994).
10. P. Ball and V. M. Braun, *Nucl. Phys.* **B543**, 201 (1999) hep-ph/9810475.
11. V.M. Braun, S.E. Derkachov, G.P. Korchemsky and A.N. Manashov, hep-ph/9902375.
12. S. J. Brodsky and G. P. Lepage, *Phys. Rev.* **D24**, 2848 (1981).
13. S. J. Brodsky and A. H. Mueller, *Phys. Lett.* **206B**, 685 (1988). L. Frankfurt and M. Strikman, *Phys. Rept.* **160** , 235 (1988); P. Jain, B. Pire and J. P. Ralston, *Phys. Rept.* **271**, 67(1996).
14. L. Frankfurt, G. A. Miller and M. Strikman, *Comments Nucl. Part. Phys.* **21**, 1 (1992).
15. S. J. Brodsky, C.-R. Ji, and G. P. Lepage, *Phys. Rev. Lett.* **51**, 83 (1983).
16. S. J. Brodsky and H. J. Lu, *Phys. Rev.* **D51**, 3652 (1995) hep-ph/9405218.
17. S. J. Brodsky, G. T. Gabadadze, A. L. Kataev and H. J. Lu, *Phys. Lett.* **B372**, 133 (1996) hep-ph/9512367.
18. S. J. Brodsky and J. Rathsman, hep-ph/9906339.
19. S. J. Brodsky, C. Ji, A. Pang and D. G. Robertson, hep-ph/9705221.
20. S. J. Brodsky and G. R. Farrar, *Phys. Rev.* **D11**, 1309 (1975).
21. V. A. Matveev, R. M. Muradian and A. N. Tavkhelidze, *Nuovo Cim. Lett.* **7**, 719 (1973).
22. S. J. Brodsky, L. Frankfurt, J. F. Gunion, A. H. Mueller and M. Strikman, *Phys. Rev.* **D50**, 3134 (1994) hep-ph/9402283.
23. J.C. Collins, L. Frankfurt and M. Strikman, hep-ph/9709336.
24. C. E. Carlson and A. B. Wakely, *Phys. Rev.* **D48**, 2000 (1993); A. Afanasev, C. E. Carlson and C. Wahlquist, *Phys. Lett.* **B398**, 393 (1997), hep-ph/9701215 and *Phys. Rev.* **D58**, 054007 (1998), hep-ph/9706522.
25. S. J. Brodsky, M. Diehl, P. Hoyer and S. Peigne, *Phys. Lett.* **B449**, 306 (1999) hep-ph/9812277.

26. J. F. Gunion, S. J. Brodsky and R. Blankenbecler, *Phys. Rev.* **D6**, 2652 (1972); R. Blankenbecler and S. J. Brodsky, *Phys. Rev.* **D10**, 2973 (1974).
27. S. J. Brodsky and D. S. Hwang, *Nucl. Phys.* **B543**, 239 (1999) hep-ph/9806358.
28. S. D. Drell and T. M. Yan, *Phys. Rev. Lett.* **24**, 181 (1970).
29. G. B. West, *Phys. Rev. Lett.* **24** 1206 (1970).
30. S. J. Chang, R. G. Root and T. M. Yan, *Phys. Rev.* **D7**, 1133 (1973).
31. M. Burkardt, *Nucl. Phys.* **A504**, 762 (1989); *Nucl. Phys.* **B373**, 613 (1992); *Phys. Rev.* **D52**, 3841 (1995).
32. K. Hornbostel, S. J. Brodsky, and H. C. Pauli, *Phys. Rev.* **D41** 3814 (1990).
33. F. Antonuccio and S. Dalley, *Phys. Lett.* **B348**, 55 (1995); *Phys. Lett.* **B376**, 154 (1996); *Nucl. Phys.* **B461**, 275 (1996).
34. H. C. Pauli and S. J. Brodsky, *Phys. Rev.* **D32**, 1993 (1985); *Phys. Rev.* **D32**, 2001 (1985).
35. A. Szczepaniak, E. M. Henley, and S. J. Brodsky, *Phys. Lett.* **B243** 287 (1990).
36. A. Szczepaniak, *Phys. Rev.* **D54** 1167 (1996).
37. P. Ball and V. M. Braun, *Phys. Rev.* **D58**, 094016 (1998) hep-ph/9805422.
38. S. J. Brodsky, G. P. Lepage and P. B. Mackenzie, *Phys. Rev.* **D28**, 228 (1983).
39. P.V. Landshoff, *Phys. Rev.* **D10**, 1024 (1974).
40. V. Chernyak, hep-ph/9906387.
41. M.R. Adams *et al.* [E665 Collaboration], *Z. Phys.* **C74**, 237 (1997).
42. G. Bertsch, S. J. Brodsky, A. S. Goldhaber, and J. F. Gunion, *Phys. Rev. Lett.* **47**, 297 (1981).
43. L. Frankfurt, G. A. Miller, and M. Strikman, *Phys. Lett.* **B304**, 1 (1993), hep-ph/9305228.
44. L. Frankfurt, G. A. Miller and M. Strikman, hep-ph/9907214.
45. P. Kroll and M. Raulfs, *Phys. Lett.* **B387**, 848 (1996).
46. I. V. Musatov and A. V. Radyushkin, *Phys. Rev.* **D56**, 2713 (1997).
47. T. Feldmann, hep-ph/9907226.
48. A. Schmedding and O. Yakovlev, hep-ph/9905392.
49. P. Stoler, *Few Body Syst. Suppl.* **11**, 124 (1999).
50. S. J. Brodsky and B. T. Chertok, *Phys. Rev.* **D14**, 3003 (1976).
51. R. J. Holt, *Phys. Rev.* **C41**, 2400 (1990).
52. C. Bochna *et al.* [E89-012 Collaboration], *Phys. Rev. Lett.* **81**, 4576 (1998) nucl-ex/9808001.

53. G. R. Farrar, K. Huleihel and H. Zhang, *Phys. Rev. Lett.* **74**, 650 (1995).
54. S. J. Brodsky and M. Karliner, *Phys. Rev. Lett.* **78**, 4682 (1997) hep-ph/9704379.
55. S. J. Brodsky and G. F. de Teramond, *Phys. Rev. Lett.* **60**, 1924 (1988).
56. B. Melic, B. Nizic and K. Passek, hep-ph/9903426.
57. A. Szczepaniak, A. Radyushkin and C. Ji, *Phys. Rev.* **D57**, 2813 (1998) hep-ph/9708237.
58. C. J. Bebek *et al.*, *Phys. Rev.* **D13**, 25 (1976).
59. N. Isgur and C. H. Llewellyn Smith, *Phys. Lett.* **B217**, 535 (1989).
60. A. V. Radyushkin, *Phys. Rev.* **D58**, 114008 (1998) hep-ph/9803316.
61. J. Bolz and P. Kroll, *Z. Phys.* **A356**, 327 (1996) hep-ph/9603289.
62. A.S. Kronfeld and B. Nizic, *Phys. Rev.* **D44**, 3445 (1991).
63. S. J. Brodsky, F. E. Close and J. F. Gunion, *Phys. Rev.* **D6**, 177 (1972).
64. For reviews and further references see S. J. Brodsky and G. P. Lepage, SLAC-PUB-4947. Published in 'Perturbative Quantum Chromodynamics', Ed. by A.H. Mueller, World Scientific Publ. Co. (1989), p. 93-240 (QCD161:M83); V. L. Chernyak and A. R. Zhitnitsky, *Phys. Rept.* **112**, 173 (1984).
65. S.J. Brodsky and P. Huet, *Phys. Lett.* **B417**, 145 (1998) hep-ph/9707543.

QUARK ORBITAL ANGULAR MOMENTUM AND OFF-FORWARD PARTON DISTRIBUTION

XIANGDONG JI
Department of Physics, University of Maryland
College Park, Maryland 20742-4111, USA
E-mail: xji@physics.umd.edu

Since the EMC publication of its polarized deep-inelastic scattering data on protons [?], the problem of the nucleon spin structure has been with us for more than a decade. A couple of years ago, I introduced the off-forward parton distribution and deeply-virtual Compton scattering as by-products of a study of the proton spin problem [?]. I am happy to see that much progress has been made in these directions and, in particular, that many talks in this workshop are related to this interesting subject. In my presentation, I will stick to the original motivation and discuss the progress on the angular momentum distribution and its relation to the off-forward parton distributions.

In a 1990 paper, Jaffe and Manohar pointed out that there are numerous sources for the spin of the proton in quantum chromodynamics (QCD) [?],

$$\vec{J} = \int d^3\vec{x} \left[\frac{1}{2}\bar{\psi}\vec{\gamma}\gamma_5\psi + \psi^\dagger \vec{x} \times (-i\vec{\nabla})\psi + \vec{E} \times \vec{A} + E_i(\vec{x} \times \vec{\nabla})A_i \right] . \tag{1}$$

In this form, the physical meaning of the individual terms seems obvious: The first term is the quark spin, the second the quark orbital angular momentum, the third the gluon spin, and the final the gluon orbital angular momentum. According to the above equation, one can write down a sum rule for the nucleon spin,

$$\frac{1}{2} = \frac{1}{2}\Delta\Sigma(\mu^2) + L_q'(\mu^2) + \Delta g(\mu^2) + L_g'(\mu^2) , \tag{2}$$

where, for instance,

$$\Delta g(\mu^2) = \langle ps| \int d^3\vec{x}(\vec{E} \times \vec{A})^z |ps\rangle . \tag{3}$$

According to a recent analysis of the global deep-inelastic scattering data [?], $\Delta\Sigma = 0.2 \sim 0.3$, with a small dependence on the probing scale. From their definitions, it is clear that L'_q, Δg and L'_g are gauge, and hence frame, dependent.

Nonetheless, $\Delta g(\mu^2)$ in the infinite momentum frame and light-like gauge ($A^+ = 0$) can be related to an observable in polarized high-energy scattering,

$$\Delta g(\mu^2) = \int_0^1 \Delta G(x, \mu^2) dx \ , \tag{4}$$

where $\Delta G(x, \mu^2)$ is the polarized gluon distribution and is shown to be measurable in direct photon production at polarized RHIC and open charm production at COMPASS. One might be surprised at the notion of measuring a gauge-dependent quantity in experiment. What one actually measures here is the matrix element of a gauge-invariant, nonlocal operator

$$O_g = \int_{-\infty}^{\infty} dx \frac{n^-}{2} \int_{-\infty}^{\infty} \frac{d\lambda}{2\pi} e^{i\lambda x} F^{+\alpha}(\lambda n) e^{-ig\int_0^\lambda dy n\cdot A(yn)} \tilde{F}_\alpha{}^+(0) \ . \tag{5}$$

However, nobody knows the physical significance of O_g in general. In $A^+ = 0$ gauge, the operator simplifies to a local operator

$$O_g \to \vec{E} \times \vec{A} \ , \tag{6}$$

which one recognizes as the gluon spin operator. Conversely one can say O_g is a gauge-invariant extension of the gluon spin. Consequently, the measured Δg can be *interpreted* as the gluon spin in the $A^+ = 0$ gauge. It must be pointed out, though, that the gluon spin in other gauges (for instance, the covariant gauge) will be different from Δg, and there is no known method to measure it. On the other hand, the measured Δg cannot be considered as gluon spin in any other gauge.

Because of the gauge-dependence, there has been little follow-up study along Jaffe and Manohar's suggestion. There is an exception. In Ref. [?], Hoodbhoy, Tang and I studied the question how an individual contribution to the spin sum rule changes with the probing scale μ^2. For $\Delta\Sigma$ and Δg, there is the Altarelli and Parisi equation [?],

$$\frac{\partial}{\partial \ln \mu^2} \begin{pmatrix} \Delta\Sigma(\mu^2) \\ \Delta g(\mu^2) \end{pmatrix} = \frac{\alpha_s(\mu^2)}{2\pi} \begin{pmatrix} 0 & 0 \\ \frac{3}{2}C_F & \frac{\beta_0}{2} \end{pmatrix} \begin{pmatrix} \Delta\Sigma(\mu^2) \\ \Delta g(\mu^2) \end{pmatrix} \ . \tag{7}$$

We derived the evolution of the orbital angular momenta,

$$\frac{d}{d \ln \mu^2} \begin{pmatrix} L'_q \\ L'_g \end{pmatrix} = \frac{\alpha_s(\mu^2)}{2\pi} \begin{pmatrix} -\frac{4}{3}C_F & \frac{n_f}{3} \\ \frac{4}{3}C_F & -\frac{n_f}{3} \end{pmatrix} \begin{pmatrix} L'_q \\ L'_g \end{pmatrix} + \frac{\alpha_s(\mu^2)}{2\pi} \begin{pmatrix} -\frac{2}{3}C_F & \frac{n_f}{3} \\ -\frac{5}{6}C_F & -\frac{11}{2} \end{pmatrix} \begin{pmatrix} \Delta\Sigma \\ \Delta G \end{pmatrix} \ . \tag{8}$$

If one knows the decomposition of the spin of the nucleon at one perturbative scale, one can solve from the above equation the decomposition at any other perturbative scale. As $\mu^2 \to \infty$, one has the following asymptotic solution,

$$J_q(\infty) = \frac{1}{2}\frac{3n_f}{16+3n_f} ,$$
$$J_g(\infty) = \frac{1}{2}\frac{16}{16+3n_f} . \tag{9}$$

Nearly 1/2 of the proton spin is carried by gluons in the asymptotic limit. Unfortunately, we discovered recently that the above evolution equations are incomplete [?], although the conclusion about the asymptotic limit remains correct. We will come back to this important point later.

In Ref. [?], I proposed an alternative way to write the QCD angular momentum operator,

$$\vec{J} = \int d^3\vec{x} \left[\frac{1}{2}\bar{\psi}\vec{\gamma}\gamma_5\psi + \psi^\dagger(\vec{x}\times(-i\vec{D}))\psi + \vec{x}\times(\vec{E}\times\vec{B}) \right] . \tag{10}$$

As before, the first term is the quark spin. The second term, in which the covariant derivative is $\vec{D} = \vec{\partial} + ig\vec{A}$, is the canonical orbital angular momentum of quarks. The last term is the angular momentum of the gluons, as is clear from the appearance of the Poynting vector. According to the above, we can write down a gauge-invariant spin sum rule,

$$\frac{1}{2} = \frac{1}{2}\Delta\Sigma(\mu^2) + L_q(\mu^2) + J_g(\mu^2) , \tag{11}$$

where the second and third terms are quark orbital and gluon contributions, respectively. I introduced the sum of the first and second terms as $J_q(\mu^2)$, representing the total quark contribution. Although $\Delta\Sigma(\mu^2)$ is affected by the axial anomaly, $J_q(\mu^2)$ is anomaly-free [?]. In this formalism, the individual contributions to the spin are independent of the Lorentz frame, i.e., the momentum of the nucleon [?].

By examining carefully the definition of the matrix elements,

$$J_{q,g}(\mu^2) = \left\langle p\frac{1}{2}\left| \int d^3x (\vec{x}\times\vec{T}_{q,g})^z \right| p\frac{1}{2}\right\rangle , \tag{12}$$

I suggested [?] that they can be extracted from the form factors of the quark and gluon parts of the QCD energy-momentum tensor, $T^{\mu\nu}_{q,g}$. Using Lorentz symmetry, we can write down the forward matrix elements of $T^{\mu\nu}_{q,g}$,

$$\langle p'|T^{\mu\nu}_{q,g}|p\rangle = \bar{u}(p')\Big[A_{q,g}(\Delta^2)\gamma^{(\mu}\bar{P}^{\nu)} + B_{q,g}(\Delta^2)\bar{P}^{(\mu}i\sigma^{\nu)\alpha}\Delta_\alpha/2M$$

$$+C_{q,g}(\Delta^2)(\Delta^\mu\Delta^\nu - g^{\mu\nu}\Delta^2)/M + \bar{C}_{q,g}(\Delta^2)g^{\mu\nu}M\Big]u(p)\ , \quad (13)$$

where $\bar{p}^\mu = (p^\mu + p^{\mu\prime})/2$, $\Delta^\mu = p^{\mu\prime} - p^\mu$, and $u(p)$ is the nucleon spinor. Taking the forward limit in the $\mu = 0$ component and integrating over 3-space, one finds that $A_{q,g}(0)$ gives the momentum fractions of the nucleon carried by quarks and gluons, $(A_q(0) + A_g(0) = 1)$. On the other hand, substituting the above into the nucleon matrix element of Eq. (??), one finds [?],

$$J_{q,g} = \frac{1}{2}\left[A_{q,g}(0) + B_{q,g}(0)\right]\ . \quad (14)$$

There is an analogy for this. If one knows the Dirac and Pauli form factors of the electromagnetic current, $F_1(Q^2)$ and $F_2(Q^2)$, the magnetic moment of the nucleon, which is defined as the matrix element of $(1/2)\int d^3x(\vec{x}\times\vec{j})^z$, is just $F_1(0) + F_2(0)$.

How do we measure the form factors of the energy momentum tensor? If one has two vector currents which are separated along the light-cone, it is known from the operator product expansion that,

$$TJ_\alpha(z)J_\beta(0) \to ... + C_{\alpha\beta\mu\nu}(z^2)T^{\mu\nu} + ... \quad (15)$$

Thus to get the matrix element $\langle p'|T^{\mu\nu}|p\rangle$, we need $\langle p'|TJ_\alpha(z)J_\beta(0)|p\rangle$, i.e., a Compton scattering amplitude. To ensure the separation of the two currents is along the light cone, we let one of the photon momenta approach the Bjorken limit. Then it is easy to show that the Compton scattering is dominated by the single quark process. I call such a scattering process: deeply-virtual Compton scattering (DVCS) [?]. A first systematic study of the process was published in Ref. [?].

What does one learn from DVCS? An analysis shows that one learns about the off-forward parton distributions (OFPDs), which are defined through the following light-cone correlations,

$$\begin{aligned}
\int \frac{d\lambda}{2\pi}e^{i\lambda x}\langle p'|\bar{\psi}(-\lambda n/2)\ \ \gamma^\mu\psi(\lambda n/2)|p\rangle &= H(x,\Delta^2,\xi)\bar{u}(p')\gamma^\mu u(p) \\
&\quad + E(x,\Delta^2,\xi)\bar{u}(p')\frac{i\sigma^{\mu\nu}\Delta_\nu}{2M}u(p) + ... \\
\int \frac{d\lambda}{2\pi}e^{i\lambda x}\langle p'|\bar{\psi}(-\lambda n/2)\ \ \gamma^\mu\gamma_5\psi(\lambda n/2)|p\rangle &= \tilde{H}(x,\Delta^2,\xi)\bar{u}(p')\gamma^\mu\gamma_5 u(p) \\
&\quad + \tilde{E}(x,\Delta^2,\xi)\bar{u}(p')\frac{\gamma_5\Delta^\mu}{2M}u(p) + ... \quad (16)
\end{aligned}$$

where I have neglected the gauge link, and the dots denote higher-twist distributions. ¿From the definition, H and $\tilde{H}$ are nucleon helicity-conserving amplitudes and E and $\tilde{E}$ are helicity-flipping ones. Such distributions have been considered in the literature before [?]. More recent works include those of Radyushkin [?].

The off-forward parton distributions have the characters of both ordinary parton distributions and nucleon form factors. In fact, in the limit of $\Delta^\mu \to 0$, we have

$$H(x,0,0) = q(x) \ , \quad \tilde{H}(x,0,0) = \Delta q(x) \ , \tag{17}$$

where $q(x)$ and $\Delta q(x)$ are quark and quark helicity distributions. On the other hand, forming the first moment of the new distributions, one gets the following sum rules,

$$\int_{-1}^{1} dx H(x,\Delta^2,\xi) = F_1(\Delta^2) \ ,$$
$$\int_{-1}^{1} dx E(x,\Delta^2,\xi) = F_2(\Delta^2) \ . \tag{18}$$

where F_1 and F_2 are the Dirac and Pauli form factors. The most interesting sum rule relevant to the nucleon spin is,

$$\int_{-1}^{1} dx x[H(x,\Delta^2,\xi) + E(x,\Delta^2,\xi)] = A_q(\Delta^2) + B_q(\Delta^2) \ , \tag{19}$$

where the ξ dependence, or $C_q(\Delta^2)$ contamination, drops out. Extrapolating the sum rule to $\Delta^2 = 0$, the total quark (and, hence, quark orbital) contribution to the nucleon spin is obtained. By forming still higher moments, one gets form factors of various high-spin operators.

Last year, Hagler and Schaefer published a paper [?] which goes further along the direction of Jaffe and Manohar. They generalized the gauge-dependent orbital angular momentum (OAM) into an OAM distribution $L'_q(x)$ by introducing a set of local operators in light-cone gauge

$$\psi^\dagger \vec{r} \times (-i\vec{\partial})(\partial^+)^n \psi \ . \tag{20}$$

The first moment of $L'_q(x)$ is defined to give L'_q. They also used Ji, Tang, and Hoodbhoy's method to derive $L'_q(x)$ evolution. Again, as pointed out lately in Ref. [?], their evolution equations are actually incomplete.

Now, I would like to turn to the recent work of Bashinsky and Jaffe [?]. In the paper, they recognized that $L'_q(x)$ defined by Hagler and Schaefer is not gauge invariant and then they claim to define a gauge-invariant angular momentum distribution, $L_q^{BJ}(x)$, which (i) in the light-cone gauge $A^+ = 0$ reduces to that of Hagler and Schaefer and (ii) has the same evolution equation as derived in Refs. [?,?]. Although the paper is somewhat difficult to understand, the basic point is clear. In a fixed gauge, nothing is not gauge covariant. For instance, A^μ is gauge covariant as defined in $A^+ = 0$ gauge because $A^\mu = (1/\partial^+)F^{+\mu}$; ∂^μ is gauge covariant because $\partial^\mu = D^\mu - (1/\partial^+)F^{+\mu}$, etc. In the following, I shall call this the gauge invariance of the second kind because it *is* different from the gauge symmetry discussed in ordinary context.

The second-kind gauge-invariant operators will in general mix with an infinite number of other second-kind gauge-invariant operators. Going back to the orbital angular momentum evolution derived in Ref. [?], there exists actually additional operator mixing. Indeed, the operator $\psi^\dagger \vec{x} \times \vec{A}\psi$ has the same quantum number and dimension as $\psi^\dagger \vec{x} \times i\vec{\partial}\psi$ and therefore will mix in through evolution. Not only that, the light-cone power counting indicates that the operator of the type

$$f(\partial^+)\psi^\dagger \vec{x} \times i\vec{\partial}\psi\ , \tag{21}$$

will also mix, where f is an arbitrary function! Therefore, it is impossible to write down a complete evolution equation for these operators. Some additional comments are as follows: First, it is impossible to find a hard process to measure matrix elements of type $L_q^{BJ}(x)$. Second, even if there were processes to measure $L_q^{BJ}(x)$, it is impossible to relate data from one μ^2 to another. Finally, L_q^{BJ} *is* just $L'_q(x)$ defined by Hagler and Schaefer, and therefore it is plagued by all of the problems of a gauge-dependent quantity.

The simple gauge-invariant approach presented in Ref. [?] has met some resistance. Critics say that while $\Delta\Sigma$ has a simple parton interpretation, the gauge-invariant orbital angular momentum clearly does not. Well, not every interesting physical observable has a simple parton interpretation. The nucleon mass is obviously an interesting observable, but it does not have a simple parton model interpretation. The quark orbital angular momentum has to do with transverse momentum of the partons in the nucleon. It is well known that the transverse momentum effects cannot be described in the simple parton picture. That is why the higher-twist effects are so difficult to describe in general.

Feynman used a simple example to illustrate that in gauge theories the orbital angular momentum associated with a covariant derivative has a clear physical interpretation [?]. Consider a charged particle in the neighborhood of a solenoid. The particle has charge q and is described by a quantum-mechanical wave function $\psi(\vec{r})$ at $t = 0$. Now suppose a current is established in the solenoid during a subsequent period Δt. A varying current creates a time-dependent magnetic field $\vec{B}$ within the solenoid which, according to

$$\oint \vec{A} \cdot d\vec{r} = \iint \vec{B} \cdot d\vec{S} \tag{22}$$

signals the presence of a changing vector potential, $\vec{A}(\vec{r}, t)$, outside the solenoid. Using Faraday's law, a time-dependent vector potential induces an electric field,

$$\vec{E} = -\frac{d\vec{A}}{dt} \tag{23}$$

which can be very large if Δt is short. In fact, as $\Delta t \to 0$, the electric force on the charged particle becomes infinite and then it is useful to talk about the impulse which represents the sudden change of the momentum of the particle,

$$\begin{aligned}\Delta P &= \vec{P}_f - \vec{P}_i \\ &= \int_0^{\Delta t} \vec{F} dt \\ &= -q\vec{A} \ .\end{aligned}$$

On the other hand, quantum mechanical wave function of a particle does not change suddenly in time. This means that in gauge theories the expectation value of $-i\nabla$ cannot yield the entire linear momentum of a particle. For instance, the final linear momentum of the charged particle must be calculated according to

$$\vec{P}_f = \int d^3\vec{r}\psi^\dagger \left(-i\nabla - q\vec{A}\right)\psi \ , \tag{24}$$

which naturally gives rise to the covariant derivative.

Finally, let me return to Hagler and Schaefer's question—what is the orbital angular momentum distribution $L_q(x)$? Consider the energy-momentum tensor $T^{\mu\nu} = \bar{\psi}\gamma^{(\mu}iD^{\nu)}\psi$ which gives a symmetric representation of the Lorentz group. The quark distribution $q(x)$ is related to the generalized energy-momentum tensor $T^{\mu_1\mu_2\cdots\mu_n} = \bar{\psi}\gamma^{(\mu_1}iD^{\mu_2}\dots iD^{\mu_n)}\psi$, which again gives a totally symmetric representation of the Lorentz group. The angular momentum tensor $M^{\mu\alpha\beta} = T^{\mu\alpha}x^\beta - T^{\mu\beta}x^\alpha$ belongs to a mixed symmetric representation. Therefore, it is natural to define the generalized angular momentum density $M^{\mu_1\cdot\mu_n\alpha\beta} = T^{\mu_1\cdots\mu_n\alpha}x^\beta - T^{\mu_1\cdots\mu_n\beta}x^\alpha$ as a generalized mixed symmetric representation. Using the strict definition of twist, the angular momentum density operators have twist one. In terms of light-cone power counting, their matrix elements behave like those of the ordinary twist-two operators. However, in light-cone quantization, one can see that the matrix elements have no simple parton interpretation (like twist-three). Using the generalized angular momentum density, Ref. [?] derived the orbital angular momentum distribution

$$L_q(x) = \frac{1}{2}\left[x(q(x) + E(x)) - \Delta\Sigma(x)\right] , \tag{25}$$

where $E(x)$ is the forward limit of the off-forward Pauli-like distribution and $\Sigma(x)$ is the total quark helicity distribution. Likewise, one can define the gluon orbital angular momentum distribution $L_g(x)$ [?]. The evolution equation for $L_q(x)$ and $L_g(x)$ depends on only the known twist-two evolution kernels.

I hope that study of these orbital angular momentum distributions evolves to an important direction in the future.

Acknowledgements

This work is supported in part by funds provided by the U.S. Department of Energy (D.O.E.) under cooperative agreement DOE-FG02-93ER-40762.

References

1. J. Ashman et al, Nucl. Phys. **B328**, 1 (1989).
2. X. Ji, Phys. Rev. Lett. **78**, 610 (1997).
3. R. L. Jaffe and A. Manohar, Nucl. Phys. **B337**, 509 (1990).
4. K. Abe et al, Phys. Rev. D **58**, 112003 (1998).
5. X. Ji, J. Tang, and P. Hoodbhoy, Phys. Rev. Lett. **76** 740 (1996).
6. G. Altarelli and G. Parisi, Nucl. Phys. **B126**, 278 (1977).
7. P. Hoodbhoy, X. Ji, and Wei Lu, Phys. Rev. D **59**, 014013 (1999).
8. X. Ji, Phys. Rev. D **58**, 056003 (1998).
9. X. Ji, Phys. Rev. D **55**, 7114 (1997).
10. F. M. Dittes, D. Muller, D. Robsschik, B. Geyer, and J. Horejsi, Phys. Lett. **B209**, 325 (1988); Fortschr. Phys. **42**, 101 (1994).
11. A. V. Radyushkin, Phys. Lett. **B385**, 333 (1996); Phys. Rev. D **56**, 5524 (1997).
12. P. Hagler and A. Schafer, Phys. Lett. B**430**, 179 (1998).
13. S. V. Bashinsky and R. L. Jaffe, Nucl. Phys. B**536**, 303 (1998).
14. R. P. Feynman, R. B. Leighton and M. Sands, *The Feynman Lectures on Physics, Vol. III: Quantum Mechanics* (Addison-Wesley, Reading, Mass., 1965).
15. P. Hoodbhoy, X. Ji, and Wei Lu, Phys. Rev. D **59**, 074010 (1999).

FACTORIZATION FOR HARD EXCLUSIVE ELECTROPRODUCTION

John C. Collins
Penn State University, 104 Davey Lab,
University Park PA 16802, U.S.A.

This talk summarized the proof of hard-scattering factorization for hard exclusive electroproduction processes: deeply virtual Compton scattering and exclusive meson production.

1 Introduction

The processes discussed in this talk were $ep \to eMp$, $ep \to e\gamma p$, and $ep \to e\mu^+\mu^- p$, all in the deep-inelastic region where Q is large. The factorization theorems that I will discuss provide a sound basis for the phenomenology of these processes, which was discussed extensively at this workshop. They express the amplitudes for the processes in terms of operator matrix elements and perturbatively calculable coefficient functions.

As always, the fundamental problem is that we do not know how to solve QCD exactly, and we must appeal to approximations. In high energy processes, such as we consider here, large ranges of scales and of rapidities are important. The factorization theorems demonstrate that enormous simplifications occur in suitable asymptotic limits, and then enable calculations to be done in perturbation theory, where asymptotic freedom implies that hard scattering coefficients etc. can be usefully approximated by low-order Feynman graph calculations.

My presentation of the arguments tends to be graphical, since I find this is the easiest way to exhibit their structure. However they correspond to quantitative mathematical proofs. The principles are quite general; the processes under discussion turn out to provide the simplest situations for explaining how to prove factorization theorems. Although the proofs are apparently based on Feynman graph arguments, some of the key elements are actually non-perturbative, as I will emphasize.

2 The simplest case: muon-pair production

The simplest case is the deep-inelastic production of high-mass muon pairs: $ep \to e\mu^+\mu^- p$, for which the kinematic variables are defined in Fig. 1. The asymptotics can be treated by the simplest version of the light-cone expansion

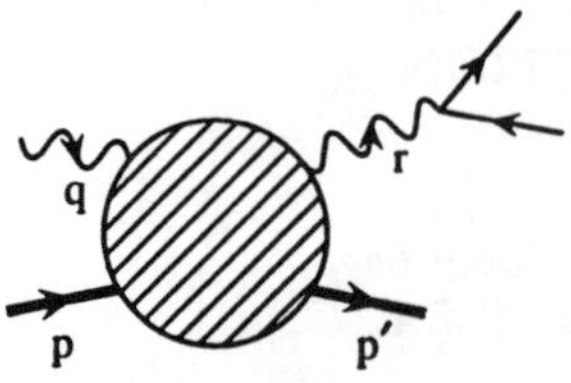

Figure 1: Deeply virtual muon-pair production.

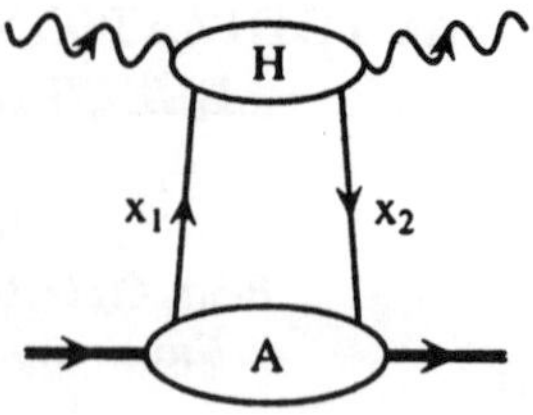

Figure 2: Leading regions for deeply virtual muon-pair production.

[1,2]. However, I will explain methods that generalize to more complicated processes, such as meson production [3]. Other relevant references are [4–6].

From the momenta defined in Fig. 1, we define a set of 4 scalar variables to specify the kinematics of the process by: $Q = \sqrt{-q^2}$, $\alpha = r^2/Q^2$, $x = Q^2/2p{\cdot}q$, and $t = (p - p')^2$. We consider the process at large Q, with x, α and t fixed. Then the leading-power part of the amplitude is given by regions symbolized in Fig. 2. There the lines in the upper bubble, H, have large virtualities, of order Q^2, and the lines in the lower bubble are approximately collinear to the proton and have relatively low virtualities. Power-counting arguments [3–6] show that only two lines connect the two bubbles, except that extra longitudinally polarized gluon lines can join the two bubbles; these can be eliminated by a gauge transformation.

The methods for the power-counting can be reduced [3] to dimensional analysis for the finite-order graphs used for calculating H, together with an analysis of the behavior of the meson and proton subgraphs under boosts from the rest frames of the hadrons. Thus the methods are actually non-perturbative as regards the parts of the amplitude that are fundamentally non-perturbative.

An expansion of H in powers of small parameters (relative to Q) gives the factorization theorem. This expansion is conveniently performed with the aid of light-front coordinates ($V^\pm = (V^0 \pm V^z)/\sqrt{2}$), so that

$$\begin{aligned}
(p^+, p^-, \mathbf{p}_T) &= (p^+,\ \text{small}, \mathbf{0}_T), \\
q^\mu &= (-xp^+, Q^2/2xp^+, \mathbf{0}_T), \\
r^\mu &= (\alpha x p^+, Q^2/2xp^+, \mathbf{0}_T) + \ \text{small}, \\
p'^\mu &= (p^+(1 - x - \alpha x),\ \text{small},\ \text{small}).
\end{aligned} \tag{1}$$

The fractional momenta of the lines joining the two bubbles in Fig. 2 are x_1 and x_2, and they obey $x_1 - x_2 = x(1 + \alpha)$. The small variables in Eq. (1) are small compared with Q, and the variables x and α have also gotten redefined by a fraction of order M^2/Q^2 to give the formulae for q^μ and r^μ.

A factorization theorem results after one sums over all possibilities for the graphs and after one neglects relatively small external momentum components for H:

$$A = \int d\xi \sum_i H_i(Q^2, x/\xi, \alpha) f_{i/p}(\xi, \xi - x(1+\alpha)) + \text{non-leading power.} \quad (2)$$

where the momentum fractions have been rewritten as $x_1 = \xi$ and $x_2 = \xi - x(1+\alpha)$. The sum is over the flavors of the partons joining the two bubbles.

In this equation, there is implicitly the usual DGLAP scale-dependence for the hard scattering and for the skewed parton densities. To get a valid perturbative calculation of the hard scattering coefficient H_i one should choose the renormalization and factorization scale $\mu_{\overline{\rm MS}}$ to be of the order of a typical scale of transverse momentum in H, i.e., of order Q. The hard scattering can be expanded in Feynman graphs for muon production from a quark or gluon target, but the higher order terms have subtractions. The subtractions have two effects: they prevent double counting of contributions to the amplitude from different orders of perturbation theory, and they cancel the collinear divergences in the unsubtracted graphs.

The skewed parton densities are matrix elements of suitable non-local gauge-invariant operators, whose details are in Refs. [2–5]. Taking into account the dependence on the polarization of the proton state, Ji [5] decomposed this parton density into scalar form factors. There are a number of different conventions on the kinematic variables and the normalization for the parton densities, but the details are not important for my purposes.

It is important that there exists an operator definition of the parton densities: (a) it permits a precise definition of the non-perturbative quantities involved, and (b) it gives a definite link to non-perturbative physics. Positivity constraints have been derived for the parton densities in Ref. [7].

Since at fixed Q and t, both the amplitude and the skewed parton density depend on two scalar variables, one can in principle unfold the skewed parton density from the amplitude, up to the usual issue of flavor dependence. This is unlike the case of deeply virtual Compton scattering or meson production, where the amplitude depends on one variable, and the unfolding is much problematic (*pace* Freund [8]).

Finally, conformal symmetry, exactly valid at lowest order, gives great simplification in the calculation of the DGLAP evolution kernels for skewed parton distributions [9].

3 Deeply virtual Compton Scattering

Deeply virtual Compton scattering ($\gamma^* + p \to \gamma p$), is intermediate between the case of high-mass muon-pair production that was treated in the previous section and deeply inelastic meson production to be treated in the next section, so I will not cover it explicitly. See Refs. [4–6].

4 Meson production

The case of meson production [3], $e + p \to e + M + p$, needs a generalization of the methods used for muon pair production. No longer does a simple light-cone expansion work, and one must use the methods summarized in Sec. 2. The reason for this is that there are two hadrons in the final state and therefore the factorization theorem not only separates the short-distance part of the scattering from the long-distance part, but also it gives separate factors for each of the two hadrons. The original discussions were for vector mesons only (ρ, ω, etc.), but Ref. [3] showed that the production of other mesons, like the π, also obey a factorization theorem.

4.1 Momentum regions

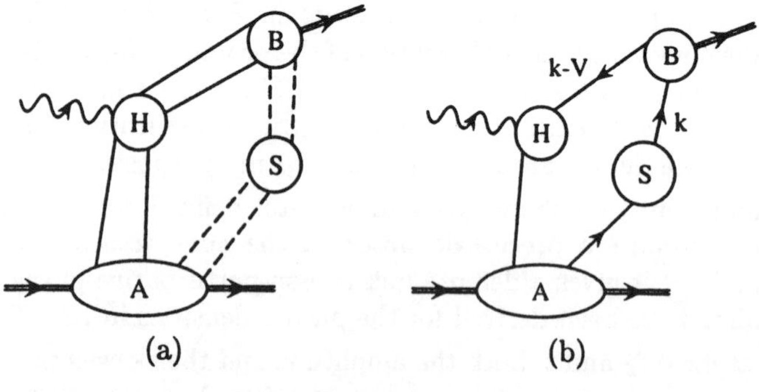

Figure 3: The two kinds of leading region for meson production. In (a), the soft subgraph is optional. In both graphs, there may be extra gluons connecting the subgraphs A and B to S and H.

The usual power-counting techniques tell us that the leading power contributions to the amplitude come from regions symbolized by Fig. 3, which has groups of lines that are: collinear to the target proton, collinear to the meson, soft, and hard. The possible soft lines joining the two collinear subgraphs are a new feature of this process compared with muon-pair production. Roughly, one can characterize the sizes of the momenta in the various subgraphs associated with the leading regions by:

$$\begin{aligned} H: &\quad (O(Q), O(Q), O(Q)) \\ \text{proton}: &\quad (O(Q), O(M^2/Q), O(M)) \\ \text{meson}: &\quad (O(M^2/Q), O(Q), O(M)) \\ \text{soft}: &\quad \text{small compared with } Q. \end{aligned} \tag{3}$$

There are two kinds of potentially leading region:

- Fig. 3(a), where two lines connect each of the meson and proton subgraphs to the hard scattering, and only gluons connect the soft subgraph to the meson and proton subgraphs.

- Fig. 3(b), where one quark connects each of the meson and proton subgraphs to the hard scattering, and where one (anti)quark plus any number of gluons connect the soft subgraph to the meson and proton subgraphs.

Relative to the first kind of region, the second kind gets an enhancement because of the lower number of external lines of the hard scattering and it gets a suppression because the external lines of the soft subgraph are quarks. Which kind of region actually is most important depends on the polarization of the meson. It turns out [3, 10], by a quite non-trivial argument, that the leading-most power only occurs for longitudinally polarized virtual photons making either pseudo-scalar mesons or longitudinally polarized vector mesons, and that only regions of type (a) contribute.

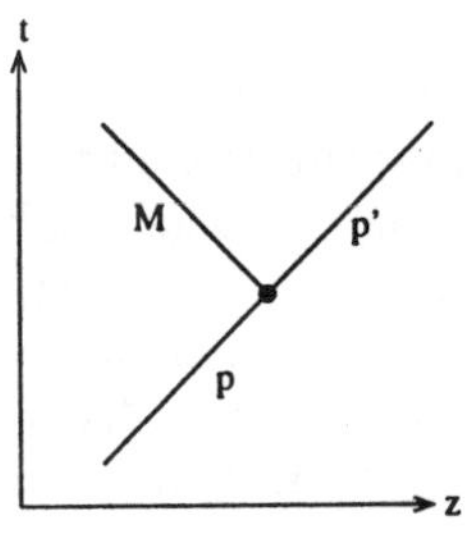

Figure 4: Space-time structure for leading regions.

It is useful to observe that any of the regions involved can be interpreted in terms of a space-time picture, Fig. 4. The incoming and outgoing hadrons move in almost light-like directions, and the hard scattering is localized close to the intersection of the world-lines of the external hadrons. What will be important for our treatment of the soft subgraph in the next section is that the lines collinear to the meson are created at the hard scattering. This implies that the soft interactions happen in the final state, and not in the initial state.

4.2 Analyticity and causality

Consider a region of the form of Fig. 3(b). The line coupling the hard scattering to the final-state meson has a factor

$$\frac{1}{(V-k)^2 - m^2 + i\epsilon} \simeq \frac{1}{V^2 - 2V^- k^+ - k_T^2 - 2\mathbf{V}_T \cdot \mathbf{k}_T - m^2 + i\epsilon}, \tag{4}$$

where the approximation is valid to the leading power of Q given that the integration momentum k is soft. Now there are two cases to consider for the relative sizes of the components of k^μ:

(i) The longitudinal components are small: $k^+k^- \ll k_T^2$. This is the region that is conventionally associated with rescattering corrections, i.e., small angle scattering of low mass objects.

(ii) The longitudinal and transverse components are comparable: $k^+k^- \sim k_T^2$; this is the conventional soft region.

In the first case, the only significant dependence on k^+ is in the propagator (4). Hence the k^+ integration can be deformed into the lower half-plane. Then the propagator (4) is far off-shell, so that in effect conventional rescattering interactions do not exist. This derivation implements the statement that, as in Fig. 4, the meson only has final-state interactions and that there is no causal connection between the oppositely moving final-state particles. The second case has some line(s) forced off-shell, by order QM, where M is a typical hadron scale.

Given that case (i) is not important, a further approximation to (4) is valid whenever k is soft:

$$\frac{1}{(V-k)^2 - m^2 + i\epsilon} \simeq \frac{1}{V^2 - 2V^- k^+ - m^2 + i\epsilon} \simeq \frac{1}{-2V^- k^+ + i\epsilon}, \tag{5}$$

which only depends on k^+, and not on k_T. The second half of the approximation depends on the non-perturbative statement that soft lines have virtualities of at least of the order of a hadron mass. The same approximation applies to all the soft lines coupling to the meson subgraph in Fig. 3. The resulting power counting shows [3] that of the regions in Fig. 3 only (a) is leading; the soft lines connecting to the meson are all gluons and they have a polarization that permits certain Ward identity arguments to be used — see Sec. 4.4. Only the first part of the approximation is needed for the Ward identity arguments; these give the factorization theorem.

The above arguments are quite general. As stated earlier, the results of the power counting encompass relevant non-perturbative effects. In addition, the analyticity arguments go beyond perturbation theory; they are similar in style to the arguments of DeTar, Ellis and Landshoff [11], and of Landshoff and Polkinghorne [12]. So we should expect that the proof is valid beyond perturbation theory, and, in particular, as regards final-state interactions.

4.3 "Endpoint" contribution

The soft contributions can also be viewed as the endpoint contributions of the situations where lines joining the proton to the hard scattering have small momentum fractions. Hence the treatment of the soft contributions has some notable implications for the analyticity properties of the hard-scattering coefficients — see the papers of Radyushkin [4] and of Collins and Freund [6] for details.

4.4 Actual leading contributions

We have now found that the actual leading-power contributions with a non-trivial soft subgraph have the form of Fig. 3(a), where the only soft lines joining to the meson subgraph are gluons, and where the soft momenta put lines collinear to the meson off-shell by order QM. Effectively the soft gluons are on the same footing as the extra collinear longitudinal gluons joining A to H.

A Ward identity argument [3] valid to the leading power shows that all effects of these gluons either cancel or are effectively in the skewed parton distributions. This implements the statement that the "color of the meson is zero". The final result is as if the soft subgraph S is not present, and we obtain a factorization theorem of the form:

$$\begin{aligned} \text{Amplitude} &= H \otimes \text{skewed pdf} \otimes \text{meson distribution amplitude} \\ &\quad + \text{non-leading power}. \end{aligned} \tag{6}$$

As usual, the hard scattering H is perturbatively calculable in powers of $\alpha_s(Q)$, the skewed parton distribution is the same as in other processes, and is computed at a scale Q (or of that order), while the distribution amplitude of the meson is the same as in other exclusive processes and is computed at a scale Q.

At small x (e.g., for ρ production at the HERA collider) additional steps are used to discuss the Regge limit and to relate the skewed parton density to the ordinary gluon density [13].

For π production the skewed parton densities have a relation to the ordinary polarized parton density, since the operators are the same, and there are some interesting selection rules.

5 Summary

The full proof of hard-scattering factorization provides a sound basis for the phenomenology of the processes discussed, since it follows from the QCD Hamiltonian. There are many non-perturbative elements in the proof, so that it goes beyond mere perturbation theory. This includes definite operator definitions of the parton densities.

Part of the theorem is that amplitudes involving transverse polarization are power suppressed. More work is needed here to prove an effective generalized factorization theorem for this case.

Acknowledgments

This work was supported in part by the U.S. Department of Energy.

References

1. S.A. Anikin and O.I. Zavialov, *Ann. Phys. (NY)* **116** 135, (1978).
2. D. Müller, D. Robaschik, B. Geyer, F.M. Dittes and J. Hořejši, *Fortsch. Phys.* **42**, 101 (1994).
3. J.C. Collins, L. Frankfurt, and M. Strikman, *Phys. Rev. D* **56**, 2982 (1997).
4. A.V. Radyushkin, *Phys. Rev.* **D56**, 5524 (1997).
5. X.-D. Ji, *Phys. Rev.* **D55**, 7114 (1997).
6. J.C. Collins and A. Freund, *Phys. Rev.* **D59**, 074009 (1999).
7. B. Pire, J. Soffer, and O. Teryaev, *Eur. Phys. J.* **C8**, 103 (1999).
8. Andreas Freund, "Extraction of Skewed Parton Distributions from Experiment", hep-ph/9905233; "On the Extraction of Skewed Parton Distributions from Experiment", hep-ph/9903488.
9. E.g., A.V. Belitsky, D. Muller, and A. Freund, "Reconstruction of Non-forward Evolution Kernels", hep-ph/9904477, and references therein.
10. L. Mankiewicz, G. Piller and T. Weigl, *Eur. Phys. J.* **C5**, 119 (1998); M. Diehl, Th. Gousset and B. Pire, *Phys. Rev.* **D59**, 034023 (1999); J.C. Collins and M. Diehl, "Transversity Distribution Does Not Contribute to Hard Exclusive Electroproduction of Mesons", hep-ph/9907498.
11. C. DeTar, S.D. Ellis and P.V. Landshoff, *Nucl. Phys.* **B87**, 176 (1975).
12. P.V. Landshoff and J.C. Polkinghorne, *Phys. Rept.* **5**, 1 (1972).
13. S.J. Brodsky, L. Frankfurt, J.F. Gunion, A.H. Mueller, and M. Strikman, *Phys. Rev. D* **50**, 3134 (1994).

SKEWED PARTON DISTRIBUTIONS AND REAL AND VIRTUAL COMPTON SCATTERING

P. KROLL

Fachbereich Physik, Universität Wuppertal
Gaußstrasse 20, D-42097 Wuppertal, Germany

The soft physics approach to Compton scattering at moderately large momentum transfer is reviewed. It will be argued that in that approach the Compton cross section as well as other exclusive observables exhibit approximate scaling in a limited range of momentum transfer.

1 Introduction

QCD provides three valence Fock state contributions to nucleon form factors, real (RCS) and virtual (VCS) Compton scattering at large momentum transfer: a soft overlap term with an active quark and two spectators (see Fig. 1), the asymptotically dominant perturbative contribution where by means of the exchange of two hard gluons the quarks are kept collinear with respect to their parent nucleons (see Fig. 1) and a third contribution that is intermediate between the soft and the perturbative contribution where only one hard gluon is exchanged and one of the three quarks acts as a spectator. Both the soft and the intermediate terms represent power corrections to the perturbative contribution. Higher Fock state contributions are suppressed. The crucial question is what is the relative strengths of the three contributions at experimentally accessible values of momentum transfer, i.e. at $-t$ of the order of 10 GeV^2? The pQCD followers (cf. Ref.[1] for recent calculations of RCS) assume the dominance of the perturbative contribution and neglect the other two contributions while the soft physics community presumes the dominance of the overlap contribution. Which group is right is not yet fully decided although comparison with the pion case[2] seems to favour a strong overlap contribution.

In this talk I am going to report on results for Compton scattering obtained within the soft physics approach[3]. It has been shown recently[3,4] that at moderately large momentum transfer Compton scattering off protons approximately factorises into a hard photon-parton subprocess and soft proton matrix elements described by new form factors specific to Compton scattering. These form factors, as the ordinary electromagnetic ones, represent moments of skewed parton distributions (SPD)[5]. The soft physics approach bears resemblance to the fixed-pole model advocated for in[6] long time ago.

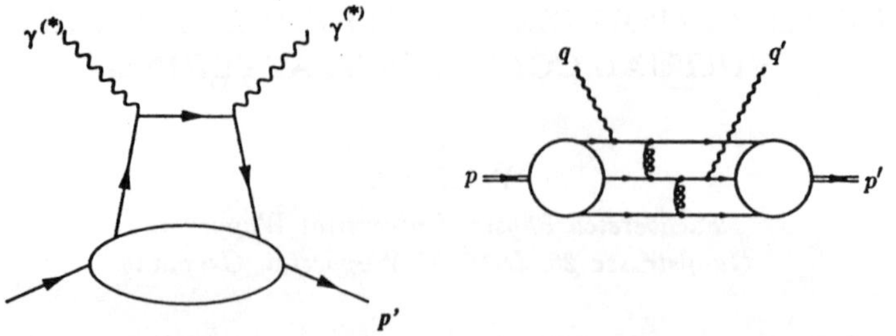

Figure 1: The handbag diagram (left) and an example of a Feyman graph for Compton scattering within pQCD (right). The momenta of the incoming and outgoing protons (photons) are denoted by p (q) and $p' = p + \Delta$ ($q' = q - \Delta$), respectively.

2 The soft physics approach to Compton scattering

For Mandelstam variables, s, t and u, that are large on a hadronic scale the handbag diagram shown in Fig. 1 describes RCS and VCS. To see this it is of advantage to choose a symmetric frame of reference where the plus and minus light-cone components of Δ are zero. This implies $t = -\Delta_\perp^2$ as well as a vanishing skewdness parameter $\zeta = -\Delta^+/p^+$. To evaluate the SPD appearing in the handbag diagram one may use a Fock state decomposition of the proton and sum over all possible spectator configurations. The crucial assumption is then that the soft hadron wave functions are dominated by virtualities in the range $|k_i^2| \lesssim \Lambda^2$, where Λ is a hadronic scale of the order of 1 GeV, and by intrinsic transverse parton momenta, $k_{\perp i}$, defined with respect to their parent hadron's momentum, that satisfy $k_{\perp i}^2/x_i \lesssim \Lambda^2$. Under this assumption factorisation of the Compton amplitude in a hard photon-parton amplitude and $1/x$-moments of SPDs is achieved [3].

As a consequence of this result the Compton amplitudes conserving the proton helicity are given by

$$\mathcal{M}_{\mu'+,\,\mu+} \;=\; 2\pi\alpha_{\rm em}\left[\mathcal{H}_{\mu'+,\,\mu+}\,(R_V + R_A) \,+\, \mathcal{H}_{\mu'-,\,\mu-}\,(R_V - R_A)\right]. \qquad (1)$$

Proton helicity flip is neglected. μ and μ' are the helicities of the incoming and outgoing photon in the photon-proton cms, respectively. The photon-quark subprocess amplitudes, $\mathcal{H}$, are calculated for massless quarks in lowest order QED. The form factors in Eq. (1), R_V and R_A, represent $1/x$-moments of SPDs at zero skewedness parameter. R_V is defined by

$$\sum_a e_a^2 \int_0^1 \frac{{\rm d}x}{x}\, p^+ \int \frac{{\rm d}z^-}{2\pi}\, e^{i\,xp^+z^-} \langle p'|\,\overline{\psi}_a(0)\,\gamma^+\,\psi_a(z^-) - \overline{\psi}_a(z^-)\,\gamma^+\,\psi_a(0)\,|p\rangle$$

$$= R_V(t)\,\bar{u}(p')\,\gamma^+ u(p) + R_T(t)\,\frac{i}{2m}\bar{u}(p')\sigma^{+\rho}\Delta_\rho u(p)\,, \qquad (2)$$

where the sum runs over quark flavours a $(u,\ d,\ \ldots)$, e_a being the electric charge of quark a in units of the positron charge. R_T being related to nucleon helicity flips, is neglected in (1). There is an analogous equation for the axial vector nucleon matrix element, which defines the form factor R_A. Due to time reversal invariance the form factors R_V, R_A etc. are real functions.

As shown in [3] form factors can be represented as generalized Drell-Yan light-cone wave function overlaps. Assuming a plausible Gaussian $k_{\perp i}$-dependence of the soft Fock state wave functions, one can explicitly carry out the momentum integrations in the Drell-Yan formula. For simplicity one may further assume a common transverse size parameter, $\hat{a}$, for all Fock states. This immediately allows one to sum over them, without specifying the x_i-dependence of the wave functions. One then arrives at [3,4]

$$F_1(t) = \sum_a e_a \int \mathrm{d}x \exp\left[\frac{1}{2}\hat{a}^2 t \frac{1-x}{x}\right] \{q_a(x) - \bar{q}_a(x)\}\,,$$

$$R_V(t) = \sum_a e_a^2 \int \frac{\mathrm{d}x}{x} \exp\left[\frac{1}{2}\hat{a}^2 t \frac{1-x}{x}\right] \{q_a(x) + \bar{q}_a(x)\}\,, \qquad (3)$$

and the analogue for R_A with $q_a + \bar{q}_a$ replaced by $\Delta q_a + \Delta\bar{q}_a$. q_a and Δq_a are the usual unpolarized and polarized parton distributions, respectively. The result for F_1 can also been found in [7].

The only parameter appearing in (3) is the effective transverse size parameter $\hat{a}$; it is known to be about 1 GeV^{-1} with an uncertainty of about 20%. Thus, this parameter only allows some fine tuning of the results for the form factors. Evaluating, for instance, the form factors from the parton distributions derived by Glück et al. (GRV) [8] with $\hat{a} = 1\,GeV^{-1}$, one already finds good results. Improvements are obtained by treating the lowest three Fock states explicitly with specified x-dependencies. For the valence Fock distribution amplitude, for instance, a form proposed in Ref. [9] was used

$$\Phi_{123}^{BK} = 60\,x_1 x_2 x_3\,(1 + 3x_1)\,, \qquad (4)$$

valid at a factorisation scale of 1 GeV. All higher Fock states were still treated in the global way, using the parton distributions of Ref. [8] as input. Results for $t^2 F_1$ and $t^2 R_V$ obtained that way are displayed in Fig. 2. Both the scaled form factors, as well as $t^2 R_A$, exhibit broad maxima and, hence, mimic dimensional counting rule behaviour in the t-range from about 5 to 15 GeV^2. For very large momentum transfer the form factors turn gradually into the soft physics

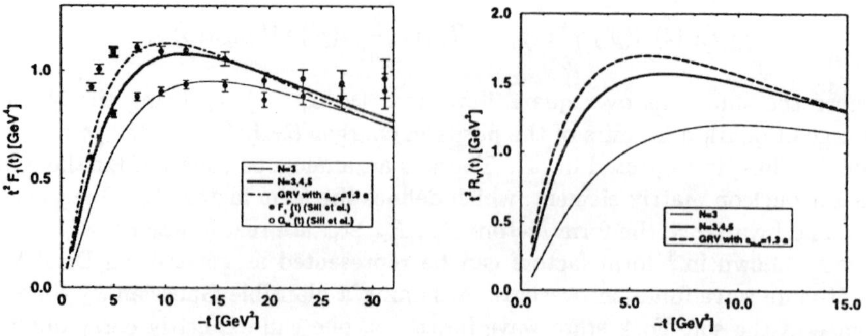

Figure 2: The Dirac (left) and the vector Compton (right) form factor of the proton as predicted by the soft physics approach[3,9]. Data are taken from[10]. The data on the magnetic form factor, G_M, are shown in order to demonstrate the size of spin-flip effects.

asymptotics $\sim 1/t^4$. This is the region where the perturbative contribution ($\sim 1/t^2$) takes the lead.

It is to be stressed that the perturbative contribution to the proton form factor evaluated from the distribution amplitude (4) within the modified perturbative approach[9,11] is as small as only a few % of the experimental value. A similarly small value is expected for Compton scattering. Therefore, the analysis performed in Ref. [3] is to be regarded as a consistent calculation in which the soft contribution clearly dominates for experimentally accessible values of momentum transfer.

3 Results on Compton scattering

The amplitude (1) leads to the RCS cross section

$$\frac{d\sigma}{dt} = \frac{d\hat{\sigma}}{dt}\left[\frac{1}{2}(R_V^2(t) + R_A^2(t)) - \frac{us}{s^2+u^2}\left(R_V^2(t) - R_A^2(t)\right)\right]. \qquad (5)$$

It is given by the Klein-Nishina cross section

$$\frac{d\hat{\sigma}}{dt} = \frac{2\pi\alpha_{em}^2}{s^2}\frac{s^2+u^2}{-us} \qquad (6)$$

multiplied by a factor that describes the structure of the proton in terms of two form factors. Evidently, if the form factors scale as $1/t^2$, the Compton cross section would scale as s^{-6} at fixed cm scattering angle θ. In view of the above discussion (see also Fig. 2) one therefore infers that approximate

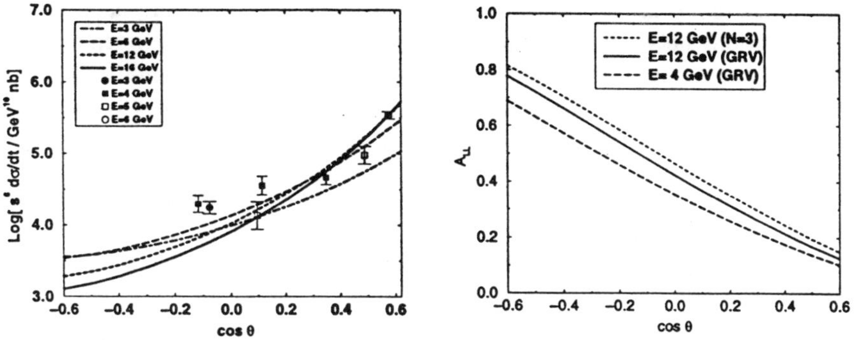

Figure 3: The Compton cross section, scaled by s^6, (left) and the initial state helicity correlation $A_{\rm LL}$ (right) as predicted by the soft physics approach [3]. Data taken from [12].

dimensional counting rule behaviour holds in a limited range of energy. The magnitude of the Compton cross section is quite well predicted as is revealed by comparison with the admittedly old data [12] measured at rather low values of s, $-t$ and $-u$ (see Fig. 3). Cross sections of similar magnitude have been obtained within the perturbative approach [1] (evaluated from very asymmetric distribution amplitudes) and within the diquark model [13]. The latter model is a variant of the standard perturbative approach [14] in which diquarks are considered as quasi-elementary constituents of the proton [15].

The soft physics approach also predicts characteristic spin dependencies of the Compton process. Of particular interest is the initial state helicity correlation

$$A_{\rm LL}\frac{d\sigma}{dt} = \frac{2\pi\alpha_{\rm em}^2}{s^2}\,R_V(t)R_A(t)\left(\frac{u}{s}-\frac{s}{u}\right). \tag{7}$$

Approximately, $A_{\rm LL}$ is given by the corresponding subprocess helicity correlation $\hat{A}_{\rm LL} = (s^2-u^2)/(s^2+u^2)$ multiplied by the dilution factor $R_A(t)/R_V(t)$. Thus, measurements of both the cross section and the initial state helicity correlation allows one to isolate the two form factors R_V and R_A experimentally [16]. In Fig. 3 predictions for $A_{\rm LL}$ are shown. Interestingly, the diquark model predicts the opposite sign for $A_{\rm LL}$. For the final state helicity correlation, $C_{\rm LL}$, and for the helicity transfer from the incoming photon to the outgoing proton, $K_{\rm LL}$, one finds

$$C_{\rm LL} = K_{\rm LL} = A_{\rm LL}, \tag{8}$$

while the helicity transfer from the initial to the final state photon , $D_{\rm LL}$, is unity since the photon helicity is conserved. Since in the soft physics approach photon helicity flips are zero and proton helicity flips are neglected many other polarization observables are zero, e.g. all single spin asymmetries or correlations between longitudinal and sideway polarizations (normal to the particle's momentum and in the scattering plane). In the diquark model [13], on the other hand, the photon and proton helicity flip amplitudes are non-zero although suppressed by either $1/\sqrt{s}$ or $1/s$; there are even perturbatively generated phase differences between the helicity amplitudes. Therefore, the diquark model predicts small deviations from zero for the latter polarization observables.

The VCS contribution to the unpolarized $ep \to ep\gamma$ cross section is decomposed as (see e.g. [17])

$$\frac{\mathrm{d}\sigma}{\mathrm{d}s\mathrm{d}Q^2\mathrm{d}\varphi\mathrm{d}t} \sim \frac{\mathrm{d}\sigma_T}{\mathrm{d}t} + \varepsilon\frac{\mathrm{d}\sigma_L}{\mathrm{d}t} + \varepsilon\cos 2\varphi\frac{\mathrm{d}\sigma_{TT}}{\mathrm{d}t} + \sqrt{2\varepsilon(1+\varepsilon)}\cos\varphi\frac{\mathrm{d}\sigma_{LT}}{\mathrm{d}t}\,, \quad (9)$$

where ε denotes the ratio of longitudinal and transverse photon flux in the Compton process and φ is the azimuthal angle between the electron and hadron planes. The partial cross sections in (9) refer to the scattering of transverse and longitudinal photons and to the transverse-transverse and longitudinal-transverse interference terms. The VCS partial cross sections have also been calculated in Ref. [3]. Comparing with the only other available results, namely those from the diquark model [17], one observes that the transverse cross section in both approaches comes out rather similar, while the other three cross sections are generally larger and with a smoother Q^2-dependence in the soft physics approach than in the diquark model. In contrast to the diquark model the transverse-transverse interference term, $\mathrm{d}\sigma_{TT}/\mathrm{d}t$ is strictly zero in the limit $Q^2 = 0$. In addition to VCS the full $ep \to ep\gamma$ cross section receives substantial contributions from the Bethe-Heitler process, in which the final state photon is radiated by the electron. Dominance of the VCS contribution requires high energies, small values of $|\cos\theta|$ and an out-of-plane experiment, i.e. an azimuthal angle larger than about 60°. The relative importance of the VCS and of the Bethe-Heitler contributions is rather similar in both the approaches the soft physics approach [3] and the diquark model [17].

In Ref. [17] the relevance of the beam asymmetry for $ep \to ep\gamma$

$$A_{\mathrm{L}} = \frac{\mathrm{d}\sigma(+) - \mathrm{d}\sigma(-)}{\mathrm{d}\sigma(+) + \mathrm{d}\sigma(-)}\,, \quad (10)$$

where the labels + and − denote the lepton beam helicity, has been pointed out. It is sensitive to the imaginary part of the longitudinal-transverse interference in the Compton process, while $\mathrm{d}\sigma_{\mathrm{LT}}/\mathrm{d}t$ measures its real part. In the soft physics approach, A_{L} is zero since all amplitudes are real within the accuracy of the calculation. In the diquark model [17], on the other hand, A_{L} is non-zero due to the perturbatively generated phases of the VCS amplitudes. In regions of strong interference between the Compton and the Bethe-Heitler amplitudes the beam asymmetry is even spectacularly enhanced. In the standard perturbative approach [14] a non-zero value of A_L is also to be expected.

4 Scaling - evidence for pQCD?

It is particularly interesting that the soft physics approach can account for the experimentally observed approximate scaling, i.e dimensional counting rule behaviour, at least for Compton scattering and for form factors. One may object that the perturbative explanation (leaving aside the logarithms from the running of α_s and from the evolution) works for many exclusive reactions, while in the soft physics approach the approximate counting rule behaviour is accidental, depending on specific properties of a given reaction. It however seems that the approximate counting rule behaviour is an unavoidable feature of the soft physics approach. As discussed in Ref. [3] similar arguments as for Compton scattering lead to a factorisation of baryon-baryon and meson-baryon amplitudes into hard scatterings of spin 1/2 partons and soft hadronic matrix elements which are represented by form factors analogue to the electromagnetic or Compton ones. In elastic pp scattering, for instance, the form factor

$$F_V^{pp}(t) = \sum_a \int \mathrm{d}x \exp\left[\frac{1}{2}\hat{a}^2 t \frac{1-x}{x}\right] \{q_a(x) + \bar{q}_a(x)\} \qquad (11)$$

appears. All these form factors are smooth functions of the momentum transfer and, if scaled by t^2, exhibit a broad maximum in the $-t$-range from about 5 to 15 GeV^2, set by the transverse hadron size, i.e. by a scale of order 1 GeV^{-1}. The position, t_0, of the maximum of $t^2 F_i$, where F_i is any of the soft form factors, is determined by the solution of the implicit equation

$$-t = 4\hat{a}^{-2} \left\langle \frac{1-x}{x} \right\rangle_{F_i,t}^{-1} . \qquad (12)$$

The mean value $\langle\frac{1-x}{x}\rangle$ comes out around 0.5 at $t = t_0$, hence, $t_0 \simeq 8\hat{a}^2$. Since both sides of Eq. (12) increase with $-t$ the maximum of the scaled form factor, F_i, is quite broad. The scaling behaviour of the form factors lead to approximative s^{-10} (s^{-8}) scaling of $\mathrm{d}\sigma/\mathrm{d}t$ in baryon (meson)-baryon scattering around $s = 10\,\mathrm{GeV}^2$. For elastic proton-proton scattering, shown in Fig. 4, fair agreement with experiment is obtained. The proton-proton data [18] show fluctuations superimposed to the s^{-10} behaviour. These fluctuations, if a real dynamical feature, tell us that there still is another momentum scale relevant

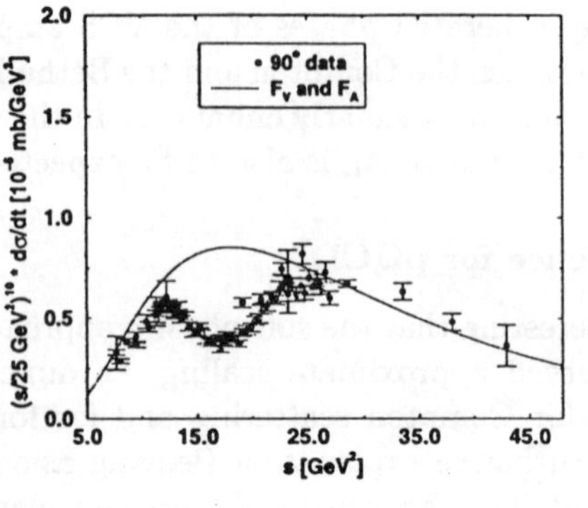

Figure 4: The cross section for elastic proton-proton scattering, scaled by s^{10}, at an cm scattering angle of 90° vs. s. The solid line is the soft physics approach prediction obtained in Ref. [3]. Data taken from [18].

in that kinematical region, contradicting the very idea of dimensional scaling. Theoretical interpretations of these fluctuations have been attempted in [19].

5 Summary

The soft physics approach leads to detailed predictions for RCS and VCS as well as for form factors. The predictions exhibit interesting features and characteristic spin dependences with marked differences to other approaches. Dimensional counting rule behaviour for form factors, Compton scattering and perhaps for other exlusive observables is mimicked in a limited range of momentum transfer. This tells us that it is premature to infer the dominance of perturbative physics from the observed scaling behaviour . The soft contributions although formally representing power corrections to the asymptotically leading perturbative ones, seem to dominate form factors and Compton scattering for momentum transfers around 10 GeV^2. However, a severe confrontation of this approach with accurate large momentum transfer data on RCS and VCS is still pending.

Acknowledgments

It is a pleasure to thank Carl Carlson and Anatoly Radyushkin for organising a most interesting meeting. This work has been supported in part by the European TMR network ERB 4061 Pl 95 0115.

References

1. A. Kronfeld and B. Nižić, Phys. Rev. **D44, 3445** (1991); Erratum *ibid.* **D46**, 2272 (1992); M. Vanderhaeghen, P.A.M. Guichon and J. Van de Wiele, Nucl. Phys. **A622**, 144c (1997).
2. R. Jakob and P. Kroll, Phys. Lett. **B315**, 463 (1993), Erratum *ibid.* **B319**, 545 (1993); P. Kroll and M. Raulfs, Phys. Lett. **B387**, 848 (1996); V. Braun and I. Halperin, Phys. Lett. **B328**, 457 (1994); L.S. Kisslinger and S.W. Wang, Nucl. Phys. **B399**, 63 (1993).
3. M. Diehl, T. Feldmann, R. Jakob and P. Kroll, Eur. Phys. J. **C8**, 409 (1999) and hep-ph/9903268, to be published in Phys. Lett. B.
4. A.V. Radyushkin, Phys. Rev. **D58**, 114008 (1998).
5. D. Müller *et al.*, Fortschr. Physik **42**, 101 (1994), hep-ph/9812448; X. Ji, Phys. Rev. Lett. **78**, 610 (1997); Phys. Rev. **D55**, 7114 (1997); A.V. Radyushkin, Phys. Rev. **D56**, 5524 (1997).
6. S.J. Brodsky, F.E. Close and J.F. Gunion, Phys. Rev. **D6**, 177 (1972).
7. V. Barone *et al.*, Z. Phys. **C58**, 541 (1993); A.V. Afanasev, hep-ph/9808291.
8. M. Glück, E. Reya and A. Vogt, Z. Phys. **C67**, 433 (1995); Eur. Phys. J. **C5**, 461 (1998); M. Glück, E. Reya, M. Stratmann and W. Vogelsang, Phys. Rev. **D53**, 4775 (1996).
9. J. Bolz and P. Kroll, Z. Phys. **A356**, 327 (1996).
10. A.F. Sill *et al.*, Phys. Rev. **D48**, 29 (1993).
11. J. Bolz, R. Jakob, P. Kroll, M. Bergmann and N.G. Stefanis, Z. Phys. **C66**, 267 (1995).
12. M.A. Shupe *et al.*, Phys. Rev. **D19**, 1921 (1979).
13. P. Kroll, M. Schürmann and W. Schweiger, Intern. J. Mod. Phys. **A6**, 4107 (1991).
14. G.P. Lepage and S.J. Brodsky, Phys. Rev. **D22**, 2157 (1980).
15. M. Anselmino, P. Kroll and B. Pire, Z. Phys. **C36**, 89 (1987).
16. A.M. Nathan, hep-ph/9807397 and these proceedings.
17. P. Kroll, M. Schürmann and P.A.M. Guichon, Nucl. Phys. **A598**, 435 (1996).
18. C.W. Akerlof *et al.*, Phys. Rev. **159**, 1138 (1967); J.V. Allaby *et al.*, Phys. Lett. **25B**, 156 (1967).
19. B. Schrempp and F. Schrempp, Phys. Lett. **55B**, 303 (1975); B. Pire and J.P. Ralston, Phys. Lett. **117B**, 233 (1982).

POWER CORRECTIONS TO DVCS AND HARD MESON ELECTROPRODUCTION

M. VANDERHAEGHEN
Institut für Kernphysik, University of Mainz, D-55099 Mainz, Germany

P.A.M. GUICHON
Service de Physique Nucléaire, CEA/Saclay, F-91191 Gif-sur-Yvette, France

M. GUIDAL
Institut de Physique Nucléaire - IN2P3, F-91406 Orsay, France

We report on predictions for the leading order amplitudes for deeply virtual Compton scattering and hard meson electroproduction reactions at large Q^2 in the valence region in terms of skewed quark distributions. We furthermore give estimates for the power corrections to these leading order amplitudes. In particular, we outline examples of experimental opportunities to access the skewed parton distributions at the current high-energy lepton facilities : JLab, HERMES and COMPASS.

1 Introduction

In recent years, a unified theoretical description of a wide variety of exclusive processes in the Bjorken regime has emerged through the formalism introducing new generalized parton distributions, the so-called skewed parton distributions (SPD's). It has been shown that these distributions, which parametrize the structure of the nucleon, allow to describe in leading order perturbative QCD (PQCD), various exclusive processes such as, in particular, deeply virtual Compton scattering (DVCS) and longitudinal electroproduction of vector and pseudoscalar mesons (see e.g. Refs.[1,2,3,4] and references therein).

2 Leading order hard electroproduction amplitudes and skewed parton distributions

The leading order PQCD diagrams for DVCS and hard meson electroproduction are of the type as shown in Fig. 1. It has been proven [1,2] that the leading order DVCS amplitude in the forward direction can be factorized in a hard scattering part (which is exactly calculable in PQCD) and a soft, nonperturbative nucleon structure part as is illustrated on the left panel of Fig. 1. The nucleon structure information can be parametrized, at leading order, in terms of 4 generalized structure functions. In the notation of Ji, these functions are the off-forward parton distributions (OFPD's) denoted as $H, \tilde{H}, E, \tilde{E}$ which

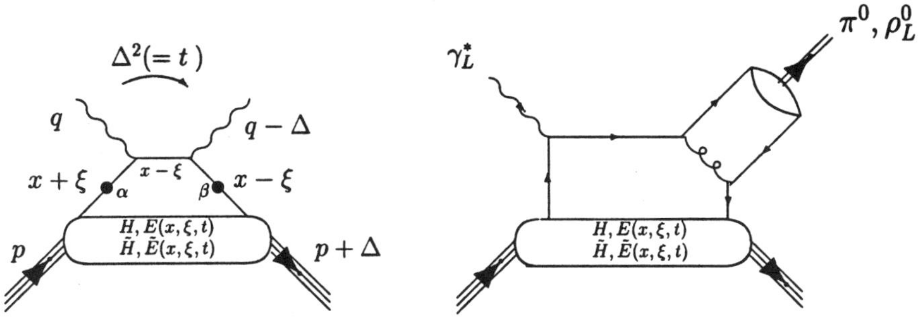

Figure 1: Leading order diagrams for DVCS (left) and for longitudinal electroproduction of mesons (right).

depend upon three variables : x, ξ and t. The light-cone momentum fraction x is defined by $k^+ = xP^+$, where k is the quark loop momentum and P is the average nucleon momentum (using the definition $a^\pm \equiv 1/\sqrt{2}(a^0 \pm a^3)$). The skewedness variable ξ is defined by $\Delta^+ = -2\xi\, P^+$, where Δ is the overall momentum transfer in the process and where $2\xi \to x_B/(1-x_B/2)$ in the Bjorken limit. Furthermore, the third variable entering the OFPD's is given by the Mandelstam invariant $t = \Delta^2$. In Fig. 1, the variable x runs from -1 to 1. As noted by Radyushkin [2], one can identify two regions for the SPD's. In the regions where $x > \xi$ or $x < -\xi$, the SPD's are the generalizations of the usual parton distributions from DIS. Actually, in the forward direction, the SPD's H and $\tilde{H}$ respectively reduce to the quark density distribution $q(x)$ and quark helicity distribution $\Delta q(x)$, obtained from DIS. In the region $-\xi < x < \xi$, the SPD's behave as a "meson-like" distribution amplitude and contain new information about nucleon structure [5].

To provide estimates for electroproduction observables, we need a model for the SPD's. The following calculations were performed using ξ-dependent SPD's based on a product ansatz (for the double distributions [6]) of a quark distribution (we use the MRST98 quark distributions [7] as input) and an asymptotic "meson-like" distribution amplitude (see Ref. [9] for more details). The t-dependence is given by the corresponding form factors (Dirac form factor for H, axial form factor for $\tilde{H}$), so that the first moments of the SPD's are satisfied by construction. We show on Fig. 2 the d-quark SPD using this ansatz.

Besides the DVCS process, a factorization proof was also given for the leading order meson electroproduction amplitude in the valence region at large Q^2 [3], which is shown on the right panel of Fig. 1. This factorization theorem only applies when the virtual photon is longitudinally polarized. The

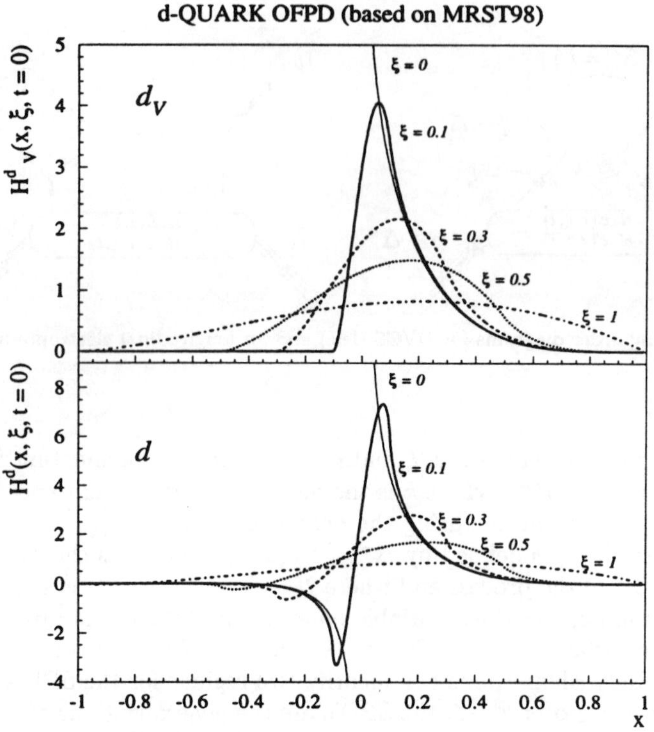

Figure 2: ξ dependence of the OFPD H^d at $t = 0$ using the ansatz (based on the MRST98 [7] quark distributions). Upper panel : valence down quark OFPD, lower panel : total down quark OFPD. The thin lines ($\xi = 0$) correspond with the ordinary d-quark distributions (MRST98 parametrization at $Q^2 = 2$ GeV^2).

leading order longitudinal amplitude for meson electroproduction behaves as $1/Q$. This leads to a $1/Q^6$ scaling behavior for the longitudinal cross section $d\sigma_L/dt$ at large Q^2, which provides an experimental signature of the leading order mechanism. Besides the dependence on the SPD's, the meson electroproduction amplitudes require the additional non-perturbative input from the meson distribution amplitudes, for which we take asymptotic forms in the calculations. Furthermore, it was shown in Ref. [3] that electroproduction of vector (pseudoscalar) mesons accesses the unpolarized (polarized) SPD's H and E ($\tilde{H}$ and $\tilde{E}$) respectively. According to the produced meson (ρ^0, $\rho^\pm$, ω, ϕ, π^0, $\pi^\pm$, η,...), the SPD's for the different quark flavors enter in different combinations due to the different quark charges and isospin factors, allowing to perform a flavor separation.

When measuring hard electroproduction reactions in the region $Q^2 \approx 1-20$ GeV^2, the question arises how important are power corrections to the leading order amplitudes, i.e. how fast one approaches the scaling regime predicted by the leading order amplitudes. A first estimate of these corrections has been made in Ref. [9], to which we refer for details. The correction due to the intrinsic transverse momentum dependence is known to be important to get a successful description of the $\pi^0\gamma^*\gamma$ transition form factor, for which data exist in the range $Q^2 \approx 1 - 10$ GeV^2. For the pion electromagnetic form factor in the transition region before asymptotia is reached, the power corrections due to both the transverse momentum dependence and the soft overlap mechanism are quantitatively important. We took these form factors as a guide to estimate the corrections due to the parton's intrinsic transverse momentum dependence in the DVCS and hard meson electroproduction amplitudes. In addition, for the meson electroproduction amplitude, an estimate is given for competing soft overlap mechanism.

3 Results for DVCS and hard meson electroproduction observables

We show here some results for DVCS and meson electroproduction observables using the ξ-dependent ansatz for the SPD's described previously. For more results, we refer to our works in Refs.[4,8,9]. In Fig. 3, the ρ^0_L, π^+ and γ cross sections are compared as function of the beam energy at fixed Q^2 in the valence region. Going up in energy, the increasing virtual photon flux factor boosts the meson leptoproduction cross sections and the DVCS part of the γ leptoproduction cross section. Comparing the different channels, it is clear on this picture that the ρ^0_L channel is very favorable as it depends on the unpolarized SPD's. For the γ channel, the contaminating Bethe-Heitler (BH) process is hardly influenced by the beam energy and therefore overwhelms the DVCS cross section at low beam energies. Although Fig. 3 shows that a high energy such as planned at COMPASS is preferable, one can try to undertake a preliminary study of the hard electroproduction reactions using the existing facilities such as HERMES or JLAB, despite their low energy. Recently, an experiment has been approved at JLAB [10] to investigate (the onset of) the scaling behavior for ρ^0_L electroproduction in the valence region ($Q^2 \approx 3.5$ GeV^2, $x_B \approx 0.3$).

Although at present, no experimental data for the ρ^0_L electroproduction at larger Q^2 exist in the valence region ($x_B \approx 0.3$), the reaction $\gamma^* p \longrightarrow \rho^0_L p$ has been measured at smaller values of x_B. We therefore compare our results in Fig. 4 to see how these data approach the valence region, where one is sensitive to the quark SPD's. For the purpose of this discussion, we call

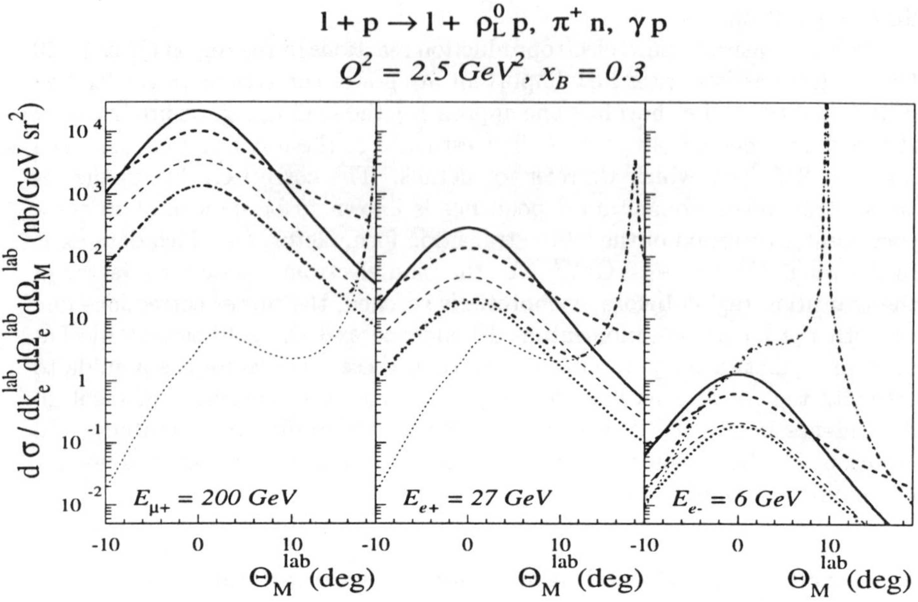

Figure 3: Comparison between the angular dependence of the leading order predictions for the ρ^0_L (full lines), π^+ (thick, upper dashed lines) leptoproduction and DVCS (thick dotted lines) in-plane cross sections at $Q^2 = 2.5$ GeV^2, $x_B = 0.3$ and for different beam energies : $E_{\mu^+} = 200$ GeV (COMPASS), $E_{e^+} = 27$ GeV (HERMES), $E_{e^-} = 6$ GeV (JLab). The BH (thin dotted lines) and total γ (dashed-dotted lines) cross sections are also shown. For the π^+, the result excluding the pion pole is also shown (thin, lower dashed lines).

the mechanism which proceeds through the quark SPD's, the Quark Exchange Mechanism (QEM). Besides the QEM, ρ^0 electroproduction at large Q^2 and small x_B proceeds predominantly through a perturbative two-gluon exchange mechanism (PTGEM) as studied in Ref. [11]. To compare to the data at intermediate Q^2, we implemented [9] in both mechanisms the power corrections due to the parton's intrinsic transverse momentum dependence, which give a significant reduction at the lower Q^2 values. Comparing our results with the data in Fig. 4, one sees that the PTGEM explains well the fast increase at high c.m. energy (W) of the cross section but substantially underestimates the data at lower energies. This is where the QEM is expected to contribute since x_B is then in the valence region. The QEM describes well the change of behavior of the data at lower W. This has also been confirmed by recent HERMES data (around $W \approx 5$ GeV) [12] to which we compared our calculations.

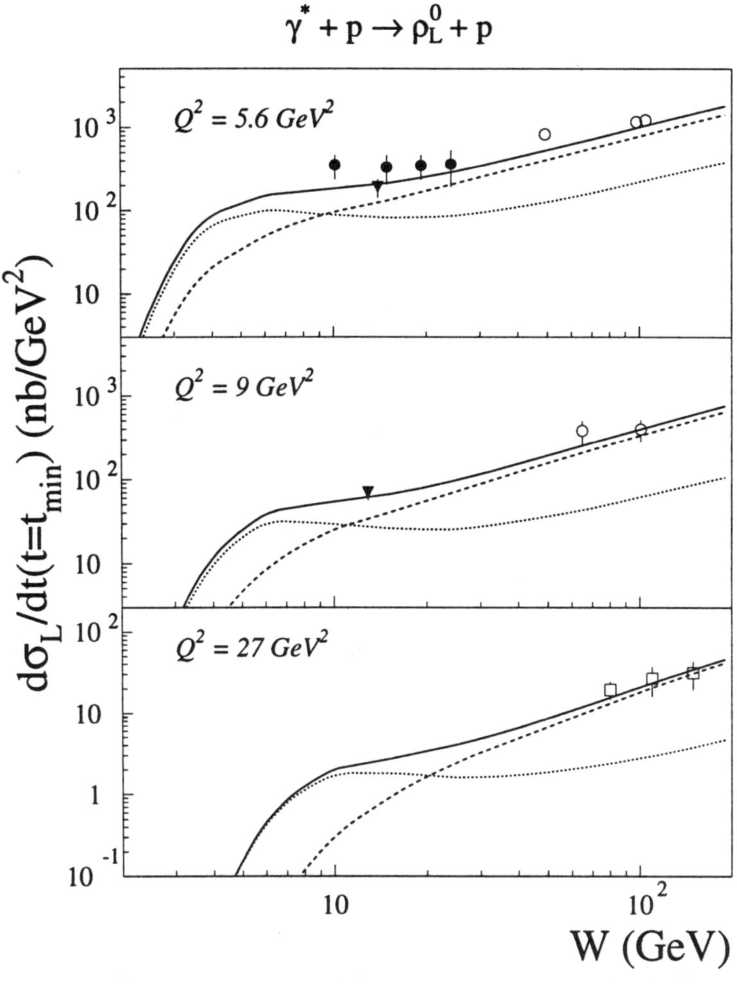

Figure 4: Longitudinal forward differential cross section for ρ_L^0 electroproduction. Calculations compare the quark exchange mechanism (dotted lines) with the two-gluon exchange mechanism (dashed lines) and the sum of both (full lines). Both calculations include the corrections due to intrinsic transverse momentum dependence. The data are from NMC (triangles), E665 (solid circles) and ZEUS (open circles and squares). References to the data can be found in Ref. [9].

In Fig. 5, we compare our calculations for $\gamma_L^* p \rightarrow \pi^+ n$ with the few data at larger Q^2. Our results are compatible with those of Ref. [13] where it was

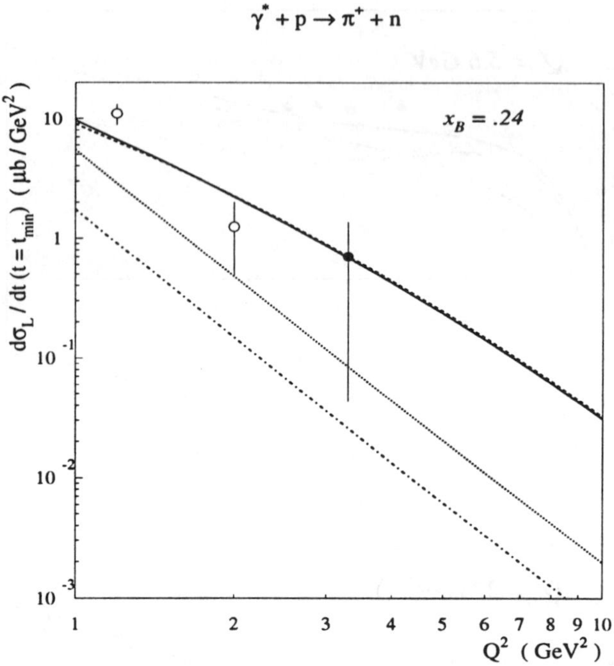

Figure 5: Q^2-dependence for the longitudinal forward differential cross section $d\sigma_L/dt$ for $\gamma^* p \rightarrow \pi^+ n$ at $x_B = 0.24$. The leading order pion pole contribution (dotted line) and $\tilde{H}$ contribution (dashed-dotted line) are shown using a running coupling constant with $\Lambda_{QCD} = 0.2$ GeV. Also shown is the calculation including the power corrections for the pion pole contribution (dashed line) as well as for the sum of the pion pole and $\tilde{H}$ contribution (full line). The data are from Ref. [14].

also found that in the valence region, the pion pole dominates the $\gamma_L^* p \rightarrow \pi^+ n$ cross section. We furthermore show in Fig. 5 our predictions including the power corrections. For the power corrections to the pion pole contribution we include in the pion electromagnetic FF the transverse momentum dependence and the soft overlap contribution. These power corrections enhance the leading order predictions for the cross sections substantially, which is also indicated by the sparse data. It will be most valuable to have more accurate data for the longitudinal cross sections for pion electroproduction to extract information on the pion pole contribution. Of particular interest should also be the ratio of π^+ versus π^0, to get information on the non-pole ($\tilde{H}$) contribution.

4 Conclusions

It is clear that at present new and accurate data are needed for these exclusive channels at large Q^2 in the valence region, where the quark exchange mechanism dominates. The fundamental interest of the SPD's justifies an effort towards their experimental determination. Although these exclusive experiments at large Q^2 are quite demanding, we think that both the scaling region and the onset of the scaling region promise to be sufficiently rich to motivate an extensive experimental investigation at different facilities.

Acknowledgments

This work was supported in part by the Deutsche Forschungsgemeinschaft (SFB 443), by the French Commissariat à l'Energie Atomique (CEA), and by the French CNRS/IN2P3.

References

1. X. Ji, Phys. Rev. Lett. **78**, 610 (1997); Phys. Rev. D **55**, 7114 (1997).
2. A.V. Radyushkin, Phys. Lett. B **380**, 417 (1996); Phys. Rev. D **56**, 5524 (1997).
3. J.C. Collins, L. Frankfurt and M. Strikman, Phys. Rev. D **56**, 2982 (1997).
4. P.A.M. Guichon and M. Vanderhaeghen, Prog. Part. Nucl. Phys. **41**, 125 (1998).
5. M.V. Polyakov and C. Weiss, hep-ph/9902451.
6. A.V. Radyushkin, Phys. Rev. D **59**, 014030 (1999).
7. A.D. Martin, R.G. Roberts, W.J. Stirling, R.S. Thorne, Eur. Phys. J. C **4**, 463 (1998).
8. M. Vanderhaeghen, P.A.M. Guichon and M. Guidal, Phys. Rev. Lett. **80**, 5064 (1998).
9. M. Vanderhaeghen, P.A.M. Guichon and M. Guidal, Phys. Rev. D **60** (1999), in press; hep-ph/9905372.
10. JLab exp. E98-107/E99-105, spokespersons M. Guidal, C. Marchand and E. Smith.
11. L. Frankfurt, W. Koepf and M. Strikman, Phys. Rev. D **54**, 3194 (1996).
12. G. van der Steenhoven, private communication.
13. L. Mankiewicz, G. Piller and A. Radyushkin, hep-ph/9812467.
14. C.J. Bebek et al., Phys. Rev. D **17**, 1693 (1978).

POLARISATION IN DEEPLY VIRTUAL MESON PRODUCTION

M. DIEHL
Deutsches Elektronen-Synchroton DESY, 22603 Hamburg, Germany

T. GOUSSET
SUBATECH, B.P. 20722, 44307 Nantes, France

B. PIRE
Centre de Physique Théorique, École Polytechnique, 91128 Palaiseau, France

We discuss two aspects of polarisation in hard exclusive meson production: the leading-twist selection rule for the meson helicity, and the different partial waves of a $\pi\pi$-pair which may or may not be due to the decay of a ρ.

1 Helicity in vector meson production

The first half of this talk is about exclusive electroproduction of a ρ (or any other light vector meson), $ep \to ep + \rho$, in the Bjorken limit where the photon virtuality $Q^2 = -q^2$ becomes large, while $x_B = Q^2/(2p \cdot q)$ and the invariant momentum transfer $t = (p - p')^2$ remain fixed. There is a factorisation theorem[1] stating that in this limit the $\gamma^* p$ amplitude factorises into a hard photon-parton scattering and non-perturbative quantities, namely skewed quark and gluon distributions in the proton and a $q\bar{q}$ distribution amplitude of the meson. This is shown in Fig. 1, where also four-momenta are defined.

The factorisation theorem provides a solid basis for extracting information on the quark and gluon structure of hadrons from this process, in particular for measuring skewed parton distributions. The description of the amplitude

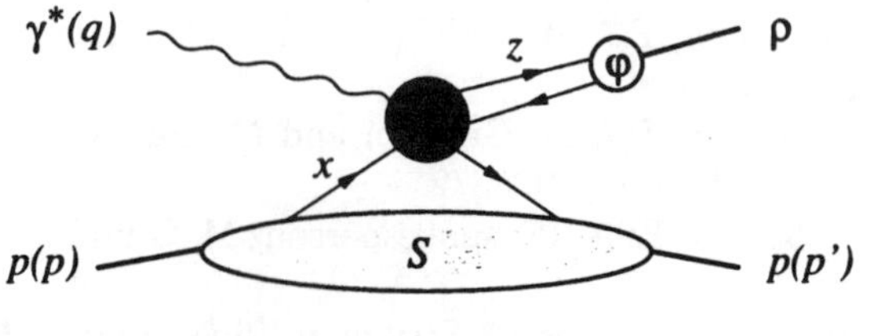

Figure 1: Factorisation of exclusive ρ-production into a hard-scattering coefficient H, a skewed quark distribution S and a meson distribution amplitude φ. There are also diagrams with the skewed gluon instead of the quark distribution. x and z denote momentum fractions in the proton and the ρ, respectively.

in terms of these quantities is of course only accurate up to power corrections in $1/Q$, and in a data analysis it is essential to investigate how close one is to the asymptotic regime at a given value of Q^2. It is at this point that helicity selection rules are of great value.

An essential part of the factorisation theorem is that the description of Fig. 1 only holds if the photon is longitudinally polarised in the $\gamma^* p$ c.m. The corresponding amplitude has a scaling behaviour $\mathcal{A}(\gamma_L^* p) \sim \mathrm{const}/Q$ in the Bjorken limit. For a transverse photon factorisation cannot be established since the loop integrals in the relevant Feynman graphs are sensitive to dangerous infrared regions, but the theorem predicts the corresponding amplitude to be suppressed by at least one power, $\mathcal{A}(\gamma_T^* p) \sim \mathrm{const}/Q^2$.

It turns out that there is a second selection rule, stating that to leading power in $1/Q$ one can only produce a longitudinal vector meson, while transverse ρ production is again power suppressed. Collins, Frankfurt and Strikman[1] observed that a transition $\gamma_L^* p \to \rho_T p$ at order $1/Q$ must involve the so-called chiral-odd quark distribution in the proton and the chiral-odd distribution amplitude of the ρ. This was promising given that experimental information on these quantities is very scarce. Unfortunately, the corresponding hard-scattering coefficient H was found to be zero to lowest order in the strong coupling.[2] A general symmetry argument shows that it is in fact zero to all orders in α_s.[3]

Let us see what goes into this argument. The first ingredient is that the leading power behaviour of the diagrams in Fig. 1 is obtained by replacing the relative transverse momenta of the partons with zero when evaluating the hard scattering H. The hard subprocess is therefore collinear, say along the z-axis. The loop integral over the transverse momenta is performed in the soft quantities S and φ alone, and from rotation invariance it is easy to see that in φ the helicities of the quark and antiquark must add up to the helicity of the meson. For a transverse ρ this gives a chiral-odd distribution amplitude, i.e., one where q and $\bar{q}$ have opposite chirality and thus equal helicity. The second ingredient is that when calculating to leading-power accuracy one can set the quark mass to zero in H. Then the hard scattering conserves quark helicity and thus selects the chiral-odd quark distribution in the proton target; chiral-even quark distributions and the gluon distribution do not contribute. When one follows the flow of helicity from the incoming to the outgoing quark lines in the Feynman graphs for H one finds that the angular momentum along the z-axis changes by two units, a mismatch that cannot be compensated by the photon polarisation. Thus one finds $H = 0$.

The previous argument did not take into account that at higher orders in α_s the Feynman diagrams for H have infinities, so that one must first regularise them, then perform appropriate subtractions of these divergences, and finally remove the regularisation. Now, as soon as one regularises the theory, chirality is no longer conserved. Chirality breaking terms can survive even after the regularisation has been removed again, which gives for instance rise to the axial anomaly in QCD. Hoodbhoy and Lu[4] have calculated the one-loop diagrams for H and found indeed a non-zero result.

Their result is however incomplete since it misses the subtractions of collinear divergences that have to be made in the calculation of a hard-scattering coefficient H; if these are included one will find $H = 0$ again. One can in fact give a general proof[5] that to any finite order in perturbation theory the hard-scattering coefficient conserves the chirality of its external quark lines, so that the symmetry argument sketched above is valid. The idea of the proof is to regularise the theory by going to $4 - \epsilon$ dimensions, perform all subtractions of divergences and then set $\epsilon = 0$, leaving the helicities of the external quarks in H unspecified. The final result for H lives in 4 dimensions, where chirality conservation it ensured by having an odd number of Dirac matrices γ^μ along each quark line. The latter is a consequence of the Feynman rules of massless QCD and is not invalidated by any of the intermediate steps in $4 - \epsilon$ dimensions.

An alternative would be to use Pauli-Villars regularisation to render the loop integrals finite. Then chirality remains conserved along massless fermion lines. Only in internal quark loops do massive regulator fermions occur whose chirality is not conserved; this leads for instance to the axial anomaly in the triangle diagram of two vector and one axial current. It does, however, not disturb the fact that chirality is conserved for the external quark lines of H, which remain massless.

It is important to note that the argumentation leading to our helicity selection rule is very similar to the one that establishes hadron helicity conservation for exclusive processes at leading twist.[6] The proof that a hard-scattering coefficient in perturbative QCD conserves chirality carries over to this case.

In summary, we have that the only leading-twist helicity amplitude is the one for $\gamma^*_L p \to \rho_L p$. All others must be due to power corrections. As the angular distributions in $ep \to ep + \rho \to ep + \pi^+\pi^-$ contain detailed information on the various helicity transitions, their study can give valuable information on how far one is from the asymptotic regime where the helicity selection rules apply.

Beyond this they may guide theory in describing the physics of power corrections itself, a field where we are far yet from a systematic theory. A crucial ingredient in deriving our selection rule was the collinearity of the hard scattering subprocess, and as soon as one keeps transverse parton momenta (and hence t) finite there, all helicity combinations appear. The relevance of the transverse momentum distribution of partons within hadrons has been emphasised in several papers dealing with the helicity structure of meson production in the small-x_B regime,[7,8,9] where increasingly accurate data are becoming available.[10] Another important question is that of perturbative control over infrared regions in the loop integrals, to which the different helicity amplitudes are sensitive to different degrees (remember what we said about γ_L^* and γ_T^* in the beginning).

2 From ρ to $\pi\pi$

So far we have looked at $\gamma^* p \to \rho p$ and its description by the factorised diagrams of Fig. 1, having in the back of our minds that in practice one observes the ρ via its decay channel $\rho \to \pi^+\pi^-$. In the second part of this talk we will take a different point of view on the same process and directly describe $\gamma^* p \to \pi^+\pi^- + p$ in a factorised way. That is, we replace in Fig. 1 the distribution amplitude $\varphi(z)$ of the ρ by a generalised distribution amplitude (GDA), which describes the non-perturbative transition from a $q\bar{q}$-pair, produced in the hard scattering, to the final state $\pi^+\pi^-$.[11,12] The proof of factorisation carries over including its fine print,[13] and the amplitude for $\gamma^* p \to \pi^+\pi^- + p$ has the same scaling behaviour and photon helicity selection rule as the one for $\gamma^* p \to \rho p$. To formulate the second selection rule we replace the helicity of the ρ by the projection l_z of the $\pi\pi$ angular momentum along the z-axis (defined as opposite to the momentum p' in the $\pi\pi$ c.m.): To leading power in $1/Q$ one has $l_z = 0$.

It is very natural that such a generalised factorisation holds: the transition from $q\bar{q}$ to the pion pair is all long-range physics, and the formation and decay of a ρ resonance are just different parts of this. Maybe even more important is that the transition also receives contributions from the nonresonant $\pi\pi$ continuum. In the context of factorisation between short and long distances (and thus for instance in the study of skewed parton distributions) it is not necessary to separate a resonance "signal" from a continuum "background". In fact one need not even require the invariant mass M of the pion pair to be close to the ρ mass; all that counts for factorisation is that M^2 be small compared to the large scale Q^2.

Factorisation not only extends to invariant masses off the ρ peak, but also to $\pi\pi$ partial waves other than the P-wave corresponding to ρ-decay (each in its $l_z = 0$ projection according to our selection rule). To discuss this in more detail we list the kinematical variables on which a GDA Φ depends: There is the invariant mass M, the quark momentum fraction z with respect to the total momentum of the pair (in a frame where the pair moves fast), and the polar angle θ of the π^+ in the $\pi\pi$ c.m. (with the z-axis defined as above). Expanding

$\Phi(z, \cos\theta, M^2)$ in Legendre polynomials $P_l(\cos\theta)$ one obtains the partial wave expansion of the pion system. Odd partial waves ($l = 1, 3, \ldots$) have negative charge conjugation parity C, just as mesons of the ρ family, while the even ones ($l = 0, 2, 4, \ldots$) have positive C-parity and thus the quantum numbers of f-mesons. It is useful to project out the C-even and odd parts of Φ according to $\Phi^{\pm}(z, \cos\theta, M^2) = \frac{1}{2}[\,\Phi(z, \cos\theta, M^2) \pm \Phi(z, -\cos\theta, M^2)]$.

Just as ordinary distribution amplitudes, Φ also depends on a factorisation scale μ, and this dependence is described by the usual ERBL evolution equations.[14] For pion pairs in the ρ-channel they have the asymptotic solution

$$\Phi^- \overset{\mu\to\infty}{\sim} \text{const}\; z(1-z)\, F_\pi(M^2)\, \beta \cos\theta, \qquad (1)$$

where $\beta = \sqrt{1 - 4m_\pi^2/M^2}$ is the velocity of the pions in the $\pi\pi$ c.m. and $F_\pi(M^2)$ the timelike pion form factor, measured in $e^+e^- \to \pi^+\pi^-$. Note that even in the completely asymptotic regime the mass distribution is governed by the pion form factor, which contains the ρ resonance *and* the $\pi\pi$ continuum. While there are certainly non-factorising contributions to the amplitude that lead to a distortion of the mass spectrum away from a resonance shape, the converse is not true: factorisation does *not* predict the $\pi\pi$ mass spectrum to become a pure Breit-Wigner form for the ρ.

While the asymptotic form (1) of Φ^- is a pure P-wave, the asymptotic form of Φ^+ has a z-dependence $z(1-z)(2z-1)$ and contains an S- and a D-wave, each multiplied with an M^2-dependent form factor. Higher partial waves, $l \geq 3$, occur only to the extent that the GDAs deviate from their asymptotic forms.[15]

In the process $\gamma^* p \to \pi^+\pi^- + p$ both C-even and odd pion pairs can be produced, so that in general one has a coherent superposition of pairs with different quantum numbers, even when M is close to or on the ρ mass peak. Observable consequences of the presence of f-type pairs are that

- the angular distribution of $\pi^+\pi^-$ is not the one of a pure P-wave. For analyses of the helicity density matrix in ρ-production it is essential to know how important contributions from other partial waves, e.g. the S- and D-waves are, since the usual extraction methods for the ρ helicity density matrix assume a pure P-wave.
  It is worth noting that the interference between even and odd partial waves gives a contribution to the $\pi^+\pi^-$ angular distribution that is odd under the exchange of the four-momenta of π^+ and π^-. It should be easy to test whether such C-odd terms are present in the angular distribution, e.g. by taking moments of C-odd functions like $\cos\theta$ or $\sin\theta\cos\varphi$.
- not only $\pi^+\pi^-$ but also $\pi^0\pi^0$-pairs are produced. In fact, isospin invariance tells us that the GDA for $\pi^0\pi^0$ is equal and opposite to Φ^+ for $\pi^+\pi^-$.

The production of $\pi\pi$-pairs with different C-parity involves exchanges with different quantum numbers in the t-channel, namely C-plus exchange for C-odd pairs and C-minus exchange for C-even pairs. In the Bjorken limit this means different combinations of skewed parton distributions. For ρ-type pairs one is sensitive to the quark and the gluon distributions, while for f-type pairs quarks contribute but gluons do not: C-minus exchange requires at least three gluons in the t-channel, which gives an amplitude that is power suppressed relative to the leading $1/Q$ of quark-antiquark or two-gluon exchange.[1] It is beyond the scope of our investigation to estimate how important this nonleading-twist contribution becomes when Q^2 is not so large.

To leading-twist accuracy and leading order in α_s the ratio of the amplitudes for f- and ρ-channel pion pairs is

$$\frac{\mathcal{A}(C\text{-even pairs})}{\mathcal{A}(C\text{-odd pairs})} = \frac{\int_0^1 dz\, \frac{2z-1}{z(1-z)}\, \Phi_u^+}{\int_0^1 dz\, \frac{1}{z(1-z)}\, \Phi_u^-}$$
$$\times \frac{\int_0^1 dx\, \frac{1}{(x-\xi+i\epsilon)(x+\xi)} \left(\frac{2}{3}\xi(H_u - H_{\bar{u}}) - \frac{1}{3}\xi(H_d - H_{\bar{d}})\right) + \dots}{\int_0^1 dx\, \frac{1}{(x-\xi+i\epsilon)(x+\xi)} \left(\frac{2}{3}x(H_u + H_{\bar{u}}) + \frac{1}{3}x(H_d + H_{\bar{d}}) + \frac{3}{4}H_g\right) + \dots} \tag{2}$$

in the case where the proton helicity is not flipped. The dots stand for terms with the skewed distributions E_q, $E_{\bar{q}}$, E_g; they are the only ones that contribute in the case of proton helicity flip. We use skewed quark distributions $H_q(x,\xi,t)$ as defined by Ji,[16] antiquark distributions $H_{\bar{q}}(x,\xi,t) = -H_q(-x,\xi,t)$, and $H_g(x,\xi,t) = 2xH_g^{\mathrm{Ji}}(x,\xi,t)$ for the gluons. The non-skewed limits of these functions are $H_q(x,0,0) = q(x)$, $H_{\bar{q}}(x,0,0) = \bar{q}(x)$ and $H_g(x,0,0) = xg(x)$. In (2) we have neglected the formation of a pion pair from $s\bar{s}$, and used isospin invariance to relate the GDAs for d- and u-quarks, $\Phi_d^{\pm} = \pm\,\Phi_u^{\pm}$.

A number of theory predictions for skewed parton distributions can be found in the literature, and one may use one's favourite model in order to evaluate the ratio of integrals over skewed parton distributions. For a very crude order-of-magnitude estimate we replace the second line of Eq. (2) with the ratio

$$\frac{\frac{2}{3}(u-\bar{u}) - \frac{1}{3}(d-\bar{d})}{\frac{2}{3}(u+\bar{u}) + \frac{1}{3}(d+\bar{d}) + \frac{3}{4}g} \tag{3}$$

of the usual parton densities, evaluated at a momentum fraction of order $\xi = x_B/(2-x_B)$. As a numerical example we take the GRV LO parametrisation[17] at a factorisation scale $\mu^2 = 4\ \mathrm{GeV}^2$ and find that the ratio (3) changes from 0.15 to 0.5 for momentum fractions from 0.1 to 0.4. We thus expect the second line of Eq. (2) to be small at values of x_B where gluons dominate over quarks, whereas for x_B in the valence region it may be of order 1.

To estimate the ratio of integrals over GDAs in Eq. (2) we take the asymptotic solution (1) in the ρ-channel, which is completely determined given our knowledge of the pion form factor. The asymptotic GDA in the f-channel involves two form factors which are unknown at the values of M where we need them. Indirect information on them can be obtained from crossing symmetry, which relates GDAs to the parton distributions in the pion. Using the results of Polyakov[15] and making a number of approximations[18] we obtain for M below 1 GeV

$$\frac{\int_0^1 dz \frac{2z-1}{z(1-z)} \Phi_u^+}{\int_0^1 dz \frac{1}{z(1-z)} \Phi_u^-} \approx -\frac{\frac{5}{3}R_q[\,(3-\beta^2)\exp(i\delta_S) - \beta^2\,(3\cos^2\theta - 1)\exp(i\delta_D)\,]}{6\beta\,\cos\theta\exp(i\delta_P)\,|F_\pi(M^2)|}, \quad (4)$$

where R_q denotes the fraction of the pion momentum carried by quarks and is between 0.6 and 0.5 for the GRS LO parton distributions in the pion[19] at a factorisation scale μ^2 between 1 and 20 GeV^2. δ_S, δ_P and δ_D are the phase shifts for $\pi\pi$ elastic scattering in the appropriate partial waves, which are rather well known for $\pi\pi$ invariant masses below 1 GeV. A determining factor in (4) is $|F_\pi(M^2)|$, which has a value around 1.4 at $M = 400$ MeV and around 1.8 at $M = 1000$ MeV, but is as large as 6.7 at the ρ mass peak.

In conclusion, we estimate that the relative contribution from f-channel pion pairs in electroproduction should be rather small for invariant masses M around the ρ-peak. A few 100 MeV off peak, however, it may well be of importance in the x_B-region where gluon exchange does not dominate over quarks, and especially the interference between f- and ρ-channel pairs may be visible. We also note that in (4) not only $|F_\pi|$ has a strong dependence on M^2 but also the $\pi\pi$ phase shifts, so that the different contributions to the pion angular distribution will have a marked M^2-behaviour.

We finally remark that there are other pion pair production processes where the $\pi\pi$ formation is described by GDAs. One example is $\gamma^*\gamma \to \pi\pi$ with the photon virtuality much larger than the $\pi\pi$ invariant mass; in this case Φ^+ but not Φ^- contributes.[12] Another is photoproduction, $\gamma p \to \pi^+\pi^- + Y$, where the proton dissociates into a hadronic system Y with invariant mass much smaller than the γp c.m. energy, and where the momentum transfer t between p and Y is large. Data for this reaction are beginning to come in.[20] At high energies it is dominated by two-gluon exchange in the case where the pion pair is due to the decay of a ρ. A fair amount of theory has been worked out for this.[21] It is important to realise that the production of f-type pion pairs through three-gluon exchange has the *same* scaling behaviour in t as the production of ρ-type pairs, i.e., it is not power suppressed as in electroproduction at small t.[22] Away from the mass peak of the ρ the contribution from f-channel pairs may therefore be important.

Acknowledgements

This work has been partially funded by the TMR programme of the European Union, Contracts No. FMRX-CT96-0008 and No. FMRX-CT98-0194. SUBATECH is Unité mixte 6457 de l'Université de Nantes, de l'Ecole des Mines de Nantes et de l'IN2P3/CNRS. Centre de Physique Théorique is Unité mixte C7644 du CNRS.

References

1. J.C. Collins, L. Frankfurt and M. Strikman, *Phys. Rev.* D **56**, 2982 (1997);
   A.V. Radyushkin, *Phys. Rev.* D **56**, 5524 (1997).
2. L. Mankiewicz, G. Piller and T. Weigl, *Eur. Phys. J.* C **5**, 119 (1998).
3. M. Diehl, T. Gousset and B. Pire, *Phys. Rev.* D **59**, 034023 (1999).
4. P. Hoodbhoy and W. Lu, hep-ph/9902286.
5. J.C. Collins and M. Diehl, hep-ph/9907498.
6. S.J. Brodsky and G.P. Lepage, *Phys. Rev.* D **24**, 2848 (1981).
7. A.D. Martin, M.G. Ryskin and T. Teubner, *Phys. Rev.* D **55**, 4329 (1997);
   D.Yu. Ivanov and R. Kirschner, *Phys. Rev.* D **58**, 114026 (1998).
8. E.V. Kuraev, N.N. Nikolaev and B.G. Zakharov, *JETP Lett.* **68**, 696 (1998), hep-ph/9809539; N.N. Nikolaev, these proceedings.
9. I. Royen and J.-R. Cudell, *Nucl. Phys.* B **545**, 505 (1999).
10. H1 Collaboration, hep-ex/9902019;
    ZEUS Collaboration, hep-ex/9908026.
11. D. Müller *et al.*, *Fortsch. Phys.* **42**, 101 (1994), hep-ph/9812448.
12. M. Diehl *et al.*, *Phys. Rev. Lett.* **81**, 1782 (1998).
13. A. Freund, hep-ph/9903489.
14. G.P. Lepage and S.J. Brodsky, *Phys. Lett.* B **87**, 359 (1979);
    A.V. Efremov and A.V. Radyushkin, *Phys. Lett.* B **94**, 245 (1980);
    M.K. Chase, *Nucl. Phys.* B **174**, 109 (1980).
15. M.V. Polyakov, hep-ph/9809483.
16. X. Ji, *J. Phys.* G **24**, 1181 (1998).
17. M. Glück, E. Reya and A. Vogt, *Eur. Phys. J.* C **5**, 461 (1998).
18. M. Diehl, T. Gousset and B. Pire, in preparation.
19. M. Glück, E. Reya and I. Schienbein, hep-ph/9903288.
20. J.A. Crittenden, these proceedings.
21. J.R. Forshaw and M.G. Ryskin, *Z. Phys.* C **68**, 137 (1995); J. Bartels, J.R. Forshaw, H. Lotter and M. Wüsthoff, *Phys. Lett.* B **375**, 301 (1996);
    I.F. Ginzburg and D.Yu. Ivanov, *Phys. Rev.* D **54**, 5523 (1996).
22. S.J. Brodsky *et al.*, *Phys. Lett.* B **449**, 306 (1999).

HARD EXCLUSIVE ELECTROPRODUCTION OF PSEUDOSCALAR MESONS AND QCD AXIAL ANOMALY

M. I. EIDES
Department of Physics, Pennsylvania State University, University Park, PA 16802, USA
E-mail: eides@phys.psu.edu

L. L. FRANKFURT
School of Physics and Astronomy, Raymond and Beverly Sackler Faculty of Exact Sciences,
Tel Aviv University, Ramat Aviv 69978, Israel
frankfur@lev.tau.ac.il

M. I. STRIKMAN
Department of Physics, Pennsylvania State University, University Park, PA 16802, USA
E-mail: strikman@phys.psu.edu

We consider hard exclusive electroproduction of π^0, η, η' mesons by the longitudinal photons. On the basis of the QCD factorization theorem we show that at sufficiently large virtualities Q^2 amplitudes of hard exclusive processes may be calculated in terms of the target parton distributions and the minimal Fock component of the pseudoscalar meson light cone wave function at zero interquark distances. We obtain the ratios of the $\pi^0 : \eta : \eta'$ yields for the electroproduction processes off nucleons and deuterons from the low energy QCD chiral dynamics, and the QCD anomaly.

1 Introduction

Recently a number of "two-body processes off nucleons, nuclei, photons..." were discovered which can be legitimately calculated in pQCD: $\pi + \mathrm{T} \to 2$ jets $+\mathrm{T}$ [1]; $\gamma_L^* N \to V(\rho, J/\Psi, \rho'..) + N$ [2]; $\gamma_L^* N \to Meson(\pi, K, \eta,)$; $\gamma_L^* \to$ [Few meson system]+Baryon[3]; $\gamma_L^* p \to forward\ \ N + \pi$; $\gamma_L^* p \to forward\ \ \Lambda + K^+$; $\gamma_L^* p \to forward\ \ \bar{p} + NN$ [4]; $\gamma^* + N \to \gamma + N$ [5,6,7,8,9,10,11,12]; and $\gamma^* + \gamma \to Meson + Meson'$ [13]. These processes allow us to investigate in a new way the parton structure of nucleons and to compare it to that of Δ-isobars, hyperons, $N\pi$. It would allow us to address a novel question of short-range parton correlations in nuclei such as: what is the probability to find a small color singlet cluster in the nucleus built of a quark-antiquark pair, three quarks or even three antiquarks? These processes also probe the minimal light-cone $q\bar{q}$ components of various mesons and few meson systems. In addition, these processes allow us to address a number of key questions of high-energy QCD including the determination of the maximum transverse color separations in high-energy strong interactions that are dominated by perturbative QCD and how far down in x one can use linear QCD evolution equations.

We will discuss below one of such processes, hard exclusive electroproduction of π^0, η, η' mesons. We will obtain a number of predictions for the ratios of the $\pi^0 : \eta : \eta'$ production cross sections for hard exclusive processes in terms of the parameters derived from the low-energy chiral dynamics [14].

2 Factorization theorem and meson distribution amplitudes

The key observation which leads to connection between the hard diffractive electroproduction and the parameters of the low-energy physics was made in [15], where the QCD factorization theorem was proven for the forward hard exclusive electroproduction of mesons by longitudinally polarized photons

$$\gamma_L^*(q) + p \to M(q+\Delta) + B'(p-\Delta) \tag{1}$$

at large Q^2, with t and $x = Q^2/2p{\cdot}q$ fixed. It asserts that the electroproduction amplitude has the form (see Fig.1)

$$\sum_i \int_0^1 dz \int d\xi f_{i/p}(\xi, \xi - x; t, \mu)\, H_{iM}(\xi/x, Q^2, z, \mu)\, \phi_M(z, Q^2, \mu^2)$$
$$+\text{power-suppressed corrections,} \tag{2}$$

where f's are skewed (off(non)-diagonal(forward)) parton helicity densities within a nucleon, ϕ is the light-cone meson distribution amplitude [16], z is the light-cone fraction of the photon momentum carried by a quark (antiquark), and H is a hard-scattering coefficient, which may be calculated in the form of the expansion in powers of the strong coupling constant $\alpha_s(Q)$. The scale μ^2 in Eq. (2) as usual for the factorization relations is $\sim Q^2$ hence there is a direct relation between meson wave functions probed in hard exclusive two-body processes and in form factor physics.

For sufficiently small $x \leq 0.1$ and large Q^2 the skewed distributions can be expressed through the diagonal distributions at large x and $Q^2 \sim$ few GeV^2 [17,18]. Relation between skewed and usual parton densities at moderately small x has been discussed in [19]. At the same time for $x \geq 0.1$ and $Q^2 \sim$ few GeV^2 calculations of the skewed distributions within the chiral model were recently performed in [20].

In spite of these theoretical developments current knowledge of skewed parton distributions is insufficient to avoid related to them uncertainties. Notice,

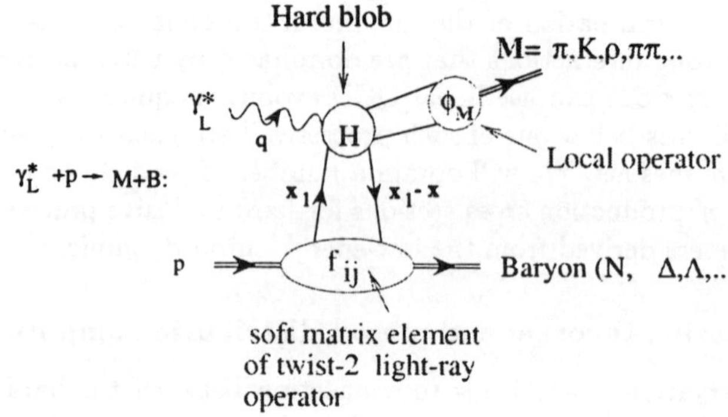

Figure 1: The block structure of the DIS exclusive process.

however, that due to negative charge parity in the crossed channel practically the same combinations of the skewed parton densities enter into amplitudes of hard exclusive electroproduction of pseudoscalar mesons, and they almost completely cancel in the ratios of cross sections. This cancellation of parton distributions is especially precise in the case of the target with vanishing isospin, namely for the deuteron target. We neglect here the contribution of the difference between the s and $\bar{s}$ distributions within a nucleon target. Thus for large Q^2 only the integral

$$\int \frac{dz}{z(1-z)} \phi_M(z, Q^2, \mu^2 = Q^2). \tag{3}$$

of the light-cone meson distribution amplitude $\phi_M(z, Q^2, \mu^2)$ enters in the expressions for the ratios of cross sections of electroproduction of pseudoscalar mesons.

As is well known the meson distribution amplitude is completely determined by the conformal invariance of QCD and the quantum numbers of the meson M and can be decomposed in a series over the Gegenbauer polynomials $C_n^{\frac{3}{2}}$ [16,21,22]. In the limit $Q^2 \to \infty$ only the leading term with $n = 0$ survives and

$$\phi_M(z, Q^2, \mu^2 = Q^2) = c_M z(1-z). \tag{4}$$

The two-gluon component of the meson distribution amplitude rapidly decreases with increase of Q^2, as follows from the calculations of eigenfunctions of the QCD evolution equation [23]. Thus c_M is scale independent. As a result in the case of pseudoscalar mesons integration over z in the expression for the amplitude becomes trivial and the amplitude in Eq. (3) is simply proportional to c_M. Hence, all nontrivial dynamical information about the amplitude is contained in this case in the normalization factor of the light-cone wave function which is equal to the matrix element of the local axial current [16,21,22]. In this respect calculation of ratios of hard diffractive pseudoscalar meson production cross sections is theoretically even more transparent than calculation of absolute cross sections at moderate Q^2, when respective integrals of the meson wave function are more complicated and do not reduce to matrix elements of a local operator.

3 Low energy QCD and $U_A(1)$ axial anomaly

Wealth of experimental data on the low-energy hadron dynamics was understood already before advent of QCD in terms of the current algebra and chiral Lagrangians (see, e.g., review in [24]). After emergence of QCD as the theory of strong interactions it turned out that predictions of the current algebra and the chiral Lagrangians could be more or less rigorously derived from the microscopic theory of strong interactions, if one assumes that the inherent chiral symmetry of QCD with massless quarks is spontaneously broken in the ground state. This bright picture was marred by the fact that naively spontaneous breaking of the chiral symmetry in QCD predicts a pattern of masses of the neutral pseudoscalar mesons π, η, η' which contradicts the experimental data [25]. This problem of the mass spectrum of the neutral pseudogoldstone bosons (and also of the large partial width of the decay $\eta \to 3\pi$) is known as the U_1 problem. Resolution of the U_1 problem was triggered by the remark by t'Hooft [26] that due to the Adler-Bell-Jackiw anomaly [27] the singlet axial current in QCD is non conserved and thus there could be a nontrivial contribution to the mass of the respective would be Goldstone boson. Such additional contribution to the boson mass could be generated for example by the instanton contribution to the axial anomaly. A practical way to obtain a heavy η' in QCD with large number of colors N_c was suggested by Witten [28]. Notice that the very existence of the axial anomaly is insufficient for resolution of the U_1 problem. Veneziano [29] has demonstrated that one has to assume existence of an additional massless pole (so called ghost pole) in the correlators of the certain gauge-noninvariant gluon currents. It was checked a long time ago that the ghost pole hypothesis is self-consistent and provides a satisfactory explanation of the spectrum of the low-lying pseudoscalar mesons, of the large width of the decay $\eta \to 3\pi$, the widths of the decays $\eta, \eta' \to 2\gamma$, and successfully describes also a number of other low-energy reactions [30]. As we will see below the ghost pole hypothesis leads to observable phenomena not only in the realm of the low-energy physics, but also in the hard exclusive electroproduction of pseudoscalar mesons. The same matrix element of the axial current determines both the amplitudes of the low-energy interaction of the neutral pseudoscalar mesons and the amplitudes of the hard diffractive electroproduction of these mesons.

4 Production of π^0, η, η' and QCD Axial Anomaly

Using the factorization theorem [15] and light cone meson wave functions Eq. (4) one may easily obtain hard electroproduction amplitudes in terms of pseudoscalar-quark-antiquark vertices. The strengths of these vertices are determined by the matrix elements of the local quark axial currents. In the naive $SU_f(3)$ limit, neglecting $U_A(1)$ axial anomaly the ratios of the π^0, η and η' yields are given by the expressions

$$\pi^0 : \eta : \eta' = \frac{1}{2}\left(\frac{2}{3}\Delta u + \frac{1}{3}\Delta d\right)^2 :$$
$$\frac{1}{6}\left(\frac{2}{3}\Delta u - \frac{1}{3}\Delta d + \frac{2}{3}\Delta s\right)^2 : \frac{1}{3}\left(\frac{2}{3}\Delta u - \frac{1}{3}\Delta d - \frac{1}{3}\Delta s\right)^2, \tag{5}$$

where we have neglected the skewedness effects. However, due to the presence of the QCD axial anomaly in the neutral channel, and due to an explicit flavor $SU(3)$ breaking by the current quark masses, the matrix elements of the axial currents in QCD have a more involved structure than in the $SU(3)$ symmetric case. The cross section ratios in the realistic case acquire the form

$$\pi^0 : \eta : \eta' = \frac{1}{2}\left(\frac{2}{3}\Delta u + \frac{1}{3}\Delta d\right)^2 :$$
$$\frac{1}{6}\left(\frac{2}{3}\cdot 1.27\Delta u - \frac{1}{3}\cdot 1.27\Delta d + \frac{2}{3}\cdot 1.14\Delta s\right)^2 :$$
$$\frac{1}{3}\left(\frac{2}{3}\cdot 0.840\Delta u - \frac{1}{3}\cdot 0.840\Delta d - \frac{1}{3}\cdot 1.697\Delta s\right)^2. \tag{6}$$

To analyze the consequences of Eq. (6) we use the standard assumption that the quark helicity distributions satisfy the relation $\Delta s = 0$, see e.g. [31].

Account for the anomaly leads to the following effects [14]:

- Enhancement of the η production rate by a factor of 1.61.
- Suppression of the η' production by a factor of 0.71 as compared to the naive $SU_f(3)$ limit, leading to $\eta : \eta' = 1 : 0.87$ (QCD) vs $1 : 2$ in naive $SU_f(3)$.

Qualitatively all this is quite natural since it means enhancement of the η

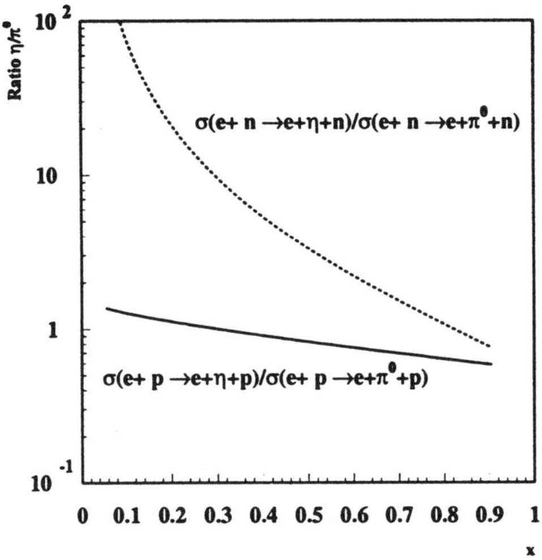

Figure 2: The x dependence of the ratio $\sigma(\gamma_L + N \to \eta + N)/\sigma(\gamma_L + N \to \pi^0 + N)$ for the proton and neutron targets calculated using parameterization of Δu_V and Δd_V from [31].

coupling with the u and d quarks, and suppression of the η' coupling with the same quarks. This may be interpreted in the sense that due to mixing with the QCD ghost pole η'-meson contents is enriched by additional glue.

In the case of the coherent scattering off the deuteron large cancellations occur since the amplitudes for scattering off proton and neutron have opposite sign due to $\Delta u \approx -\Delta d$. This indicates that coherent diffraction off the deuteron will be strongly suppressed as compared to the break up channel. There is an additional suppression due to a spin flip in the deuteron vertex. Naive $SU(3)$ symmetric result for the cross sections ratios for the coherent scattering off the deuteron is given by the relationship $\pi : \eta : \eta' = 27 : 1 : 2$, while the calculation with account of the chiral anomaly and $SU(3)$ breaking leads to the result $\pi : \eta : \eta' = 27 : 1.6 : 1.4$, which is markedly different from the case of scattering off the proton as well as off the neutron (which can be studied in the deuteron break up reaction).

For numerical studies of the implications of Eq. (6) for production of η and π^0 off nucleons we use the results of the recent global analysis of inclusive and semi-inclusive polarization data [31]. For the first estimate we neglect effects of skewedness of relevant parton densities. For certainty we fix $Q^2 = 20\ GeV^2$

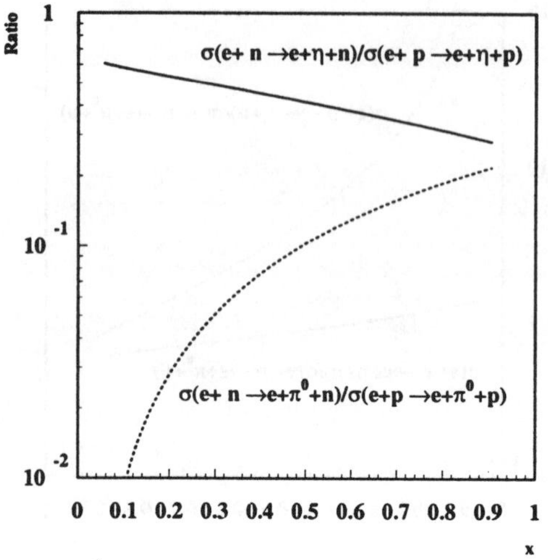

Figure 3: The x dependence of the ratio $\sigma(\gamma_L + n \to M + n)/\sigma(\gamma_L + p \to \pi^0 + p)$ for production of η, and π^0 mesons calculated using parameterization of Δu_V and Δd_V from [31].

and take the fit 1 of the set of fits presented in [31]. We checked that our results depend very weakly on Q^2 scale of the parton densities. This is due to a rather small scaling violation in the valence quark channels. Also we found that different fits of [31] lead to rather small ($\leq 10\%$) changes of the considered ratios.

In Fig.2 we present the ratio of the cross section of production of η and π^0 off the proton and neutron targets. In Fig.3 we compare yields of π^0 and η off the proton and neutron targets. One can see that the current data on Δu_V, Δd_V lead to a rich pattern of x, and target dependencies which would be worth studying experimentally. We also observe that a simplifying approximation $\Delta u_V \approx -\Delta d_V$ is inadequate for the quantitative predictions.

5 Discussion of Results

Hard exclusive electroproduction of the pseudoscalar mesons M off a nucleon $\gamma + N \to M + N$ can be described in the leading twist with the help of the perturbative quark handbag diagram. The first estimates of higher twist effects in the meson production due to renormalons were recently reported [32].

The authors find that these contributions could be strongly enhanced in the production of π^0-mesons. Further studies of this question are certainly needed. Meson production process may be also mediated by the odderon exchange [33]. Distinctive feature of the odderon exchange is that the cross section only slowly depends on the photon energy but rapidly decreases as $\alpha_s^3(Q^2)/Q^{10}$ with increase of the photon virtuality Q^2. On the other hand contribution to the cross section which is due to the exchange of $q\bar{q}$ pair discussed above behaves itself as $\alpha_s^2 x^n/(Q^6)$, i.e. this contribution decreases with energy. Here, n is equal 2 if the perturbative $q\bar{q}$ pair exchange describes amplitude at large intervals in rapidity. If the x dependence of amplitude due to the q$\bar{q}$ pair exchange is estimated assuming that the x dependence of the nonperturbative support function in the QCD evolution equation is given by the exchange by the vector meson Regge trajectory, then $n \approx 1$.

The ratio of the respective cross sections has the form $c\alpha_s(Q^2)/Q^4x^n$, where c is a model dependent numerical factor. Odderon exchange is a sum of the hard and soft QCD contributions. Even signs of these contributions cannot be derived from the general principles. Moreover, hard contribution to the odderon exchange cannot be expressed through the parton distributions of a proton since the soft (nonperturbative) end-point contribution in the integration over the fraction of the photon momentum carried by the quark lines is not suppressed in the case of the odderon exchange.

One option to suppress uncontrollable and therefore dangerous odderon contribution is to investigate the ratios at moderate x where the odderon exchange is suppressed. Another option is to consider the processes where the odderon contribution is forbidden by the quantum numbers. A reaction with Δ^+ production on the proton $\gamma_L + p \to M + \Delta^+$ is a good example. In this case the odderon exchange is suppressed by the isospin conservation. Another advantage of this reaction is that in this case the poorly understood distribution of strange quarks within a proton gives no contribution. Hence, the amplitude of this production process is expressed exclusively in terms of the valence quark distributions. The relation between the electroproduction cross sections of π, η, η' in this reaction may be easily calculated and one comes to the same ratio of the production cross sections $\eta : \eta' = 1 : 0.87$ as was obtained above.

In conclusion, we have demonstrated that the ratios of cross sections of hard exclusive electroproduction of the neutral pseudoscalar mesons may be calculated in terms of the low-energy physics. This low-energy physics crucially depends on the QCD axial anomaly and our results mean that one may explore the physics of spontaneous chiral symmetry breaking in the high-energy processes.

Acknowledgments

This work was supported by the U.S. Department of Energy Under Contract No. DE-FG02-93ER-40771, by the Israel Academy of Science under contract N 19-971, and by the NSF grant PHY-9900771.

References

1. L. Frankfurt, G.A. Miller and M. Strikman, Phys. Lett. **B304**, 1 (1993).
2. S. J. Brodsky, L. Frankfurt, J.F. Gunion, A.H. Mueller, M. Strikman, Phys.Rev. **D50**, 3134 (1994).
3. J. Collins,L. Frankfurt, and M. Strikman, Phys. Rev. **D56**, 2982 (1997).
4. L.L. Frankfurt, P.V. Pobylitsa, M.V. Polyakov and M. Strikman, Phys. Rev. **D60**:014010 (1999).
5. J. Bartels and M. Loewe, Z.Phys. **C12**, 263 (1982).
6. I.I. Balitskii, V.M. Braun, Nucl.Phys. **B311**, 541 (1989).
7. F.M. Dittes, D. Muller, D. Robaschik, B. Geyer, J. Horejsi Phys.Lett. **B209**, 325 (1988).
8. H. Abramowicz, L.L.Frankfurt, and M.Strikman, DESY preprint 95-047,Proceedings of SLAC 94 Summer School SLAC Report 484, pp.539-574, Surveys High Energy Physics **11**, 51 (1997), [hep-ph/9503437]
9. X. Ji, Phys. Rev. Lett. **78**, 610 (1997); X. Ji, Phys. Rev.**D55**, 7114 (1997).
10. A. V. Radyushkin, Phys. Lett. **B380**, 417 (1996); A. V. Radyushkin, Phys. Rev. **D56**, 5524 (1997).
11. X. Ji, J. Phys. **G24**, 1181 (1998).
12. L.L. Frankfurt et al, Phys.Lett.**B418**, 345 (1998), Erratum-ibid. Phys.Lett.**429**, 414 (1998).
13. M. Diehl, T. Gousset, B. Pire and O. Teryaev, Phys. Rev. Lett. **81**, 1782 (1998).
14. M. I. Edies, L. L. Farnkfurt, and M. I. Strikman, Phys. Rev. **D59**:114025 (1999).
15. J. C. Collins, L. Frankfurt, and M. Strikman, Phys. Rev. **D56**, 2982 (1997).
16. S. J. Brodsky and G. P. Lepage, Phys. Lett. **B87**, 359 (1979); Phys. Rev. Lett. **43**, 545 (1979); 1625(E) (1979); Phys. Rev. **D22**, 2157 (1980).
17. L. Frankfurt, A. Freund, V. Guzey, and M. Strikman, Phys.Lett. **B418**, 345 (1998); **B429**, 414(E) (1998).
18. A.D. Martin and M.G. Ryskin, Phys. Rev. **D57**, 6692 (1998).
19. A. V. Radyushkin, Phys. Rev. **D56**, 5524 (1997); *ibid*, **D59**:014030

(1999).
20. V. Yu. Petrov, P. V. Pobylitsa, M. V. Polyakov, I. Bornig, K. Goeke, and C. Weiss, Phys.Rev. **D57**, 4325 (1998).
21. A. V. Efremov and A. V. Radyushkin, Phys. Lett. **B94**, 245 (1980).
22. V. L. Chernyak and A. R. Zhitnitsky, JETP Lett. **25**, 510 (1977); Yad. Fiz. **31**, 1053 (1980).
23. M. Shifman and M. Vysotsky, Nucl.Phys. **B186**, 475 (1981); V. Baier and A. Grozin, Nucl.Phys. **B192**, 476 (1981).
24. J. F. Donoghue, E. Golowich, and B. R. Holstein, Dynamics of the Standard Model, Cambridge University Press, 1992.
25. S. Weinberg, Phys. Rev., **D11**, 3583 (1975).
26. G. t'Hooft, Phys. Rev. Lett., **37**, 8 (1976).
27. J. Schwinger, Phys. Rev., **82**, 664 (1951); S. L. Adler, Phys. Rev., **177**, 2426 (1968); J. S. Bell and R. Jackiw, Nuovo Cim., **60A**, 47 (1969).
28. E. Witten, Nucl. Phys. **B156**, 269 (1979).
29. G. Veneziano, Nucl. Phys. **B159**, 213 (1979).
30. D. I. D'yakonov and M. I. Eides, Sov. Phys. JETP, **54**, 232 (1981).
31. D. de Florian, O.A. Sampayo, and R. Sassot Phys.Rev. **D57** 5811 (1998).
32. M. Vanttinen, L. Mankiewicz and E. Stein, "Higher twist contributions in exclusive processes," hep-ph/9810527.
33. A. Schafer, L. Mankiewicz, and O. Nachtmann, preprint UFTP-291-1992, published in DESY HERA Workshop, Hamburg, v.1, 243 (1991); I. F. Ginzburg, D. Yu. Ivanov, and V. G. Serbo, Phys. At. Nucl. **56** 1474 (1993).

REAL COMPTON SCATTERING FROM THE PROTON

Alan M. NATHAN
University of Illinois at Urbana-Champaign, 1110 W. Green Street, Urbana, IL 61801, USA
E-mail: a-nathan@uiuc.edu

Real Compton Scattering on the proton in the hard scattering regime is investigated. Recent theoretical developments are reviewed. Plans for new experimental studies at Jefferson Lab are presented.

1 Introduction

Real Compton Scattering (RCS) in the hard scattering limit is a powerful probe of the short-distance structure of the nucleon. It is a natural complement to other hard exclusive reactions that are currently being pursued at Jefferson Laboratory (JLab) and elsewhere, such as high Q^2 elastic form factors, hard pion electroproduction and photoproduction, and Virtual Compton Scattering (VCS), and Deeply Virtual Compton Scattering (DVCS). The common feature of these reactions is a hard energy scale, leading to the factorization of the scattering amplitude into a part involving a hard perturbative scattering amplitude, which describes the coupling of the external particles to the active quarks, and the overlap of soft nonperturbative wave functions, which describes the coupling of the active quarks to the proton. For RCS, the hard scale is achieved when the Mandelstam variables s, $-t$, and $-u$ are all large, or equivalently when $p_\perp$ is large, on the hadronic scale.

There has been considerable theoretical effort in recent years in calculating RCS cross sections and polarization observables in the hard scattering limit, and one goal of this contribution is to reveiw this activity (Section 2). Despite this renewed theoretical interest in RCS, the only Compton scattering data available in this kinematic regime are the 20-year old Cornell data,[1] which are spase and of limited statistical precision in the theoretically interesting range of high $p_\perp$. In order to provide the high quality data necessary to discriminate among reaction mechanisms and gain new insight into the structure of the proton, a new experiment is planned at JLab (E97-108/E99-114) to measure RCS cross sections over a broad range of s and t and to have an initial look at polarization observables. A second goal of this contribution is to present an overview of the new experiment in the context of the theoretical work (Section 3). A summary is presented in Section 4.

2 Theoretical Overview

Various theoretical approaches have been applied to RCS in the hard scattering regime, and these can be distinguished by the number of active quarks participating in the hard scattering subprocess, or equivalently, by the mechanism for sharing the transferred momentum among the constituents. Two extreme pictures have emerged. In the perturbative QCD (pQCD) approach,[2,3,4] there are three active quarks which share t by the exchange of two hard gluons (see Fig. 1a). In the soft overlap approach,[5,6] the handbag diagram dominates (see Fig. 1b) in which there is one active quark and t is shared by the overlap of the high momentum components of the soft wave function. In any given kinematic regime, both mechanisms will contribute, in principle, to the cross section. It is generally believed that at sufficiently high energies, the pQCD mechanism dominates. However, the question of how high is "sufficiently high" is still an open question to be answered by more precise experiments, and it is not known with any certainty what is the dominant mechanism in the kinematic regime appropriate to JLab ($p_\perp \sim 1-2$ GeV).

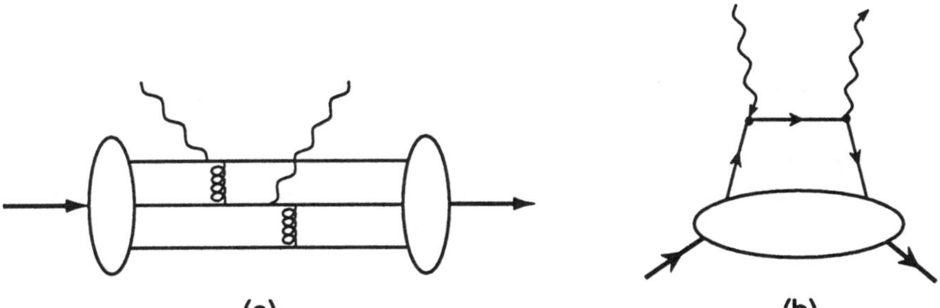

(a) (b)

Figure 1: Different hard scattering mechanisms for RCS. In the pQCD mechanism (a), the momentum is shared among the quarks by hard gluon exchange. In the handbag mechanism (b), the scattering is from a single quark and the momentum is shared by the overlap of the high momentum components of the soft wave function.

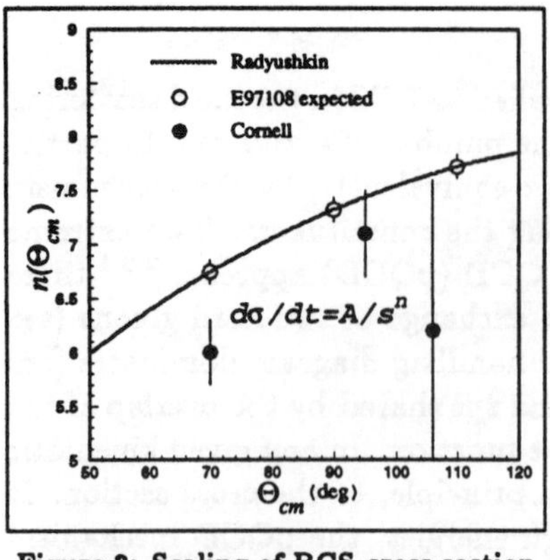

Figure 2: Scaling of RCS cross section at fixed θ_{cm}. The closed points are the Cornell data. The open points represent the projected precision from the JLab experiment. The curve is the prediction of Radyushkin, assuming dominance of the handbag diagram. For the pQCD mechanism, n=6 independent of θ_{cm}.

I next examine the two reaction mechanisms in somewhat more detail. In the pQCD approach, the exchange of two hard gluons leads naturally to the constituent counting rule and scaling [7]

$$\frac{d\sigma}{dt} = \frac{f(\theta_{cm})}{s^n}, \tag{1}$$

where n=6 for RCS. The valence Fock state dominates, since higher Fock states require additional gluon exchanges and are therefore suppressed by additional factors of $1/s$. The soft physics enters through the valence quark distribution amplitude (DA) $\phi(x_1, x_2, x_3)$. Experimentally, the Cornell data support scaling with $n \approx 6$ (Fig. 2), albeit with modest statistical precision. Nevertheless, it has been shown that the cross sections calculated using the asympotic DA badly underpredict the Cornell data.[2,4] This has led to the suggestion [5,6] that the dominant mechanism at experimentally accessible energies is the soft overlap mechanism, which I describe next.

In the soft overlap approach to hard exclusive reactions in general and to RCS in particular, the handbag diagram dominates. The hard physics is contained in the scattering from a single active quark, whereas the soft physics is contained in the wave function describing how the active quark couples to the proton. This coupling is described in terms of nonforward parton densities (ND) [8,9] which are superstructure of the nucleon from which are derived the normal parton densities, elastic form factors, and other quantities that have yet to be measured, including new form factors accessible through Compton scattering. Therefore the ND provide links among diverse physical processes, including both inclusive and exclusive reactions. For example, the dominant ND in RCS is $\mathcal{F}^a(x,t)$, which is related to both the Dirac form factor F_1 and the RCS vector form factor R_V through the expressions

$$F_1(t) = \Sigma_a e_a \int_0^1 \mathcal{F}^a(x;t)dx \qquad R_V(t) = \Sigma_a e_a^2 \int_0^1 \mathcal{F}^a(x;t)\frac{dx}{x}, \tag{2}$$

whereas the $t = 0$ limit of $\mathcal{F}^a(x,t)$ is the parton distribution function $q^a(x)$. Additional links are shown in Table 1. Despite the similarity between the (e,e)

and RCS form factors, an important distinction is the weighting by the quark charge, which is linear for (e, e) and quadratic for RCS, reflecting the one- and two-photon nature of the interaction, respectively. Thus RCS is sensitive to the flavor structure of the proton in a different way than electro-weak scattering, thereby potentially providing another tool, along with parity-violating electron scattering, for decomposing the flavor structure. In particular, RCS has a greater sensitivity to sea quarks than does (e, e).

Table 1: Nonforward parton densities (ND) and their associated electroweak scattering (EWS) and RCS form factors, where a labels the quark flavor. The last column shows the relationship with the parton distribution functions.

ND	EWS Form Factor	RCS Form Factor	$t = 0$ limit
$\mathcal{F}^a(x;t)$	$F_1(t)$	$R_V(t)$	$q^a(x)$
$\mathcal{K}^a(x;t)$	$F_2(t)$	$R_T(t)$	$2J^a(x)/x - q^a(x)$
$\mathcal{G}^a(x;t)$	$G_A(t)$	$R_A(t)$	$\Delta q^a(x)$

The relationship between the RCS cross section and the form factors has been worked out with several simplfying approximations, including the neglect of terms such as R_T which involve hadron helicity flip. This leads to the factorization of the RCS cross sections into a simple product of the Klein-Nishina (KN) cross section describing the hard scattering from a single quark and a sum of form factors depending only on t [5,6]:

$$\frac{d\sigma}{d\sigma_{\text{KN}}} = f_V R_V^2(t) + (1 - f_V) R_A^2(t) \qquad f_V = \frac{(\tilde{s} - \tilde{u})^2}{2(\tilde{s}^2 + \tilde{u}^2)}, \tag{3}$$

where $\tilde{s} = s - m^2$ and $\tilde{u} = u - m^2$. The vector and axial vector form factors R_V and R_A, respectively, have a simple physical interpretation. The combination $|R_V(t) + R_A(t)|^2$ is the probability that a photon can scatter elastically from the proton by transferring t to a single active quark whose helicity is oriented in the direction of the proton helicity. Similarly $|R_V(t) - R_A(t)|^2$ is the probability that the active quark has helicity opposite to that of the proton. In order to measure R_V and R_A, it is necessary to measure the RCS cross section at fixed t with a variable f_V in order to achieve a "Rosenbluth-like" separation. However, for the kinematics of interest, where s, $-t$, and $-u$ are all large, f_V is always close to 1. Consequently the unpolarized cross sections are largely insensitive to R_A. This leads to the very nice feature that the left-hand-side of Eq. 3 is nearly s-independent at fixed t. This is shown in Fig. 3, where one sees that the pQCD mechanism predicts a very different behavior, thereby allowing a very powerful test of the reaction mechanism as well as a precise measurement of R_V.

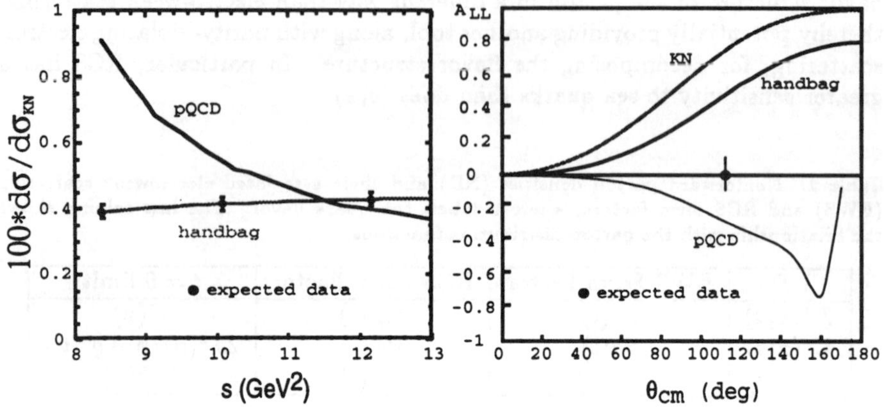

Figure 3: (Left) The ratio $d\sigma/d\sigma_{KN}$, scaled by a factor of 100, as a function of s at $-t = 5$ GeV^2. This ratio is nearly independent of s in the handbag but not in the pQCD model. (Right) The longitudinal spin transfer parameter at 4 GeV. The handbag model essentially follows the Klein-Nishina (KN) result for a structureless Dirac particle. For both plots the points show the precision of the data expected from JLab E99-114.

Simple models for the ND's have been proposed, leading to predictions for the RCS form factors and cross sections.[5,6] An interesting feature of these form factors is that they decrease approximately as $1/t^2$ in the few-10 GeV^2 range, leading to $n \approx 6$ scaling factor, in agreement with the asymptotic law (Eq. 1) but for very different reasons. However, nontrivial violations of $n = 6$ scaling are predicted in the form of an angle-dependent scaling factor $n(\theta_{CM})$, which seems to agree with the limited Cornell data (Fig. 2). Another interesting feature is that at sufficiently high $-t$, the form factors fall as $1/t^4$, leading to $n \approx 10$. Thus the handbag contribution to RCS will be asymptotically subdominant to the pQCD contribution, even though the former may dominate at experimentally accessible energies.

A measurement of polarization observables provides further tests of the reaction mechanism as well as access to additional form factors. The longitudinal polarization transfer observable A_{LL} is defined by

$$A_{LL}\frac{d\sigma}{dt} \equiv \frac{d\sigma(\uparrow\uparrow)}{dt} - \frac{d\sigma(\uparrow\downarrow)}{dt} \tag{4}$$

where the first arrow refers to the incident photon helicity and the second to the recoil proton helicity. In the handbag mechanism, this is related to the form factors by the expression [6]

$$A_{LL}\frac{d\sigma}{d\sigma_{\mathrm{KN}}} = A_{LL}^{\mathrm{KN}} R_V(t) R_A(t)\,, \tag{5}$$

where A_{LL}^{KN} is the result for a structureless Dirac particle. This is plotted in Fig. 3, where one sees that the handbag calculation essentially follows the KN result, whereas the pQCD prediction looks very different, thereby providing another stringent test of the reaction mechanism in addition to a measurement of the axial form factor R_A. One can similarly define the transverse polarization transver observable A_{LT}, which arises as an interference between proton helicity flip and non-flip amplitudes. In the strict pQCD limit, it must vanish since hadron helicity is conserved. Thus far this observable has not been calculated with either the pQCD or handbag mechanisms. However, one can anticipate that in the handbag mechanism it will be proportional to R_T, the RCS form factor that is closely related both to the Pauli elastic form factor F_2 and to the quark total angular momentum (see Table 1), both of which are topics of high current interest. The induced polarization P_N is the component of recoil polarization normal to the scattering plane and involves the imaginary part of the interference between helicity flip and nonflip amplitudes. In the handbag mechanism, it is suppressed since all amplitudes are strictly real in this model.[6] In the strict pQCD limit, it vanishes due to hadron helicity conservation. No calculation has yet been done for this quantity.

3 A New Experiment: JLab E99-114

Experiment E97-108 (recently upgraded to E99-114) at JLab plans to measure differential cross sections for Compton scattering from the proton at incident photon energies between 3 and 6 GeV (s=6-12 GeV^2) and over a wide range of CM scattering angles ($-t$=2-7 GeV^2). The goal is to achieve a statistical precision of order 5%, with systematic errors less than 6%. In addition a measurement of the components of the proton recoil polarization at $s = 8$, $-t = 4$ GeV^2 is planned, using a polarized photon beam. Both sets of measurements utilize the technique shown schematically in Fig. 4.

A high duty factor electron beam with current ≥ 10 μA is incident on a 6% copper radiator located just upstream of the scattering target. The mixed beam of electrons and bremsstrahlung photons is incident on a 15-cm LH_2 target. For incident photons near the bremsstrahlung endpoint, the recoil proton and scattered photon are detected with high angular precision in a magnetic spectrometer and photon spectrometer, respectively.

The magnetic spectrometer is one of the High Resolution Spectrometers that are part of the standard Hall A equipment, along with the cryogenic hydrogen target and bremsstrahlung radiator. The photon spectrometer is a new piece of equipment which is being constructed for this experiment. For the polarization measurements, a longitudinally polarized electron beam is used and the polarization is nearly completely transferred to the bremsstrahlung photon. The components of the polarization of the recoil proton are measured in the focal plane polarimeter (FPP) which is also part of the standard Hall A equipment. One essential feature of the experimental technique is the use of the kinematic correlation between the scattered photon and recoil proton in the RCS reaction to reduce the background of π^0 decay photons from the $p(\gamma, \pi^0 p)$ reaction, thereby placing stringent demands on the combined angular resolution of two-spectrometer system. A second essential feature is the mixed electron-photon beam, which is required in order to achieve the desired photon luminosity. On the one hand, this introduces the necessity to identify and reject electrons from *ep* elastic scattering, while on the other hand it provides a convenient tool for an *in situ* calibration of the photon spectrometer and normalization of cross sections.

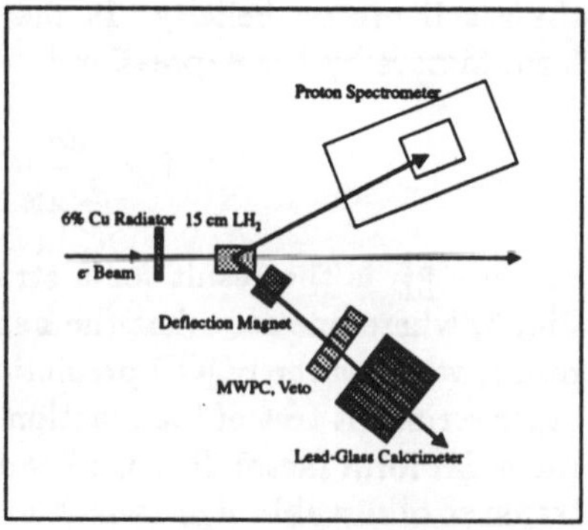

Figure 4: Schematic of the planned JLab experiment.

The principal new piece of equipment for this experiment is the photon spectrometer, which is currently under construction. The principal component is a large-area segmented Pb-glass calorimeter with excellent position resolution and modest energy resolution. In order to reduce the potential background of electrons from *ep* elastic scattering, which are kinematically indistinguishable from RCS photons, several techniques will be used. First, the number of *ep* elastic electrons will be considerably reduced by avoiding the region very close to the bremsstrahlung endpoint, where the e/γ ratio in the beam is very large. Next, electrons will be identified in a plexiglass Čerenkov veto detector that is segmented to allow for a veto that is spatially correlated with an event in the calorimeter. Finally, a magnet will be used to deflect the *ep* elastic electrons by a sufficient amount on the front face of the calorimeter to allow identification by the altered kinematic correlation with the recoil proton relative to undeflected RCS photons. The mixed $\gamma - e$ beam is advantageous in that the *ep* elastic electrons can be used to calibrate the photon spectrometer and normalize the RCS cross sections. For this purpose, several planes of

MWPC just in front of the calorimeter will be used in a separate *in situ ep* elastic scattering experiment to calibrate the position of each element of the calorimeter and veto detector, measure the position resolution, and measure the veto efficiency.

The technique outlined here is conceptually similar to that used in the Cornell experiment. However, the combined effects of a high duty factor electron beam, a state-of-the-art magnetic spectrometer, the ability to calibrate *in situ* with *ep* elastic scattering, and high segmentation in the photon detector should allow significantly better measurements in the range of s and t already covered by Cornell, as well as significant extensions beyond that. The equipment would also be suitable for measurements at higher energies, should those energies become available at JLab in the future.

4 Summary

In this contribution, I have summarized our present theoretical understanding of the RCS process in the hard scattering regime. In addition, I have presented the conceptual design of a new experiment that should extend the RCS data base in the very near future.

Acknowledgments

It is a pleasure to acknowledge stimulating discussion with M. Diehl, P. Kroll, M. Vanderhaeghen, and especially A. Radyushkin. I also thank my E99-114 colleagues C. Hyde-Wright, B. Wojtsekhowski, and F. Sabatie. This work was supported in part by the USNSF under Grant No. 94-20787.

References

1. M.A. Shupe *et al.*, Phys. Rev. **D19** (1979) 1929.
2. A. S. Kronfeld and B. Nizic, Phys. Rev. **D44** (1991) 3445.
3. G. R. Farrar and H. Zhang, Phys. Rev. Lett. **65** (1990) 1721; Glennys R. Farrar and H. Zhang, Phys. Rev. **D41** (1990) 3348; and E. Maina and G. R. Farrar, Phys. Lett. **B206** (1988) 120.
4. M. Vanderhaeghen, *et al.*, Nucl. Phys. **A622** (1997) 144c.
5. A. Radyushkin, Phys. Rev. D **58** (1998) 114008.
6. M. Diehl, Th. Feldman, R. Jakob, and P. Kroll, hep-ph/9811253, hep-ph/9903268. P. Kroll, these proceedings.
7. S.J. Brodsky and G. Farrar, Phys. Rev. Lett. **31** (1973) 1953.
8. X. Ji, Phys. Rev. Lett. **78** (1997) 610; Phys. Rev. D **55** (1997) 7114.
9. A. Radyushkin, Phys. Rev. D **56** (1997) 5524.

DEEP EXCLUSIVE PSEUDOSCALAR MESON PRODUCTION AT CLAS

A. EMPL and P. STOLER
Rensselaer Polytechnic Institute, Troy, NY 12180, USA

M. GUIDAL
Institut de Physique Nucléaire, F-91406 Orsay, France

V. D. BURKERT
Jefferson Lab, Newport News, VA 23606, USA

We discuss measurements of pseudoscalar meson electroproduction above the resonance region: $ep \to e'\pi^+n$ and $ep \to e'(\pi^0,\eta)p'$. These reactions are seen in the light of the OFPDs. The feasibility of the measurements for the JLAB-CLAS detector is shown both with simulation performing an extrapolation from 4 GeV data and data from a very short test run at 5.6 GeV.

1 Motivation

Reaction cross sections decrease rapidly with increasing beam energy and momentum transfer. In particular, there exists very little quality data on exclusive meson electroproduction at momentum transfer greater than a few $(\text{GeV/c})^2$. Only through the combination of high energy, high luminosity accelerators and large acceptance, high resolution detectors, such as CLAS at CEBAF, has the investigation of exclusive reactions at intermediate energies become possible experimentally.

Recently, there has been important theoretical progress[1,2] in the treatment of exclusive reactions in the kinematical region which is becoming available at JLAB with increasing beam energy. The techniques rely on the factorization between hard and soft physics. This factorization was demonstrated for the case of longitudinally polarized virtual photons at high enough Q^2, as illustrated in Figure 1.

The hard physics, which is specific to the reaction studied, can be treated by pQCD techniques, given a large enough momentum transfer (mostly believed to be around a Q^2 of about 10 $(\text{GeV/c})^2$). The soft physics of the nucleon states can be parametrized in the framework of Off-forward Parton Distributions (OFPDs). The OFPDs are generalizations of the deep inelastic distribution functions g(x) and Δg(x) which are measured in inclusive non-polarized and polarized scattering experiments respectively.

It was also shown[3] that in leading-order pQCD the vector meson channels

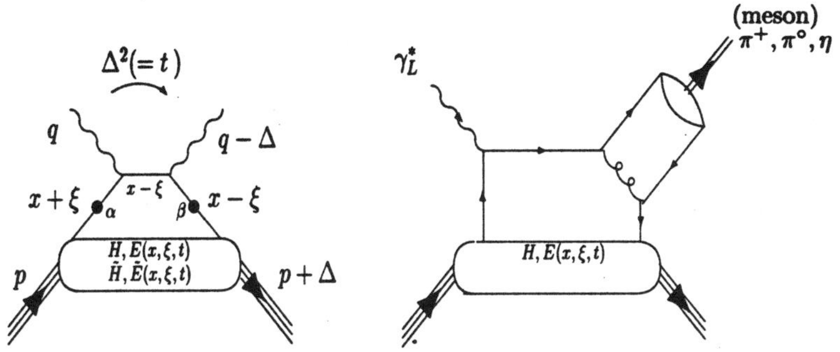

Figure 1: 'Handbag' diagrams for DVCS (left) and pseudoscalar meson electroproduction (right). The OFPDs are defined as $H, \tilde{H}, E$ and $\tilde{E}$ and depend on the momentum fraction of the struck quark x, the momentum transfer t between the virtual photon and the meson, and $\xi = (x_B/2)/(1 - x_B/2)$.

are sensitive primarily to the unpolarized OFPDs (H, E) while in contrast the pseudoscalar meson channels are sensitive primarily to the polarized OFPDs ($\tilde{H}$, $\tilde{E}$). Therefore, by selecting the nature of the outgoing meson one selects the spin-independent or spin-dependent generalized parton distributions. In case of pseudoscalar electroproduction, one can access polarized nucleon structure information without any polarization of external particles.

The OFPD formalism is valid at small momentum transfer t and large Q^2. In Figure 2a predictions of the forward cross section as a function of Q^2 are compared for different mesons [4]. The leading-order longitudinal cross section $d\sigma/dt$ for meson electroproduction shows a $1/Q^6$ behavior, analogous to the one observed in DIS and is a prediction of pQCD at large Q^2. For a first exploratory approach, the meson channels hold the best promise due to the relatively high expected cross sections.

For the charged pion electroproduction in forward direction it was shown [5,6] that there should be a pseudovector contribution (the virtual photon interacting with a quark of the nucleon, which after a hard-gluon exchange, combines with a second quark of the nucleon to form the final pion) and a pseudoscalar contribution (the virtual photon knocks out a pion from the meson cloud of the nucleon). This latter contribution is absent from the neutral pion channel. In Figure 2b the leading-order production cross section of π^0 and π^+ are compared. As a result, at $t = t_{min}$ the pseudovector contribution to the π^+ production is about 5-10 times larger than for the π^0. At $x_B \approx 0.3$ large contributions from the pion cloud, which are absent in case of the neutral pion account for much of the cross section difference. This ratio will be accessed by the proposed simultaneous measurement of π^+ and π^0 channels.

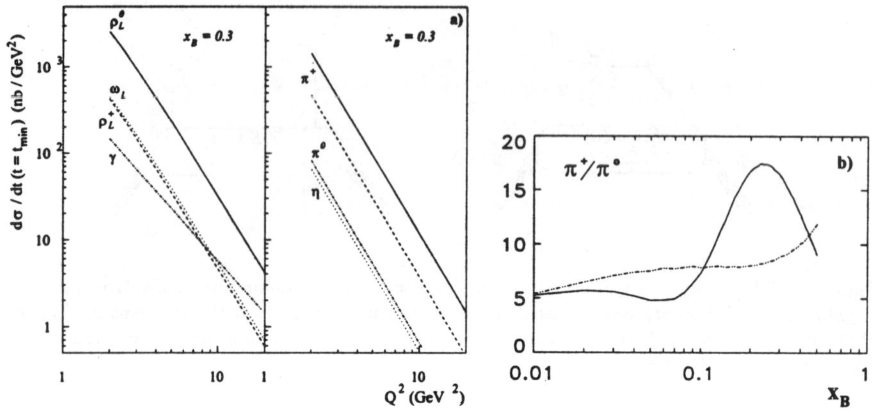

Figure 2: a) Scaling behavior of the leading-order predictions for the longitudinal forward differential electroproduction cross section $d\sigma/dt$ on the proton for vector mesons (left panel) and pseudoscalar mesons (right panel, including prediction for DVCS). b) The predicted ratio[5] of differential production cross sections for exclusive electroproduction of π^+ and π^0 through scattering of longitudinally polarized photons on protons at $t = t_{min}$ and $Q^2 = 10$ $(\mathrm{GeV/C})^2$ (dotted curve corresponds to the pseudovector contribution).

2 Experiment

As an initial investigation the CLAS collaboration has proposed[7] to investigate the exclusive pseudoscalar electroproduction above the resonance region ($W >$ 2 GeV) up to the highest Q^2 values attainable at CLAS using a 6 GeV electron beam. This will allow us to obtain unseparated cross sections to a maximum Q^2 of about 7 $(\mathrm{GeV/c})^2$, and L/T separated cross sections to about 3 $(\mathrm{GeV/c})^2$. With the increase of JLAB's beam energy to about 12 GeV the accessible range of Q^2 for separated data will greatly increase. The selected exclusive reaction channels are $ep \to e'\pi^+ n$ and $ep \to e'(\pi^0, \eta)p'$. By measuring Q^2, x_B and t distributions simultaneously one might be able to distinguish between 'soft' and 'hard' processes, the transition from meson-exchange mechanisms at low Q^2 to quark-exchange mechanisms at larger Q^2. At high Q^2 a search for the onset of the $1/Q^6$ scaling behavior of the cross sections, as predicted by OFPD/pQCD will be possible.

To study the feasibility of the experiment, existing data (eg. for 4 GeV electron beam and a short test run of a few hours at 5.6 GeV taken at CLAS) was used in conjunction with simulation techniques. Results were scaled to 40 days of beam time at a nominal luminosity of $L = 10^{34}$ cm^{-2} s^{-1}. For details on these studies see Ref. [7].

The $ep \to e'\pi^+ n$ reaction is the exclusive meson electroproduction that is most easily accessible to experiment due to its relatively high cross section, and due to the simple identification of the long-lived π^+. It is therefore an ideal candidate to be studied over a wide kinematical range, from the 'soft' to the 'hard' domain. The neutron here is reconstructed by missing mass technique as demonstrated in Figure 3, with data from the test run at 5.6 GeV. A clear peak is found in the spectrum for the neutron by requiring a scattered electron and a single π^+ in the detector, which is even enhanced when selecting events with pion momenta above 3 GeV/c.

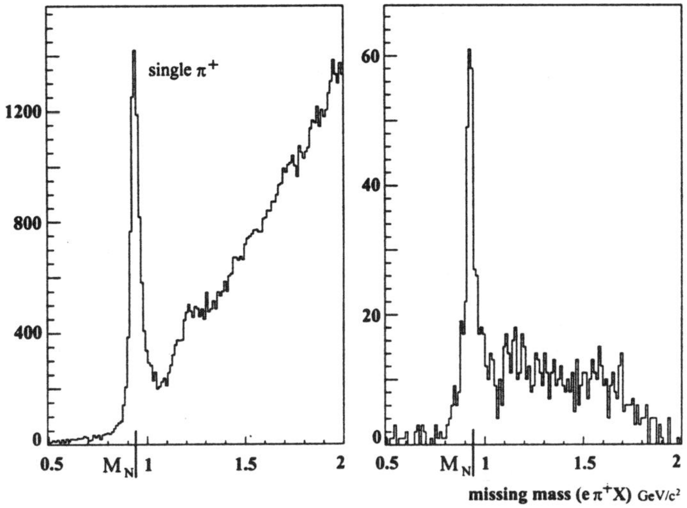

Figure 3: Missing mass distribution for the reaction $ep \to e'\pi^+ X$ from a preliminary analysis of the 5.6 GeV test run data for all pion momenta (left) and pion momenta above 3 GeV/c (right).

To make the connection to the OFPDs, which rely on the factorization for the longitudinal part, it is important to separate the longitudinal response function of the reaction. The objective therefore is to carry out the Rosenbluth L/T separation above the resonance region. The general form of the unpolarized differential cross section for the reaction $ep \to e'\pi^+ n$ is given by $\sigma_{\gamma^* p} = \sigma_T + \epsilon\,\sigma_L$. In order to separate σ_T and σ_L, it is necessary to vary ϵ, which can be achieved by varying the beam energy while keeping Q^2 and x_B

fixed. Three beam energies, 4, 5 and 6 GeV[a] were considered and the study was restricted to the region of $Q^2 > 1.5$ (GeV/c)^2 and $-t < 1.5$ (GeV/c)^2.

A model [8] that gives a good description of all the existing charged-pion electroproduction data was adapted and used to generate events which were then subjected to the resolution and acceptance of the CLAS detector. As an example, Figure 4a displays the reconstructed cross section as a function of ϵ,

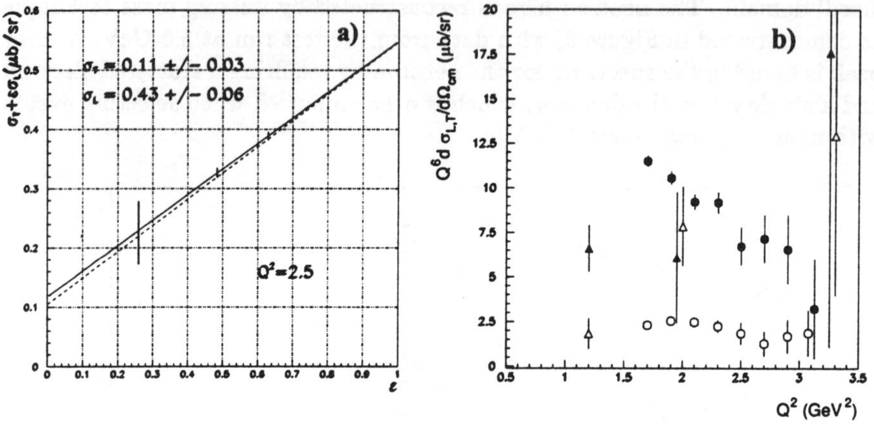

Figure 4: a) Simulated reconstruction of $d\sigma/dt$ as a function of ϵ for $< x_B >= 0.35$ and $t - t_{min} = 0.1$ (GeV/c)^2 at $Q^2 = 2.5$ (GeV/c)^2. The dashed line corresponds to the input model and the solid line is the fit result. b) Separated differential cross section ($\times Q^6$) as a function of Q^2. The existing data[10] are displayed as triangles and the circles are the result of the simulation (open σ_T and filled σ_L symbols respectively).

from simulation, at $Q^2 = 2.5$ (GeV/c)^2 for $< x_B >= 0.35$ and $t - t_{min} = 0.1$ (GeV/c)^2. From a linear fit one extracts the intercept σ_T and the slope σ_L. The fit result (solid line, purely statistical error bars) is compared to the input model (dashed line). The extracted σ_T and σ_L cross sections for various bins in Q^2 are plotted in Figure 4b and compared to the existing data [9]. This shows, that in terms of statistics, an accurate measurement of the separated cross sections is possible up to $Q^2 \approx 3.1$ (GeV/c)^2. Given the statistics, the final error bars will be driven by the systematic error at low Q^2, which was estimated to be of the order of 4%[b].

Neutral pseudoscalar electroproduction is naturally complementary to the

[a]Equal amounts of beam time corresponding to 40 days of data taking were assumed for 4 and 6 GeV beam energy, compared to 20 days for 5 GeV. Bins in (GeV/c)^2 of $\Delta x_B = 0.1$, $\Delta t = 0.1$ and $\Delta Q^2 = 0.2$ were used.

[b]Please see Ref. [7] for more details on the estimate of the systematic error.

charged pion study. Recent articles point to the richness of the physics that can be accessed through the neutral pseudoscalar channels [10]. Because the pseudoscalar pion cloud term does not contribute to π^0 electroproduction, the x_B dependence of the π^+/π^0 ratio is expected to exhibit a strong enhancement in the valence region which should be investigated.

Shown in Figure 5a is the missing-mass squared spectrum below 0.6 GeV^2 obtained from the short test run at 5.6 GeV beam energy[c]. Only events with a mass of the recoiling hadron system $W > 1.8$ GeV for virtual photons of $Q^2 > 2$ (GeV/c)^2 were selected. Cuts to suppress elastic, radiative and multi-pion events were applied. This is the first time that deep exclusive π^0 and η production have been observed in the high W, high Q^2 kinematical region.

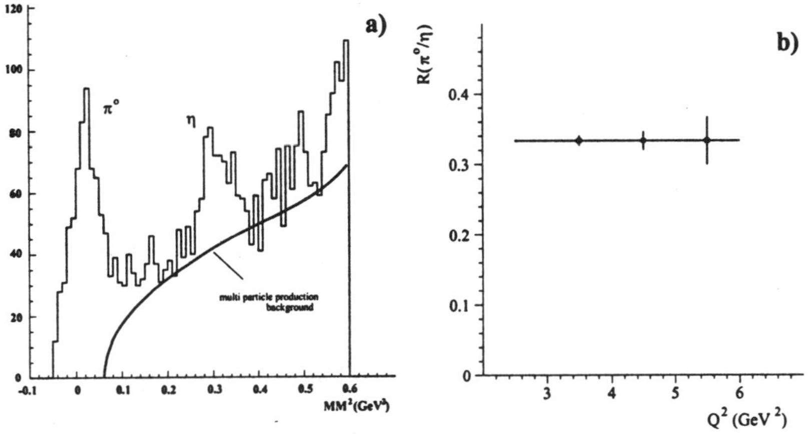

Figure 5: a) Missing mass squared spectrum for $ep \to e'p'X$ for $Q^2 > 2$ (GeV/c)^2 (not acceptance corrected). b) Simulation results for several Q^2 bins of the ratio $R(\pi^0/\eta)$, corresponding to the low t regions of Figure 6. The error bars correspond to the statistical accuracy of accepted data including the multi-pion continuum underlying the η missing-mass peak. The solid curve is the naive prediction.

In order to simulate the expected exclusive π^0 and η production, the differential cross section was modeled as: $d\sigma = \sigma_0 f(t) g(Q^2)$. σ_0 was obtained using comparable data that is only available at low Q^2 from a measurement at DESY [11]. The Q^2 dependence was assumed to follow $g(Q^2) \sim 1/Q^6$. From the 4 GeV data, we extracted the angular distribution of the produced π^0, with respect to the direction of the virtual photon, in the rest frame of the hadronic system and the distribution of the momentum transfer t to the meson. A

[c]The statistics for this test run is equivalent to approximately 2 hours of data taking at about half of the nominal luminosity.

rough geometrical acceptance for the CLAS detector, extracted from the test run data at 5.6 GeV, was applied to the results of the simulation. In Figure 6 the resulting t distributions for several bins in Q^2 are displayed indicating the expected number of events. The above findings apply similarly in the case of η production.

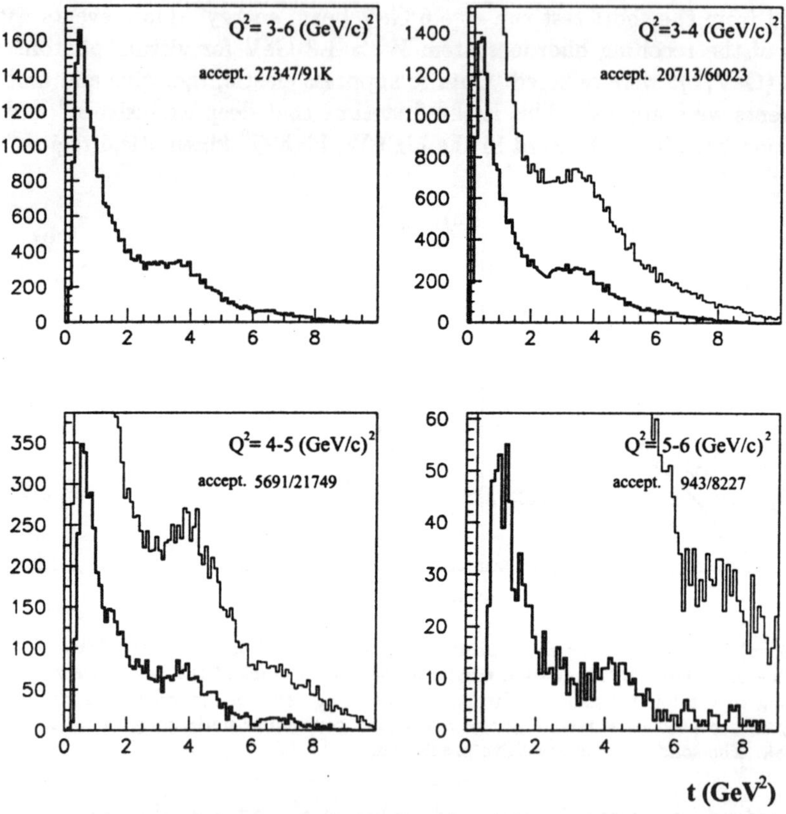

Figure 6: Simulated t distributions for several Q^2 bins at a beam energy of 6 GeV. In each individual graph the distributions for both the Monte Carlo generated and accepted events are displayed. The statistics correspond to what is expected in CLAS for one month of beam time.

The comparison of the π^0 and η (along with the η') cross sections allows one to perform a flavor separation of the polarized OFPDs ($\tilde{H}^u$ and $\tilde{H}^d$) as these OFPDs enter with a different combination for the different channels. In the naive quark model the ratio of the cross sections is predicted to [10]:

$\pi^0 : \eta : \eta' = 1 : 3 : 6$. In Figure 5b we indicate the accuracy for a measurement of the cross section ratio $\eta : \pi^0$ including events at low t only.

3 Conclusion

The study of deep exclusive pseudoscalar meson production, together with the exclusive vector meson production is a field which will become increasingly feasible with increased electron beam energies. The accessible kinematics at JLab energies (moderate x_B) will allow us to investigate the quark currents as illustrated in Figure 1. The kinematic region appropriate to study gluonic currents, (small x_B) can only be accessed at much higher energies. Specific to pseudoscalar mesons, there is considerable theoretical activity to calculate the possible experimental manifestations. We have shown that the current configuration of CLAS is nicely matched to complement this theoretical drive by supplying high quality cross sections for pseudoscalar electroproduction up to a Q^2 of about 7 (GeV/c)^2 and a separated longitudinal cross section measurement for the charged-pion production up to a Q^2 of 3.1 (GeV/c)^2. With an increased JLAB energy, and upgraded version of CLAS, separated cross sections at higher Q^2 will be obtainable.

References

1. X. Ji, *Phys. Rev. Lett.* **78**, 610 (1997), *Phys. Rev.* D **55**, 7114 (1997).
2. A.V. Radyushkin, *Phys. Lett.* B **380**, 417 (1996), *Phys. Rev.* D **56**, 5524 (1997).
3. J.C. Collins, L. Frankfurt and M. Strikman, *Phys. Rev.* D **56**, 2982 (1997).
4. M. Vanderhagen, P.A.M. Guichon and M. Guidal, *Phys. Rev.* D **60**, 094017 (1999).
5. L. Mankiewicz, G. Piller and A. Radyushkin, Preprint hep-ph/9903531, (1999).
6. L. Frankfurt, P. Pobylitsa. M. Poliakov and M. Strikman, Preprint hep-ph/9901429, (1999).
7. V. Burkert *etal*, Electroproduction of Hadrons at 6 GeV using the CEBAF Large Acceptance Spectrometer, May 1999.
8. M. Vanderhagen, M. Guidal and J.M. Laget *Phys. Rev.* C **57**, 1454 (1998).
9. C.J. Bebek *et al.*, *Phys. Rev.* D **13**, 1693 (1978).
10. M.I. Eides, L. Frankfurt and M. Strikman, *Phys. Rev.* D **59**, 114025 (1999).
11. F.W. Brasse *et al.*, *Phys. Lett.* B **58**, 467 (1975).

EXCLUSIVE MESON PRODUCTION AT COMPASS

Josef Pochodzalla[a], Lech Mankiewicz[b], Murray Moinester[c], Gunther Piller [b], Andrzej Sandacz[d], Marc Vanderhaeghen[e]

[a] *Max-Planck-Institut für Kernphysik, 69029 Heidelberg, Germany*

[b] *Physics Department, Technical University Munich, D-85747 Garching, Germany*

[c] *School of Physics and Astronomy, R. and B. Sackler Faculty of Exact Sciences, Tel Aviv University, 69978 Ramat Aviv, Israel*

[d] *Soltan Institute for Nuclear Studies, 00681 Warsaw, Poland*

[e] *University Mainz, D-55099 Mainz, Germany*

We explore the feasibility to study exclusive meson production (EMP) in hard muon-proton scattering $\mu\, p \rightarrow \mu\, p\, M$ at the COMPASS experiment. These measurements constrain the off-forward parton distributions (OFPD's) of the proton, which are related to the quark orbital contribution to the proton spin.

1 Introduction

The internal structure of the proton has been a subject of high interest for many decades. In general, the spin of the proton resides in its quark and gluon constituents and in their orbital angular momenta:

$$\frac{1}{2}\langle \Delta q_v \rangle + \frac{1}{2}\langle \Delta S \rangle + \langle \Delta G \rangle + L_q + L_g = \frac{1}{2} \tag{1}$$

Here, $\frac{1}{2}\langle \Delta q_v \rangle$ denotes the spin in the valance quarks, $\frac{1}{2}\langle \Delta S \rangle$ is the total spin carried by the sea quarks and antiquarks, $\langle \Delta G \rangle$ represents the spin residing in the gluonic field, L_q and L_g represent the orbital momenta of the quarks and gluons, respectively. A main goal of the COMPASS experiment at CERN is the determination of the gluon contribution ΔG. In COMPASS, ΔG will be studied via the photon-gluon fusion process, tagged either by charm-anticharm production, or by high-p_T hadron pair production [1].

A factorization theorem for deeply-virtual Compton scattering (DVCS) was provided by Ji [2] and Radyushkin [3,4]. For exclusive meson production by longitudinally polarized photons, a factorization theorem was proven recently by Collins, Frankfurt, and Strikman [5]. The theorem allows one to separate the meson production amplitudes into a perturbatively calculable component, describing the interaction of the virtual photon with quarks and gluons of the target, and a matrix element component which contains all information about the long-distance non-perturbative strong interaction dynamics in the produced meson and the nucleon target. The latter component can be parameterized, at leading order PQCD, in terms of 4 generalized structure functions,

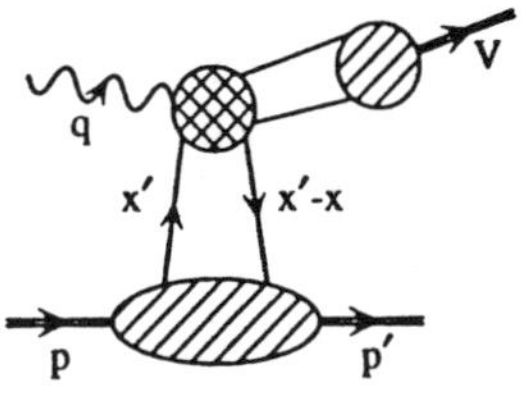

Figure 1: Illustration of the factorization theorem for exclusive meson production [5]

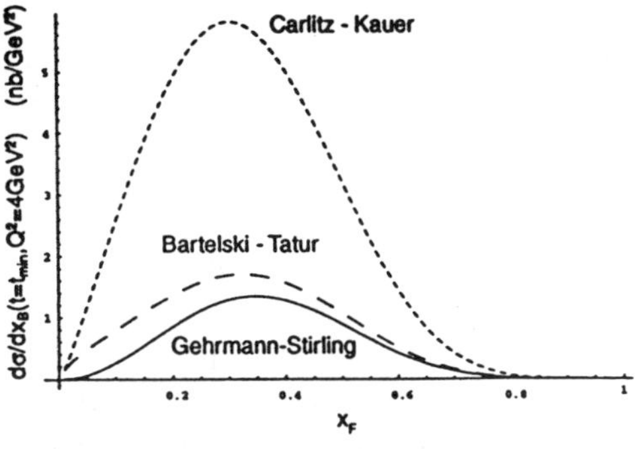

Figure 2: Differential cross section for π^0 production at $t = t_{min}$, $Q^2 = 4\ GeV^2$, as a function of x_B, in units of nb/GeV^2, for the three different valence quark parametrizations [15,16,17,18]

the so called *skewed* or *off-forward parton distributions* (OFPD's). Ji showed [2] that the second moment of these OFPD's determines the total quark angular momentum, the sum of the intrinsic quark spins and their orbital angular momentum. The OFPD's differ from the usual parton distributions probed in inclusive reactions by having a non-zero momentum transfer between the proton in the initial and final state. Therefore, the OFPD's are not accessible in standard inclusive measurements. They can, however, be measured in deeply-virtual Compton scattering or in hard exclusive lepto-production of mesons (EMP). Thus, exclusive meson production provides a unique tool to explore the contribution of orbital angular momentum and the valance quark polarization to the nucleon spin.

EMP are important not only because they relate to the quark orbital momentum contribution to the proton spin. They should test our understanding of QCD generally, and of OFPD's in particular. COMPASS may lead in the study of these new processes, that are only now calculable in QCD.

In principle, Deeply Virtual Compton Scattering represents the most elementary process to access the OFPD's. This reaction has been considered for COMPASS [6]. It was, however, realized that this channel may suffer from potentially large backgrounds. Detailed understanding of (exclusive) meson production is therefore a precondition for studying DVCS. In turn, DVCS may represent an important complement for the Exclusive Meson Production study, which we discuss below.

With the high muon beam energy and high luminosity in COMPASS [7] experiment, large Q^2 values can be reached with reasonable rates. For ρ_0 for example, a Q^2 range between 2 and 20 (80) GeV^2 is accessible for incident 100 GeV/c (200 GeV/c) muon beam. Thus, this COMPASS program extends the approved activity of the CLAS collaboration at JLAB[8], which will explore the OFPD in a limited Q^2 range up to about 4 GeV^2. We distinguish between EMP channels in which the final state meson (ρ^0, π^+[9], ϕ, etc.) can be detected via magnetic spectrometers, and those which require electromagnetic calorimetry. The charged cases can be detected with higher resolution and more easily without EM calorimetry.

As another important aspect of this study, we also consider EMP of a "di-pion" (both resonant and non-resonant), such as $\pi^+\pi^-$, $\bar{K}K$, πK, as studied recently by Polyakov[10]. These works demonstrated that the shape (not absolute cross sections!) of these "di-pion" mass distributions carry important information about quark distribution in the pion (kaon), and also about the structure of resonances (ρ, f_2, K^*, etc.).

2 The Proposed Experiment

2.1 Experimental methodology and trigger

We base our considerations on the scattered muon trigger that is forseen in the COMPASS experiment to study spin dependent DIS. The trigger fires if the muon is scattered at large angle θ_μ. In our simulations we assumed a good angular acceptance for the range $10(5) < \theta_\mu < 50$ mrad for 100 (200) GeV beam energy. The trigger is based on the X-Y correlations between hodoscope elements at about 31 and 50 meters downstream of the target.

Average recoil proton energies for EMP are quite low (~30-40 MeV). Thus, the recoil proton kinematics for COMPASS do not allow installing an efficient proton recoil detector to isolate "elastic" reactions. In order to be able to make optimal use of the experimental setup employed by the main muon program, or even to run initially in parallel with the main program, we will evaluate the inelasticity $I \propto (M_x^2 - M_p^2)$ using measured momenta of the incident and final muons and produced meson[11]. The resolution will, however, not be sufficient to cleanly resolve events with a recoil proton or a recoil Δ^+ or N* in the final state. However, the OFPD's for these processes are related[12]. Thus, a comparison of the data should be made to model calculations which combine both, proton, Δ^+ and N* production.

The main goal of the proposed program is to obtain high statistics EMP data on a pure *unpolarized* proton target during dedicated data runs. This would also help to avoid "dilution" by the presence of different target nuclei in the polarized proton target. Also, with the liquid hydrogen target, the detection efficiency of produced π^0 decay γ's would increase due to the bigger total radiation length of the LH. However, in order to explore the feasibility of this program already during the initial data taking period of COMPASS, we assumed the standard polarized target for our present simulations and count rate estimates.

2.2 *Cross section for exclusive meson production (EMP)*

We estimate the dependence of the cross section for the reaction $\mu\, p \to \mu\, p\, \pi^0$ on the kinematic variables Q^2, x_B, and t, using models for off-forward parton distributions. Following [3,4], we constructed the relevant off-forward parton distributions starting from polarized valence quark distributions.

The exact QCD calculation involves time-consuming numerical integration, which is inconvenient for Monte Carlo purposes. However, neglecting QCD logarithmic evolution of off-forward parton distributions, one can obtain a simple parametrization based on a fit to the results of the exact formulae:

$$\frac{d\sigma}{dt}(\gamma_L\, p \to \pi^0\, p) = \frac{d\sigma}{dt}(x, Q^2)\bigg|_{t=t_{min}} \times e^{B(t-t_{min})}, \tag{2}$$

where:

$$\frac{d\sigma}{dt}(x, Q^2)\bigg|_{t=t_{min}} = \frac{\alpha_S{}^2(Q^2)\cdot \mathrm{PF}(x,Q^2)\cdot \mathrm{UF}(x)}{Q^2(Q^2+M^2)^2}. \tag{3}$$

Here, $\alpha_S(Q^2)$ is the one-loop QCD coupling constant, which – as usual for LO calculations – was evaluated with $\Lambda_{QCD} = 0.2$ GeV and with $NF = 4$ for the number of active flavors. In this equation, $\mathrm{PF}(x, Q^2)$ is given by:

$$\mathrm{PF}(x,Q^2) = \frac{Q^4\,(1-x)}{(-Q^2 - 0.881721\,x + Q^2\,x)^2}. \tag{4}$$

$\mathrm{UF}(x)$ is the part of the cross-section which is independent of Q^2. It would acquire a logarithmic Q^2 dependence when QCD evolution effects are included. The shape of $\mathrm{UF}(x)$ depends on the model of the off-forward parton distributions used in the actual calculation.

Figure 2 shows the differential cross section for π^0 production at $t = t_{min}$, $Q^2 = 4\ GeV^2$, as a function of Bjorken x_B, in units of nb/GeV^2, for the three different valence quark parametrizations. The solid line refers to the Gehrmann-Stirling polarized valence parametrization [15], the short-dashed line is based on the Carlitz-Kaur model [13,16,17] and the long-dashed line results from the Bartelski-Tatur polarized valence distribution [18]. We do not include twist-2 contributions to the OFPD, which according to recent calculations [20,21] may change the absolute cross sections for EMP.

2.3 Monte Carlo results

The events of exclusive π^0 production, $\mu + p \to \mu + p + \pi^0$ with the subsequent decay $\pi^0 \to \gamma\gamma$, were generated with rates proportional to the estimates of differential cross sections from Ref. 18. The directly measured cross section could be expressed in terms of the virtual photoproduction cross section:

$$\frac{d\sigma_{\mu N \to \mu N \pi^0}(\nu, Q^2, t)}{d\nu dQ^2 dt} = \Gamma(Q^2, \nu) \frac{d\sigma_{\gamma^* N \to N\pi^0}(x = Q^2/(2M_p\nu), Q^2, t)}{dt}, \quad (5)$$

where Γ is the flux of virtual photons:

$$\Gamma = \frac{\alpha(\nu - \frac{Q^2}{2M_p})}{2\pi Q^2 E_\mu^2 (1 - \epsilon)} . \quad (6)$$

Here, α is the fine structure constant, and ϵ is the virtual photon polarization.

The cross sections were estimated for proton and carbon targets. The latter one is more representative for the complex material of the COMPASS polarized target. For the proton target, $d\sigma_{\gamma^* N \to N\pi^0}(x, Q^2, t)/dt$ is given by the Eqs. (2-4) and $B = 4.5\ \mathrm{GeV}^{-2}$. For $x = 0.1$, $Q^2 = 4\,\mathrm{GeV}^2$ and $t = t_{min}$, $\frac{d\sigma}{dt}$ is equal to $0.88\ \mathrm{nb/GeV}^2$. For similar conditions, Vanderhaeghen *et al.* [13] give about 4 times larger cross section. We will therefore give counting rate estimates in this report with an uncertainty, which reflects the difference between these two theoretical estimates. The present theoretical uncertainties are mainly due to different assumptions on the shapes/magnitudes of the valence quark spin distributions. For the carbon target, the cross section was assumed to be the product of the proton cross section and a function:

$$f(t) = 8.12\, e^{-47.7(t - t_{min})} + 0.3 . \quad (7)$$

This function was obtained from the parametrization of the NMC data for exclusive ρ^0 production on a carbon target [19].

In our simulations, we considered the kinematic range of $1\ \mathrm{GeV}^2 < \mathrm{Q}^2 < \mathrm{Q}^2_{\mathrm{max}}\ \mathrm{GeV}^2$, and $10(15)\mathrm{GeV} < \nu < 85(180)\ \mathrm{GeV}$. Here, $Q^2_{max} = 4E_\mu(E_\mu - \nu) \cdot \sin^2(\theta_\mu^{max}/2)$, $\theta_\mu^{max} = 50\mathrm{mrad}$, and E_μ is the incident muon energy. The values outside (inside) the brackets correspond to the beam energy of 100 (200) GeV. The same convention will be used hereafter. The upper cut on ν was chosen to eliminate the kinematic region where the amount of radiative events is large, whereas the lower one to eliminate events with poor kinematic resolution. The value of Q^2_{max} is about 20 (90) GeV^2. The cross section integrated within this kinematic regime.

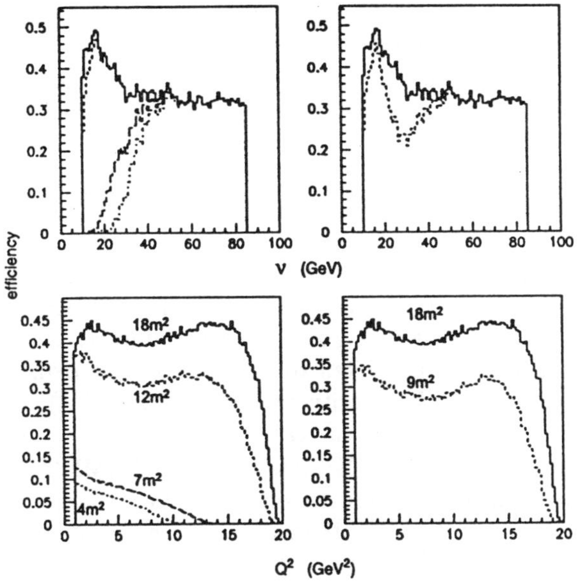

Figure 3: The total acceptance to observe exclusive π^0 events as a function of ν (top) and Q^2 (bottom) at a beam energy of 100 GeV. The different histograms correspond to different geometrical acceptances of the electro-magnetic calorimeter

The total acceptance ϵ_{π^0} to observe an exclusive π^0 event, taking into account the geometric acceptance, the absorption of the decay photons before entering the calorimeter, and the reconstruction of π^0 from the calorimetric measurements, is shown in the top parts of Fig. 3 as function of ν by the solid histograms. The main effect on ϵ_{π^0} is due to a poor reconstruction of low energy π^0 mesons from the calorimetric measurements, at small ν, and due to the absorption of photons in the polarized target, most significantly at large ν. The average total efficiency ϵ_{π^0} is 0.43 for both incident beam energies. For the kinematic range $\nu > 15$ GeV, $Q^2 > 1\,\mathrm{GeV}^2$, $0.02 < x < 0.2$, the efficiency is quite large, varying between 0.3 and 0.5.

Taking the above values of cross sections, and the expected luminosity for the COMPASS experiment[7] of 43 $\mathrm{pb}^{-1}\mathrm{day}^{-1}$, we evaluate the number of deep inelastic exclusive π^0 events per day, $N^{tot}_{\mu\pi^0 N}/\mathrm{day}(Q^2 > 1\,\mathrm{GeV}^2)$, to be about 248-992 (53-212) for 100 (200) GeV. After taking into account effects of acceptance, secondary interactions, and the efficiency of π^0 reconstruction, the number of events, $N^{cuts}_{\mu\pi^0 N}/\mathrm{day}(Q^2 > 1\,\mathrm{GeV}^2)$, is about 107-428 (23-92). The expected numbers of exclusive π^0 events for a running period of 150 days (corresponding to a 1 year period) assuming an overall (SPS and COMPASS) efficiency of 25% are 4013-16050 (855-3420). For $Q^2 > 4GeV^2$, we expect only 130 (42) counts. Adopting the Carlitz-Kauer valence quark parametrization the last numbers will increase by about a factor of 4. Thus, in 1 year of COMPASS operation we will collect at most 500 π^0 events with $Q^2 > 4GeV^2$.

Up to this point we assumed full size electro-magnetic calorimeters (total of 18 m^2) at their default positions. The different curves in Fig. 3 show the result if only limited coverage is assumed. While a reasonable efficiency may still be obtained with about half the calorimeter size (9 m^2), a further cut (e.g. down to 7 m^2) significantly reduces the π^0 efficiency. Furthermore, the count rates presented above were based on a polarized target. Assuming that the overall length of a target is limited to typically 2 m, a dedicated LH target would lead to a reduced luminosity (about a factor of 4) unless the beam intensity can be correspondingly increased.

The detailed discussion of the possibility for COMPASS experiment to measure another exclusive vector meson production channel, $\mu + N \to \mu + N + \rho^0$, is presented in Ref. [11]. Here we only briefly summarize the results of the Monte Carlo calculations for this reaction. The kinematic range is somewhat narrower than for the previous channel: $2GeV^2 < Q^2 < Q^2_{max}$ and $20(40) < \nu < 90(180)$ GeV. The estimated value of the measured cross section $\sigma^{th}_{\rho^0}$, is equal to 0.46 nb (0.5 nb) for 100 GeV (200 GeV). These values are obtained using a parametrization of the cross section measured in the NMC experiment [19]. For the given kinematic range, the average total acceptance ϵ_{ρ^0} is 0.73 (0.68). Taking into account the expected luminosity for the COMPASS experiment, and the effects of acceptance and secondary interactions, the number of events with the invariant mass of two pions in the central part of ρ^0 peak, $0.62 < m_{\pi\pi} < 0.92$ GeV,$N^{cuts}_{\mu\rho^0 N}$/day($Q^2 > 2$ GeV^2), is about 10^4 for both beam momenta. For a running period of 150 days assuming again an overall (SPS and COMPASS) efficiency of 25% this will result in about $3.7 \cdot 10^5$ events. For $Q^2 > 4GeV^2$ we expect 75 k (107 k) events.

3 Conclusions

We explored the feasibility to study exclusive meson production in hard muon-proton scattering $\mu\, p \to \mu\, p\, M$ at the COMPASS experiment. For moderate $Q^2 > 1\ GeV^2$, a significant counting rate for exclusive π^0 production can be reached. For the experimental conditions assumed in our calculations, still up to 500 events beyond a Q^2 of 4 GeV^2 can be collected during 1 year of operation.

On the other hand, COMPASS may provide already during the first main

μ runs, high statistics data on exclusive ρ^0 production even at Q^2 beyond values presently accessible at JLAB. Furthermore, data taken with the polarized target may also allow a first view at polarization phenomena.

Acknowledgements

This work was partially supported by the Polish State Committee for Scientific Research (KBN), by the U.S.-Israel Binational Science Foundation (BSF) and the Israel Science Foundation founded by the Israel Academy of Sciences and Humanities.

References

1. A. Bravar, D. von Harrach, and A. Kotzinian, *Phys. Lett.* **B421**, 349 (1998).
2. X. Ji, *Phys. Rev. Lett.***78**, 610 (1997); *Phys. Rev.* **D55**, 7114 (1997).
3. A.V. Radyushkin, Phys. Lett. B **380**, 417 (1996); *Phys. Lett.* **B385**, 333 (1996); *Phys. Rev.* **D56**, 5524 (1997); hep-ph/9805342.
4. I. V. Musatov and A. V. Radyushkin, *Phys. Rev.* **D56**, 2713 (1997).
5. J. C. Collins, L. Frankfurt, and M. Strikman, *Phys. Rev.* **D56**, 2982 (1997).
6. N. d'Hose, COMPASS collaboration meeting presentation (1998).
7. COMPASS collaboration, CERN/SPSLC 96-14, SPSC/P 297.
8. CLAS collaboration, JLab exp. E98-107/E99-105, spokespersons M. Guidal, C. Marchand and E.Smith (1998).
9. L. Mankiewicz, G. Piller, A. Radyushkin, hep-ph/9812467.
10. M.V. Polyakov, hep-ph/9809483, hep-ph/9901315, hep-ph/9906261.
11. A. Sandacz, Compass Note 1998-5.
12. L. Frankfurt, M.V. Polyakov and M. Strikman, hep-ph/9808449.
13. M. Vanderhaeghen, P.A.M. Guichon, M. Guidal, Phys.Rev.Lett. **80** (1998) 5064; P. A. M. Guichon, M. Vanderhaeghen, *Prog. Part. Nucl. Phys.* **41** 125 (1998); hep-ph/9806305.
14. L. Mankiewicz, G. Piller, and T. Weigl, *Eur. Phys. J.* **C5** 119 (1998).
15. T. Gehrmann and W. J. Stirling, *Phys. Rev.* **D53**, 6100 (1996).
16. R. Carlitz and J. Kauer, *Phys. Rev. Lett.***38**, 673 (1977).
17. M. Goshtasbpour and G.P. Ramsey, *Phys. Rev.* **D55**, 1244 (1997).
18. M. Kurzela, J. Bartelski, S. Tatur, hep-ph/9807355.
19. NMC collaboration, M. Arneodo *et al.*, *Nucl. Phys.* **B429**, 503 (1994).
20. V. Yu. Petrov *et al.*, *Phys. Rev.* **D57**, 4325 (1998).
21. M.V. Polyakov and C. Weiss, hep-ph/9902451.

SCALE DEPENDENCE OF QCD STRING OPERATORS

N. Kivel[abc], L. Mankiewicz[bd]
[a] *Alexander von Humboldt fellow,* [b] *Physik Department, Technische Universität München, D-85747 Garching, Germany*
[c] *Petersburg Nuclear Physics Institute, 188350, Gatchina, Russia*
[d] *N. Copernicus Astronomical Center, ul. Bartycka 18, PL–00-716 Warsaw, Poland*

We discuss a general solution of evolution equations for QCD twist-2 string operators in form of expansion over complete set of orthogonal eigenfunctions of evolution kernels in coordinate-space representation. In the leading logarithmic approximation the eigenfunctions can be determined using constraints imposed by conformal symmetry.

1 Introduction

Recently, there has been a renewed interest in QCD evolution of skewed parton distributions [1,2]. Skewed parton distributions play a crucial role in description of hard, exclusive QCD processes [1–7]. So far, the main effort has been devoted to studies of evolution equations for skewed parton distributions in the momentum representation [1,2,7–14]. This is certainly the most natural choice in the forward, deep inelastic scattering limit, but, as observed recently [15,16], as soon as there is a longitudinal momentum transfer between initial and final hadron states, the corresponding amplitude can be as conveniently represented in terms of momentum- as coordinate-space skewed parton distributions.

In recent papers [15,16] we have considered evolution of charge-conjugation odd and even flavor non-singlet coordinate-space skewed quark distributions in the leading-logarithmic (LO) approximation [17]. The resulting solutions have form of a Neumann series expansions [18] over a set of Bessel functions of the first kind $J_{1/2+j}$, with j assuming odd, respectively even integer values. It is interesting to note that the resulting expansions are valid uniformly for all values of the asymmetry parameter, in contrast to the situation in the momentum-space. There, one has to distinguish between cases where momentum fraction is smaller, respectively larger than the asymmetry parameter. A compact formula, based on the expansion in terms of local, multiplicatively renormalizable operators, exists only in the former case [2,9,11,12,16]

Evolution equations for twist-2 string operators have been known for a long time [17,19]. Written in the coordinate-space representation, they have been considered so far as an intermediate step in derivation of evolution equations for skewed parton distributions in the momentum space. Below, we discuss an explicit solution to the evolution equations for string operators in a form of expansion over orthogonal eigenfunctions of corresponding evolution kernels.

2 Evolution equations for QCD string operators

Let us start from a discussion of a general structure of QCD evolution equations for string operators. For simplicity, we consider here only the quark string operators defined as [17]

$$O(\alpha,\beta) = \bar{q}(\frac{\alpha+\beta}{2}z)\hat{z}\left[\frac{\alpha+\beta}{2}z, \frac{\alpha-\beta}{2}z\right] q(\frac{\alpha-\beta}{2}z)\,, \tag{1}$$

In the above formula z is a light-like vector, $z^2 = 0$. The square brackets denote the usual path-ordered exponential $[az, bz] = \mathcal{P}\exp[-ig\int_a^b z_\mu A^\mu(tz)\,dt]$ which ensures gauge-invariance of the above definitions. Note that $\alpha\, z$ describes the center of the string composed from quark fields and the gluon line between them while $\beta\, z$ corresponds to its "length", understood simply as the difference between coordinates of quark fields at the string ends.

Charge conjugation odd and even quark string operators are obtained from $O(\alpha,\beta)$ by taking its components symmetric, respectively antisymmetric in β. Denoting the former and the latter by $O^+(\alpha,\beta)$, respectively $O^-(\alpha,\beta)$ we define

$$\begin{aligned} O^+(\alpha,\beta) &= \frac{1}{2}\left(O(\alpha,\beta) + O(\alpha,-\beta)\right)\,, \\ O^-(\alpha,\beta) &= \frac{i}{2}\left(O(\alpha,\beta) - O(\alpha,-\beta)\right)\,. \end{aligned} \tag{2}$$

Similarly to local operators, bare string operators have UV-divergences and should be renormalized. In complete analogy with the former case one can define renormalized string operator through the relation:

$$O_R(\omega) = Z(\omega,\omega') \otimes O_B(\omega')\,. \tag{3}$$

Here we have introduced a compact notation $\otimes$ for the two-dimensional integral

$$\int_{-\infty}^{\infty} d\alpha \int_0^{\infty} d\beta$$

and $\omega \equiv \{\alpha,\beta\}$. Function $Z(\omega,\omega')$ corresponds to a renormalization constant in a local case.

The renormalization group equation can be derived by taking a total derivative over μ of the left- and right-hand sides of (3). In this way one arrives at an evolution equation for the renormalized string operator $O_R(\omega)$ [17,19]

$$\mu\frac{d}{d\mu}O_R(\omega) = a_s V(\omega,\omega') \otimes O_R(\omega') \,. \tag{4}$$

The evolution kernel $V(\omega,\omega')$ is defined as

$$a_s V(\omega,\omega') = \mu\frac{\mu}{d\mu}Z(\omega,\omega'') \otimes Z^{-1}(\omega'',\omega') \,. \tag{5}$$

Although it has not been indicated explicitly, $V(\omega,\omega')$ is a function of the QCD coupling constant $V(\omega,\omega') = \sum_{n\geq 0} a_s^n(\mu^2)V^{(n)}(\omega,\omega')$ except of the leading order (LO) approximation, when it is given simply by a function $V^{(0)}(\omega,\omega')$.

After these preliminaries, we shall construct now a general solution of the operator evolution equation (4). To this end, let us assume that one can find a solution of the following system of equations for eigenfunctions and corresponding eigenvalues of the evolution kernel [20]:

$$\begin{aligned} V(\omega,\omega') \otimes \varphi_i(\omega') &= \gamma_i \varphi_i(\omega') \\ \bar{\varphi}_i(\omega) \otimes V(\omega,\omega') &= \gamma_i \bar{\varphi}_i(\omega) \,. \end{aligned} \tag{6}$$

Eigenfunctions $\varphi_i(\omega)$ and $\bar{\varphi}_j(\omega)$ are orthogonal with respect to the scalar product $\otimes$, i.e.

$$\langle \bar{\varphi}_i ,\ \varphi_j \rangle = \bar{\varphi}_i(\omega) \otimes \varphi_j = \delta_{ij} N_i \ , \tag{7}$$

where N_i is some normalization constant. Note that beyond the leading order eigenfunctions $\varphi_i(\omega)$, $\bar{\varphi}_j(\omega)$ and eigenvalues γ_i depend on the running QCD coupling because the kernel $V(\omega,\omega')$ itself becomes a function of $a_s(\mu^2)$.

Solution of eq.(4) can be written in the following form:

$$O_R(\omega) = \sum_{n\geq 0} S_n \ \varphi_n(\omega) \,. \tag{8}$$

The operator-valued coefficients S_n can be obtained by taking the scalar products of both sides of the above equation with eigenfunctions $\bar{\varphi}_n$:

$$S_n = \frac{1}{N_n}\bar{\varphi}_n(\omega') \otimes O_R(\omega') \tag{9}$$

As it follows, S_n themselves are non-local operators. Moreover. eq.(4) leads to the following set of evolution equations for operators S_n:

$$\mu\frac{d}{d\mu}S_n + \sum_{m\geq 0} S_m \langle \bar{\varphi}_n ,\ \dot{\varphi}_m \rangle = a_s \gamma_n S_n , \ \ \dot{\varphi}_n \equiv \beta(\alpha_s)\frac{\partial}{\partial \alpha_s}\varphi_n \,. \tag{10}$$

The second term on the left-hand side induces mixing between operators corresponding to different orthogonal eigenfunctions. It is absent in the LO because eigenfunctions φ_n do not depend on α_s in this approximation. As a consequence operators S_n become multiplicatively renormalizable.

3 LO evolution of string operators

To the leading-order accuracy, the evolution kernel is given by $V(\omega,\omega') = V^{(0)}(\omega,\omega')$ and is therefore independent on $a_s(\mu)$. Arguments based on the conformal symmetry can be used to find explicit solutions of the eigenvalue equation (6). In [17] the nonlocal conformal operators were introduced for the first time and the conformal symmetry arguments were used to find the solution of (4) in terms of an integral representation over a complex conformal spin j. Our goal here is to represent a solution in the form of an expansion (8).

First, note that the two-dimensional eigenvalue problem (6) can be reduced to a one-dimensional one by using translational invariance to separate the dependence on the variable α, which describes the position of the center-of-mass of the string. This is done by substituting

$$\phi_j(\alpha,\beta) = e^{ik\alpha} f_j(k\beta), \;\; \bar{\phi}_j(\alpha,\beta) = e^{ik\alpha} \bar{f}_j(k\beta) \,. \tag{11}$$

In this way, $\phi_j(\alpha,\beta)$ and $\bar{\phi}_j(\alpha,\beta)$ can be interpreted as wave functions of bound systems of two particles, with a momentum k and an internal motion described by $f_j(k\beta)$ and $\bar{f}_j(k\beta)$, respectively. Inserting (11) into the eigenvalue equations (6) one arrives at the following system of equations [17]:

$$\begin{aligned} H_1 f_j &\equiv \int_0^{\rho_2} V(\rho_1,\rho_2) f_j(\rho_1) = \gamma_j f_j(\rho_2), \\ H_2 \bar{f}_j &\equiv \int_{\rho_1}^{\infty} V(\rho_1,\rho_2) \bar{f}_j(\rho_2) = \gamma_j \bar{f}_j(\rho_1) \,. \end{aligned} \tag{12}$$

where $\rho \equiv k\beta$. The reduced kernel $V(\rho_1,\rho_2)$ is given by

$$V(\rho_1,\rho_2) = -3\delta(\rho_1-\rho_2) + \frac{2\sin(\rho_1-\rho_2)}{\rho_2^2} + \frac{4\rho_1\cos(\rho_1-\rho_2)}{\rho_2(\rho_1-\rho_2)} + 4\delta(\rho_1-\rho_2)\int_0^{\rho_2}\frac{d\rho}{\rho} \,. \tag{13}$$

Equations (12) have the same form as the ERBL-equation for evolution of the pion distribution amplitude [21,22]. Conformal symmetry was shown to play a crucial role in finding solution of the LO ERBL equation a long-time ago [21–23]. In a complete analogy, we shall now demonstrate how the symmetry considerations allow to determine eigenfunctions $f_j(k\beta)$ and $\bar{f}_j(k\beta)$.

$\mathbf{J}^2$, the Casimir operator of the collinear conformal group SO(2,1), can be represented as an differential operator acting on the parameters α, β. After substitution (11) one obtains[17]:

$$\begin{aligned} \mathbf{J}^2 f_j(\rho) \equiv L_1 f_j(\rho) &= \left[\rho^2 \frac{d^2}{d\rho^2} + 4\rho\frac{d}{d\rho} + \rho^2 + 2\right] f_j(\rho) \\ \mathbf{J}^2 \bar{f}_j(\rho) \equiv L_2 \bar{f}_j(\rho) &= \left[\rho^2 \frac{d^2}{d\rho^2} + \rho^2\right] \bar{f}_j(\rho) . \end{aligned} \tag{14}$$

Using equations (12) and (14) it is easy to check by an explicit calculation that $\mathbf{J}^2$ indeed commutes with the 'Hamiltonians' $H_{1,2}$. As it follows, the eigenfunctions of $H_{1,2}$ have to be the same as the eigenfunctions of $L_{1,2}$:

$$L_1 f_j(\rho) = j(j+1) f_j(\rho), \; L_2 \bar{f}_j(\rho) = j(j+1) \bar{f}_j(\rho) , \tag{15}$$

where j are positive integers. As L_1 and L_2 are related by $L_2\rho^2 = \rho^2 L_1$ $f_j(\rho)$ and $\bar{f}_j(\rho)$ are also related to each other, $\bar{f}_j(\rho) = \rho^2 f_j(\rho)$.

Solutions of the set of differential equations (15) are given by Bessel functions:

$$f_j(\rho) = c_j(k)\rho^{-3/2} Z_{j+1/2}(\rho), \bar{f}_j(\rho) = \bar{c}_j(k)\rho^{1/2} Z_{j+1/2}(\rho) , \tag{16}$$

where $c_j(k)$, $\bar{c}_j$ are yet undetermined normalization constants. As the operator $O_R(\omega)$ has a well-defined local limit as $\beta \to 0$, the validity of expansion (8) requires that $f_j(\beta)$ are regular in this limit as well. In terms of f_j and $\bar{f}_j$ the scalar product (7) reduces to

$$\langle \bar{f}_j(\rho) f_i(\rho) \rangle = \int_0^\infty d\rho \bar{f}_j(\rho) f_j(\rho) = N_i \delta_{ij} , \tag{17}$$

and therefore f_j and $\bar{f}_j$ must be sufficiently regular such that the integral exists. These conditions are satisfied when we identify $Z_{j+1/2}(\rho)$ with $J_{j+1/2}(\rho)$, the Bessel functions of the first kind. In particular, the scalar product (17) reduces to the well known integral ($\{i,\ j\} = 1,2,3,...$)[24]:

$$\langle \bar{f}_i(\rho) f_j(\rho) \rangle = \bar{c}(k)_i c_j(k) \int_0^\infty \frac{d\rho}{\rho} J_{j+1/2}(\rho) J_{i+1/2}(\rho) = \bar{c}(k)_i c_j(k) \frac{\delta_{ij}}{2i+1} . \tag{18}$$

For reasons which will shortly become clear we set $\bar{c}_j(k) = k^{-1/2}$ and $c_j(k) = k^{3/2}$. With this choice the nonlocal operators S_j defined according to (9) assume the form:

$$S_j \equiv S(1/2+j, k; \mu^2) = \int_{-\infty}^\infty d\alpha\, e^{ik\alpha} \int_0^\infty d\beta\, \sqrt{\beta} J_{1/2+j}(|k|\beta)\, O(\alpha, \beta)_{\mu^2} . \tag{19}$$

As noted above, they are multiplicatively renormalizable[17]:

$$\mu\frac{d}{d\mu}S(1/2+j,k;\mu^2) = a_s\gamma_j S(1/2+j,k;\mu^2),$$
$$S(1/2+j,k;\mu^2) = L_j S(1/2+j,k;\mu_0^2),\ L_j \equiv \left(\frac{\log(\mu^2/\Lambda^2)}{\log(\mu_0^2/\Lambda^2)}\right)^{-\frac{\gamma(j)}{b_0}}. \tag{20}$$

Here, $\gamma(j)$ is the corresponding anomalous dimension known, e.g., from analysis of deep-inelastic scattering process.

Now we are in position to write down the decomposition (8) explicitly:

$$O(\alpha,\beta)_{\mu^2} = \int_{-\infty}^{\infty}\frac{dk}{2\pi}e^{-ik\alpha}\beta^{-\frac{3}{2}}\sum_{j=1}^{\infty}(1+2j)L_j J_{1/2+j}(|k|\beta)S(1/2+j,k;\mu_0^2) \tag{21}$$

It remains only to check the normalization. As $\mu^2 = \mu_0^2$ and therefore $L_j = 1$ for all j we should have

$$O(\alpha,\beta) = \int_{-\infty}^{\infty}d\alpha'\int_0^{\infty}d\beta'\mathcal{R}(\alpha,\beta|\alpha',\beta')\,O(\alpha',\beta') \tag{22}$$

with

$$\mathcal{R}(\alpha,\beta|\alpha',\beta') = \int\frac{dk}{2\pi}e^{ik(\alpha'-\alpha)}\beta^{-\frac{3}{2}}\beta'^{\frac{1}{2}}\sum_{j=1}^{\infty}(1+2j)J_{1/2+j}(|k|\beta)J_{1/2+j}(|k|\beta'). \tag{23}$$

To this end, recall that the series

$$\beta^{-\frac{3}{2}}\beta'^{\frac{1}{2}}\sum_{j=1}^{\infty}(1+2j)J_{1/2+j}(|k|\beta)J_{1/2+j}(|k|\beta') = \delta(\beta-\beta') \tag{24}$$

does not depend on k and represents the Neumann expansion of a δ-function in terms of Bessel functions[15,16,18]. Equation (22) follows immediately. Note that this observation explains also the choice of the coefficients c_i, $\bar{c}_i$.

Equation (21) describes solution of the LO evolution equation as an expansion in terms of orthogonal eigenfunctions of the evolution kernel. On the other hand, it can be interpreted as an expansion of a quark string operator in terms of conformal string operators. Series expansion in terms of Bessel functions is known in mathematical literature as the Neumann series, see[18] for more detailed information. Evolution of flavor non-singlet, charge-conjugation odd and even quark string operators can be obtained from (21) by taking its symmetric, respectively antisymmetric in β components.

4 Summary

We have argued that solutions of evolution equations for twist-2 QCD string operators can be written as series expansions in terms of the orthogonal sets of eigenfunctions of evolution kernels. In the LO approximation the eigenfunctions of evolution kernels can be found explicitly using conformal symmetry arguments and, at the same time, evolution equations for expansion coefficients become much simpler than in the most general case. As a result the expansion coefficients can be identified with conformal string operators.

Taking corresponding matrix elements one obtains immediately solution to the problem of scale dependence of skewed parton distributions in the coordinate representation [16]. Here, contrary to the situation in momentum space, the resulting expansion over a set of orthogonal eigenfunction is valid for all values of the asymmetry parameter.

Acknowledgments

N.K. would like to thank the Humboldt Foundation for their financial support. We gratefully acknowledge discussions with V. Braun, G. Piller, and A. Radyushkin.

References

1. X. Ji, *Phys. Rev.* D **55**, 7114 (1997).
2. A.V. Radyushkin, *Phys. Rev.* D **56**, 5524 (1997).
3. J.C. Collins, L.L. Frankfurt, M. Strikman, *Phys. Rev.* D **56**, 2982 (1997).
4. X. Ji, *J. Phys.* G **24**, 1181 (1998).
5. P.A.M. Guichon, M. Vanderhaeghen, Prog.Part.Nucl.Phys.**41**,125 (1998).
6. X. Ji, J. Osborne, *Phys. Rev.* D **57**, 1337 (1998).
7. J. Blümlein, B. Geyer,D. Robaschik, *The virtual Compton Amplitude in the generalized Bjorken region: twist-2 contributions*, DESY-99-020, hep-ph/9903250, (1999).
8. L. Frankfurt, A. Freund, V. Guzey and M. Strikman, *Phys. Lett.* B **418**, 345 (1998), Erratum-ibid.**B429**:414,1998.
9. A.V. Belitsky, B. Geyer, D. Muller, A. Schäfer *Phys. Lett.* B **421**, 312 (1998).
10. L. Mankiewicz, G. Piller, T. Weigl, *Europ. J Phys.* C **5**, 119 (1998)
11. A.V. Belitsky, D. Muller, L. Niedermeier, A. Schäfer, *Phys. Lett.* B **437**, 160 (1998);
Evolution of non-forward parton distributions in next-to-leading order:

singlet sector, hep-ph/9806232 (1998).
12. A.V. Radyushkin, *Phys. Rev.* D **59**, 014030 (1999).
13. K.J. Golec-Biernat, A.D. Martin, *Phys. Rev.* D **59**, 014029 (1999); K.J. Golec-Biernat, A.D.Martin, M.G. Ryskin, *Diagonal input for the evolution of off diagonal partons*, hep-ph/9903327, (1999).
14. A. Shuvaev, *Solution of the off forward leading logarithmic evolution equation based on the Gegenbauer moments inversion*, hep-ph/9902318, (1999).
15. N. Kivel, L. Mankiewicz, *Phys. Lett.* B **458**, 338 (1999).
16. N. Kivel, L. Mankiewicz, *Conformal string operators and evolution of skewed parton distributions*, TUM-T39-99-04, hep-ph/9903531 (1999).
17. I.I.Balitsky and V.M.Braun, *Nucl. Phys.* B **311**, 541-584 (1988/89).
18. A.Erdeylyi et al., Higher transcendental function, vol.2, McGraw-Hill, New-York,1953.
19. J. Blümlein, B. Geyer, D. Robaschik, *Phys. Lett.* B **406**, 161 (1997); I. Balitsky, A. Radyushkin, *Phys. Lett.* B **413**, 114 (1997).
20. J. Kondo, *Integral Equations*, Clarendon Press, Oxford 1991.
21. A.V. Efremov and A.V. Radyushkin, Theor.Math.Phys. **42**, 97 (1980); *ibid.* **42**, 147 (1980). A.V. Efremov and A.V. Radyushkin, *Phys. Lett.* B **94**, 245 (1980).
22. S.J. Brodsky and G.P. Lepage, *Phys. Lett.* B873791979.
23. M. Chase, *Nucl. Phys.* B **167**, 125 (1980); M. Chase, *Nucl. Phys.* B **174**, 109 (1980); Y.M. Makeenko, Sov.J.Nucl.Phys. **33**, 440 (1981); T. Ohrndorf, *Nucl. Phys.* B **198**, 26 (1982).
24. Gradshteyn and Rizhik, Table of Integrals, Series and Products, Academic Press Inc., New-York and London, 1965.

MATCHING SOFT AND HARD CONTRIBUTIONS IN FORM FACTOR

A. P. Szczepaniak
Physics Department and Nuclear Theory Center
Indiana University, Bloomington, IN 47405, USA
E-mail: aszczepa@indiana.edu

We discuss a simple approach based on QCD sum rules and light cone quantization which allows to define a unified framework for studying, soft, the wave function dependent and the hard, pQCD contributions to current matrix elements.

In deep inelastic exclusive processes the underlying quark and gluon dynamics can be successfully described in terms of the parton model with transverse motion of partons and mass effects neglected. There is however a large class of high energy process, in particular in exclusive and semi-exclusive reactions, where quark dynamics beyond the parton model seems to be relevant. For example, one of the yet unsolved issues has to do with the applicability of perturbative QCD and collinear factorization in description of far off-diagonal matrix elements of local operators, *i.e.* form factors at large momentum transfers. Unfortunately, no scheme based directly on a factorization theorem exists for taking into account transverse motion.

One possible approach would be to start from the light front Fock space expansion. The overlaps of the light front Hamiltonian energy eigenstates with the Fock space basis states give in principle a complete description of hadronic states in terms of helicities, λ_i and longitudinal, x_i and transverse $k_{\perp,i}$ momentum fractions carried by the n constituents, $i = 1 \cdots n$, $\Psi^A_n(x_i, \mathbf{k}_{\perp i}, \lambda_i)$ of the hadron A. In terms of such wave functions a typical matrix element is given by

$$\langle A|O|B\rangle = \sum_{nm} \int [dx_i \mathbf{k}_{\perp i}]_n [dy_i \mathbf{l}_{\perp i}]_m \quad \Psi^A_n(x_i, \mathbf{k}_{\perp i}, \lambda_i) O_{nm}(x_i, \mathbf{k}_{\perp i}, \lambda_i; y_i, \mathbf{l}_{\perp i}, \lambda_j) \times \Psi^B_m(y_i, \mathbf{l}_{\perp i}, \lambda_i) \tag{1}$$

where [..] stands for an invariant measure expressed in the light cone coordinates. If there are large momentum scales involved, in the light cone gauge one expects only a small number of partonic sates, $n, m = q\bar{q}, q\bar{q}g, q\bar{q}gg, \cdots$ to contribute[1]. At moderate momentum transfers, however, the number of terms contributing grows rapidly and the partonic expansion becomes less useful. In this regime the constituent basis becomes more adequate, however, the connection between the two is still not well understood. A possible scenario for matching the two regimes would be to invoke a renormalization scheme controlling the number and momentum of the partons contributing to hadronic states. Lowering the renormalization scale would have an effect of eliminating high-momentum partonic states and replacing them by effective interactions or new current matrix elements, and leaving only the low momentum part of the Fock space to contribute to the Hadronic wave function. The evolution of the Fock space under renormalization group transformations could give insight into the formation of the constituent representation. Such an approach as been recently pursued by several groups [2,3].

Here we present an alternative and simpler approach design in particular to the study of form factors [4,5]. It is based on the combination of the ideas steaming from the QCD sum rule method and from light cone quantization. The main philosophy if the approach is to calculate from factors using pQCD expansion of an appropriate Green' function and writing them as integrals over quark and gluon propagators expressed in terms of the light cone momentum components. With a properly chosen frame of reference it can be arranged that only forward (in the light-cone time) constituent propagators contribute to the hadronic Green' functions. A "smearing", similar to the Borel transformation is then introduced. It has an effect of restricting momenta in the loops over the quark and gluon propagators. With a suitably chosen smearing function, the resulting expressions can be interpreted as equivalent to an overlap of current matrix elements with hadronic wave functions.

The procedure is best illustrated on an example. Consider the pion electromagnetic form factor defined by the following matrix element,

$$(P'+P)^{\mu}F_{\pi}(Q^2)i\epsilon_{3ij} = \langle \pi^i P'|J^{\mu}_{em}(0)|\pi^j P\rangle \tag{2}$$

where $q^2 = (P'-P)^2 = -Q^2$ is the square of the energy-momentum transfer and $J^{\mu}(0) = \bar{\psi}(0)\hat{Q}\gamma^{\mu}\psi(0)$ is the quark electromagnetic current operator. In the spacelike region ($Q^2 > 0$) it is best to choose a frame with no momentum transfer in the "+" direction *i.e* with $q^+ = q^0 + q^z = 0$. This also implies $Q^2 = q_{\perp}^2$. Now consider the following correlator,

$$(p^+)^3 T(p^2, p'^2) i\epsilon_{3ij} = -\int d^4x d^4y e^{ipx - ip'y}\langle 0|T J_i^+(x) J_{em}^+(0) J_j^+(y)|0\rangle, \tag{3}$$

with $J_i^{\mu} = \bar{\psi}\frac{\tau^i}{\sqrt{2}}\gamma^{\mu}\gamma_5\psi$ used to interpolate the pion field, and allow for p^2 and p'^2 to be arbitrary. To lowest order in α_s using free quark propagators one obtains,

$$T(p^2,p'^2) = \int \frac{dxd^2\mathbf{k}_\perp}{16\pi^3} \frac{16\times 3}{\left[P^2 - \frac{\mathbf{k}_\perp^2+m_q^2}{x(1-x)}\right]\left[P'^2 - \frac{(\mathbf{k}_\perp - x\mathbf{q}_\perp)^2+m_q^2}{x(1-x)}\right]}. \tag{4}$$

Performing a double Borel transformation with respect to p^2 and p'^2, $T(p^2,p'^2) \to T(\beta)$ leads to

$$\begin{aligned} T(\beta) &= \left(\frac{1}{2\pi i}\right)^2 \oint dp^2 dp'^2 e^{-p^2/2\beta^2} e^{-p'/2\beta^2} T(p^2,p'^2) \\ &= 2f_\pi^2 \exp\left(-\frac{m_\pi^2}{\beta^2}\right) \int \frac{dxd^2\mathbf{k}_\perp}{16\pi^3} \Psi_{q\bar{q}}(x,k_\perp)\Psi_{q\bar{q}}(x,k_\perp - x\mathbf{q}_\perp) \end{aligned} \tag{5}$$

where we have introduced the $q\bar{q}$, valence wave function through

$$\Psi_{q\bar{q}}(x,k_\perp) = \frac{2\sqrt{6}}{f_\pi} \exp\left(\frac{1}{2\beta^2}\left[m_\pi^2 - \frac{\mathbf{k}_\perp^2 + m_q^2}{x(1-x)}\right]\right). \tag{6}$$

The ground state hadronic contribution to T is calculated by inserting a complete set of sates between the time-ordered operators. This results in

$$T(\beta) = 2f_\pi^2 \exp\left(-\frac{m_\pi^2}{\beta^2}\right) F_\pi(Q^2) + \cdots, \tag{7}$$

where the first term comes from two, single pion intermediate states and the ellipsis denotes contribution from higher resonances. In the standard QCD sum rule approach the contribution from higher resonances is retained (usually approximated by the perturbative one with the loop integral cutoff at small momenta). Also, on the QCD side of sum rule calculations, contributions to the corellator beyond perturbation theory are introduced and parametrized in terms of vacuum expectation values of local operators, condensates. The sum rum rule is then optimized with respect of the Borel parameter(s).

In our approach, however, we will fix the Borel parameter β to a value which we will argue makes the nonperturbative (proportional to condensates) QCD contribution to the sum rule approximately cancel the terms coming from higher resonances on the hadronic side of the calculations. In the example discussed here, $O(1)$ calculation of the pion form factor, this leads to,

$$F_\pi(Q^2) = \int [dxd\mathbf{k}_\perp]\Psi_{q\bar{q}}(x,\mathbf{k}_\perp)\Psi_{q\bar{q}}(x,\mathbf{k}_\perp - x\mathbf{q}_\perp). \tag{8}$$

Thus the procedure results in an expression identical to the standard light cone expansion of the form factor in terms of the valence $q\bar{q}$ component of the pion light cone wave function. The specific prescription used to filter out the soft contribution determines the form for the soft wave function. In the case of Borel transformation it turns out to reproduce the gaussian model.

When more loops are taken into account, nonvalence wave functions emerge. This is illustrated in Fig. 1 where it is shown how part of the $O(\alpha_s)$ correction to the electromagnetic vertex results in a contribution from a soft $q\bar{q}g$ wave function.

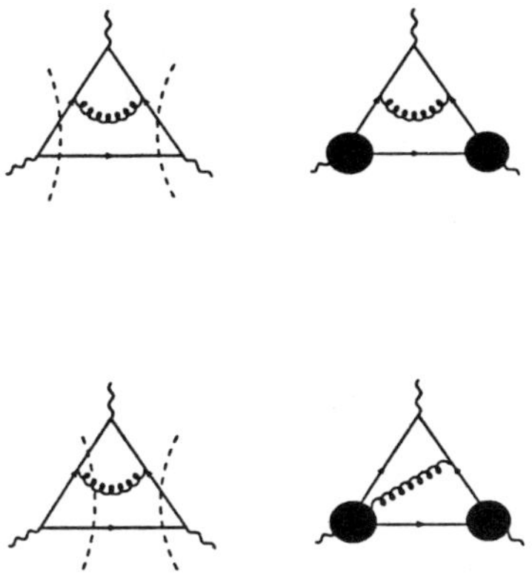

Fig 1. Soft contribution to the pion form factor. The dashed lines indicate which poles contribute to the Borel transformation and the blobs in the diagrams on the right represent the corresponding wave functions.

To $O(\alpha_s)$ the method, when applied to a matrix element of a one-body quark operator involving single meson states, results in the following, general expression,

$$\langle A|O|B\rangle = \sum_{nm} \Psi_n^A \otimes O_{nm} \otimes \Psi_n^B, \tag{9}$$

with the following states contributing : $n, m = \bar{q}q, \bar{q}qg$. In the valence sector there are two contributions to $O_{q\bar{q},q\bar{q}}$, one from the $O(1)$ original one-body operator, and the other from a two-body operator arising from one-gluon exchange. After renormalization, the sum of all contributions is both UV and IR finite, even though individual contributions are IR singular. For example the infrared singularity coming from the one gluon intermediate state in the two-body piece of $O_{q\bar{q},q\bar{q}}$ is exactly canceled by an IR singular piece from the contributions proportional to $O_{q\bar{q},q\bar{q}g}$ and $O_{q\bar{q}g,q\bar{q}}$ *i.e.* involving the nonvalence, soft gluon in the wave function. It is possible to make individual terms in the sum IR finite by subtracting (cutting-off) the IR singular piece from the valence term and adding it back to the terms containing the nonvalence wave functions. Then, it is possible to write the matrix element in a form of Eq. 9 but with the valence term containing only hard gluon exchange and the nonvalence contributions given by a new set of IR finite wave functions. This procedure introduces an IR factorization scale, μ in individual terms but there is no factorization scale dependence of the entire sum. It is interesting to study μ dependence of the one-body term with the one-gluon exchange piece which after cut-off has gluons with typical momenta $> \mu$ and will be referred to as the hard-gluon exchange. For the case of pion electromagnetic form factor this μ dependence is illustrated in Fig. 2. The contribution from the two-body current is shown together with the corresponding nonvalence contribution as a function of the momentum transfer for there values of the factorization scale, $\mu^2 = 2\beta^2$, μ=$10\beta^2$ and $\mu^2 = 0.2\beta^2$. We refer to [4] for more detail, in particular for the actual form in which the cut-off is introduced. The sum of two contributions given by the three sets of curves (solid line, circles and triangles), are identical, as a result of factorization scale independence. As $Q^2 \to \infty$ the sums approach the dashed line corresponding to the asymptotic formula $Q^2F_\pi(Q^2) = 16\pi\alpha_s f_\pi^2$ (plotted for $\alpha_s = 0.25$). The solid line approaching the asymptotic result from below is the valence, hard-gluon exchange contribution corresponding to $\mu^2 = 2\beta^2$. The circles and triangles below the asymptotic result give the hard-gluon contribution for $\mu^2 = 10\beta^2$ and $\mu^2 = 0.2\beta^2$ respectively. They approach the asymptotic limit at much higher Q^2. This is not surprising. With the factorization scale similar to the size of the wave functions (β) the nonvalence contributions are minimized, and the hard, one gluon exchange approaches the asymptotic result faster then for other choices of μ. Nevertheless the asymptotic behavior still sets in for a rather high $Q^2 \sim$ tens GeV^2.

With $\mu^2 \sim \beta^2$, to a good approximation one can ignore the contribution from nonvalence sectors and study effects of all-order resumation. For the diagrams dressing the current matrix element this results in the Sudakov suppression. In the case of the pion electromagnetic form factor, Sudakov term reduces the magnitude of the (negative) vertex correction enhancing the contribution from one-body current matrix element involving soft valence wave functions.

We have also studied matrix elements determining hadronic contributions to semileptonic and nonleptonic decays of heavy to mesons with light mesons in the final state (*e.g.* $B \to \pi$) [5]. In this case we have found that Sudakov form factor reduces the soft, valence wave function dependent contribution

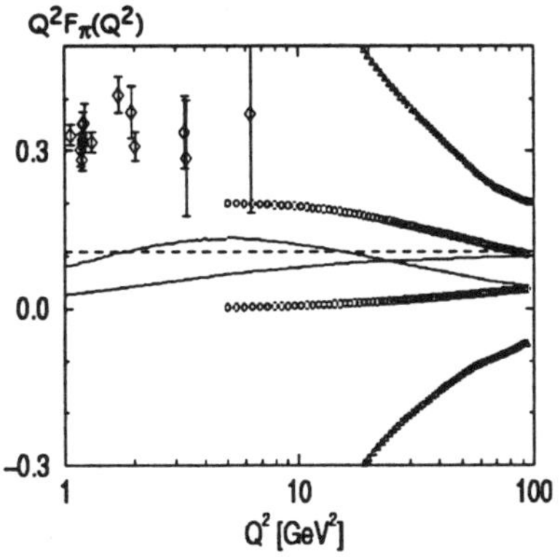

Fig 2. Factorization scale dependence of valence and nonvalence contributions to the pion from factor.

by approximately 60%. The remaining, $O(\alpha_s)$ contributions from one-gluon exchange and nonvalence wave functions amounts to $\sim$ 40% of the soft piece. Thus the net effect of pQCD corrections seems to cancel out, leaving the matrix element of the order of the soft contribution.

There are many other process were $k_\perp$ effects are potentially of interest. A simple extension of the approach presented here could be applied to study of 4-point functions. This would enable to get insight into higher twist corrections to structure functions.

Acknowledgments

It is a pleasure to thank the organizers of the conference for hosting a very stimulating meeting. This work was supported by the DOE under contract, DE-FG02-87ER40365.

References

1. G.P Lepage and S.J. Brodsky, Phys. Rev. **D** 22, 2157 (1980).
2. K.Wilson *et al.* Phys. Rev. **D** 49, 6720 (1994).

3. A.P. Szczepaniak and E.S. Swanson, Phys. Rev. **D** 55, 1578 (1997).
4. C. R. Ji, A.P. Szczepaniak and A. Radyushkin, Phys. Rev. **D 57** 2813 (1998).
5. A.P. Szczepaniak and A. Radyushkin, Phys. Rev. **D 57** 59034017, (1999).

SUM RULES FOR HIGHER-TWIST PARTON DISTRIBUTIONS

M. Burkardt
New Mexico State University, Las Cruces,
NM 88003, USA
E-mail: burkardt@nmsu.edu

In deep-inelastic scattering experiments, there is a general connection between subtractions in dispersion relations, violations of sum-rules and δ-functions in parton distribution functions. It is explained why one might expect a small violation in sum rules for the twist-3 distribution functions $g_T(x)$ and $h_L(x)$ when the sum-rules are applied to $x \neq 0$ data only. The non-perturbative predictions are studied in the context of a one-loop model.

1 Introduction

In the theoretical analysis of deep-inelastic scattering (DIS), one usually applies the operator product expansion (OPE) for $Q^2 \to \infty$ to the Compton amplitude — a procedure which implicitly involves analytic continuation of the Compton amplitude to the regime where $Q^2 > 2M\nu$. Formally this step is accomplished by invoking dispersion relations. However, at least in principle, it may happen that there appear subtractions in these dispersion relations which then manifest themselves as a violation of naive[a] sum-rules.

In this note, we study the issue of subtractions in the context of the higher-twist parton distributions $h_L(x)$ and $g_T(x)$, which are defined as correlation functions in a light-like direction [1]

$$S \cdot n\, h_L(x) \equiv \frac{1}{2M} \int \frac{d\lambda}{2\pi} e^{i\lambda x} \langle PS|\bar{\psi}(0)\sigma^{+-} i\gamma_5 \psi(\lambda n)|PS\rangle$$
$$S_\perp\, g_T(x) \equiv \frac{1}{2p^+} \int \frac{d\lambda}{2\pi} e^{i\lambda x} \langle PS|\bar{\psi}(0)\gamma_\perp \gamma_5 \psi(\lambda n)|PS\rangle, \qquad (1)$$

where $p^\pm = \frac{1}{\sqrt{2}}\left(p^0 \pm p^3\right)$ denotes the usual light-cone coordinates and the light-like vector n^μ projects out $n \cdot A = A^+$ for all 4-vectors A^μ.

Like the more familiar polarized twist-two distributions $g_1(x)$ and $h_1(x)$, these twist-three distributions are important physical quantities which summarize low energy properties of the nucleon in high-energy scattering processes [1]: $g_T(x)$ appears as a $\frac{1}{Q}$ correction in DIS [2], while $h_L(x)$ can be measured for example in the nucleon-nucleon polarized Drell-Yan process.

[a]Naive, because they are derived assuming that there are no subtractions.

Upon integrating over x in Eq. (1), comparing with similar definitions for g_1 and h_1, and using Lorentz invariance one can thus derive the sum-rules [13]

$$\int_{-1}^{1} dx h_L(x) = \int_{-1}^{1} dx h_1(x) \tag{2}$$

$$\int_{-1}^{1} dx g_T(x) = \int_{-1}^{1} dx g_1(x). \tag{3}$$

Eq. (3) is also known as the Burkhardt-Cottingham (BC) sum-rule [4]. For the parton distribution functions, defined as light-cone correlations (1), these sum-rules (2,3) are a direct consequence of Lorentz invariance, and therefore hardly anybody would question their validity (assuming the integrals converge). However, the issue here is whether these sum rules are also valid when applied to experimental data! Using the operator product expansion in the Bjorken limit, one can show that for $x \neq 0$, the parton distribution functions (defined as a light-cone correlations) agree with the corresponding experimentally measured structure functions. In DIS experiments, the point $x = 0$ must always be excluded since it corresponds to $Q^2 = 0$. In practice, DIS experiments can only measure

$$\lim_{\epsilon \to 0} \int_{\epsilon}^{1} dx \left[g_T(x) + g_T(-x)\right] \tag{4}$$

and similarly for the other distributions. Therefore, what really is being tested when one tests the above sum-rules is neither Lorentz invariance nor the OPE but whether or not Eqs. (2,3) receive a nonzero contribution from the point $x = 0$, i.e. whether or not $h_L(x)$ and $g_T(x)$ [*defined* as in Eq. (1)] contain $\delta(x)$-type singularities at the origin.

The paper is organized as follows. In the next section, we will explain the general connections between subtractions in dispersion relations, violations of sum-rules and δ-functions in parton distribution functions. The main question is whether a situation as described above does actually occur or whether it is only of academic interest. In order to understand the implications for QCD, we will use the moment relations derived from the OPE to derive relations between the small x behaviour of polarized twist-2 distributions ($g_1(x)$ and $h_1(x)$) and the coefficient of the δ-function of polarized twist-3 distributions [$h_L(x)$ and $g_\perp(x)$]. Finally, these predictions are illustrated using a one-loop model.

2 Real and Imaginary parts of the Compton Amplitude in DIS

In this section, general connections between subtraction constants, violation of sum-rules and $\delta(x)$-terms in parton distributions will be discussed. For this purpose, we will denote $G(x,Q^2)$ and $T(x,Q^2)$ the real and imaginary part respectively of some generic forward Compton amplitude, where $x = \frac{Q^2}{2M\nu}$. The Compton amplitude is an analytic function of x, except along a cut from $0 < x < 1$ (the cut appears along the physical region for DIS!). Therefore, it should be possible to relate the real and imaginary part using a dispersion relation

$$T(x',Q^2) = \frac{1}{\pi}\int_0^1 dx \frac{x'}{x'^2 - x^2} G(x,Q^2) + p\left(\frac{1}{x'},Q^2\right), \tag{5}$$

where $p(z,Q^2)$ is a polynomial in z and where $|x'| > 1$. In the language of dispersion theory, the 'polynomial subtraction' is related to 'J=0 fixed poles'.

In the theoretical analysis of DIS one usually applies the operator product expansion (OPE) for $|x'| > 1$, yielding

$$T(x',Q^2) = \sum_{n=0,2,..} \frac{1}{x'^{n+1}} a_n \tag{6}$$

where the a_n can be expressed as matrix elements of the form

$$a_n = \langle P,S|\bar{\psi}\Gamma D^{n-1}\psi|P,S\rangle, \tag{7}$$

which appear in the OPE of $TJ_\mu(\xi)J_\nu(0)$. Here, Γ is some Dirac matrix which depends on the particular structure function (e.g. polarized or unpolarized DIS) and D^n denotes an $n-th$ order covariant derivative. In order to keep the discussion as general as possible, Γ will not be specified here any further.

In the most simple case, i.e. when there is no subtraction in Eq. (5) and $p(z) \equiv 0$, the a_n can be expressed through the moments of G by expanding the geometric series in Eq. (5), yielding

$$a_n = \frac{1}{\pi}\int_0^1 dx x^{n-1} G(x,Q^2). \tag{8}$$

However, in the following we want to investigate what happens if $p(z) \neq 0$. As a specific example, let us assume that $p\left(\frac{1}{x'}\right) = \frac{c}{x'}$, where c is some (possibly Q^2 dependent) constant. Such a scenario has several important consequences:

- First of all, the simple relation between the a_n and $\int_0^1 dx x^{n-1} G(x,Q^2)$ (8) is spoiled for the lowest moment and instead one finds

$$a_n = \begin{cases} \frac{1}{\pi}\int_0^1 dx x^{n-1} G(x,Q^2) & n = 3,5,.. \\ \frac{1}{\pi}\int_0^1 dx x^{n-1} G(x,Q^2) \quad + \quad c & n = 1 \end{cases} \tag{9}$$

- As a consequence, the naive sum-rule $a_1 = \frac{1}{\pi}\int_0^1 dx G(x,Q^2)$ is of course violated if $c \neq 0$

- Suppose that one defines a parton distribution function $g(x)$ through the moments, i.e. by requiring that

$$a_n \stackrel{!}{=} \frac{1}{\pi}\mathcal{M}_n[g] \equiv \frac{1}{\pi}\int_{-1}^{1} dx x^{n-1} g(x,Q^2) \qquad n \geq 0 \tag{10}$$

then $g(x,Q^2)$ differs from the experimentally measured $G(x,Q^2)$ by a δ function at the origin, i.e.

$$g(x,Q^2) + g(-x,Q^2) = G(x,Q^2) + c\pi\delta(x). \tag{11}$$

This result should be intuitively clear since the above subtraction affects only the lowest moment, which means that there must be a δ-function present.

To summarize the above discussion, we emphasize that there is no problem with the OPE if a sum-rule derived for parton distributions fails when applied to experimental data. All it means is that the corresponding dispersion relation has a subtraction and the parton distribution function (defined as a generalized function through the light-cone moments) has a δ-function at the origin, which is not present in the experimental DIS data. Therefore, an experiment which measures $\lim_{\epsilon\to 0}\int_\epsilon^1 dx G(x,Q^2)$ would miss the δ function and hence a "violation" of the sum-rule would be observed.

There exist a number of toy models where such $\delta(x)$ terms have been observed in parton distribution for models in 2 and 4 dimensions. Because of lack of space, the reader is referred to these references [5] and we focus in the following on the most important question, namely whether such δ functions occur in QCD_{3+1}.

3 OPE, Moments and δ-Functions in QCD

In this section we start from the relation among the moments of $h_L(x)$, $h_1(x)$ and $g_1(x)$ [1] (note that $\mathcal{M}_n[\tilde{h}_L] = 0$ for $n \leq 2$).

$$\mathcal{M}_n[h_L] = \begin{cases} \mathcal{M}_n[h_1] & n = 1 \\ \frac{2}{n+1}\mathcal{M}_n[h_1] + \mathcal{M}_n\left[\tilde{h}_L\right] + \frac{m_q}{M}\frac{n-1}{n+1}\mathcal{M}_{n-1}[g_1] & n \geq 2 \end{cases}, \tag{12}$$

where $\tilde{h}_L$ is the interaction-dependent twist-three part of h_L. Upon inverting the moment relation, one finds [6]

$$h_L(x,\mu^2) = 2x\int_x^1 dy \frac{h_1(y,\mu^2)}{y^2} + \tilde{h}_L(x,\mu^2) + \frac{m_q}{M}\left[\frac{g_1(x,\mu^2)}{x} - 2x\int_x^1 dy \frac{g_1(y,\mu^2)}{y^3}\right] \tag{13}$$

$$h_L(x,\mu^2) = -2x\int_{-1}^x dy \frac{h_1(y,\mu^2)}{y^2} + \tilde{h}_L(x,\mu^2) + \frac{m_q}{M}\left[\frac{g_1(x,\mu^2)}{x} + 2x\int_{-1}^x dy \frac{g_1(y,\mu^2)}{y^3}\right] \tag{14}$$

for $x > 0$ and $x < 0$ respectively. Now we multiply Eq. (13) by x^β, integrate from 0 to 1 and let $\beta \to 0$, yielding

$$\int_{0+}^1 dx h_L(x,\mu^2) = \int_{0+}^1 dx \left(h_1(x,\mu^2) + h_L^3(x,\mu^2)\right) + \frac{m_q}{2M}\lim_{\beta\to 0}\beta\int_0^1 dx x^{\beta-1} g_1(y,\mu^2), \tag{15}$$

while multiplying Eq. (14) by $|x|^\beta$ and integration from -1 to 0 yields

$$\int_{-1}^{0-} dx h_L(x,\mu^2) = \int_{-1}^{0-} dx \left(h_1(x,\mu^2) + h_L^3(x,\mu^2)\right) - \frac{m_q}{2M}\lim_{\beta\to 0}\beta\int_{-1}^0 dx |x|^{\beta-1} g_1(y,\mu^2). \tag{16}$$

Adding Eqs. (16) and (15) and

- assuming that there is no $\delta(x)$ in h_1, i.e. assuming that

$$\int_{-1}^{0-} dx h_1(x) + \int_{0+}^1 dx h_1(x) = \int_{-1}^1 dx h_1(x) \tag{17}$$

- assuming that there is no $\delta(x)$ in h_L^3 either and using that (from OPE), the lowest moment of h_L^3 vanishes identically i.e. assuming that

$$\int_{-1}^{0-} dx h_L^3(x) + \int_{0+}^1 dx h_L^3(x) = \int_{-1}^1 dx h_L^3(x) = 0 \tag{18}$$

- using that

$$\lim_{\beta\to 0}\beta\int_0^1 dx x^{\beta-1} g_1(y,\mu^2) = g_1(0+,\mu^2) \tag{19}$$

one thus finds

$$\begin{aligned}\int_{0+}^1 dx\left[h_L^q(x,\mu^2) - h_L^{\bar{q}}(x,\mu^2)\right] &\equiv \int_{-1}^{0-} dx h_L(x) + \int_{0+}^1 dx h_L(x) \\ &= \int_{-1}^1 dx h_1(x,\mu^2) + \frac{m_q}{2M}\left[g_1(0+) - g_1(0-)\right].\end{aligned} \tag{20}$$

Since the OPE also tells us that the first moments of h_L and h_1 ought to be the same (if 0 is included in the integration), i.e. $\int_{-1}^{1} dx h_L(x,\mu^2) = \int_{-1}^{1} dx h_1(x,\mu^2)$, we thus conclude

$$h_L(x,\mu^2) = h_L^{reg}(x,\mu^2) - \frac{m_q}{2M}\left[g_1(0+,\mu^2) - g_1(0-,\mu^2)\right]\delta(x). \quad (21)$$

A similar analysis applied to $g_T(x)$ yields [7]

$$g_T(x) = \int_x^1 dy \frac{g_1(y)}{y} + \frac{m}{M}\left[\frac{h_1(x)}{x} - \int_x^1 dy \frac{h_1(y)}{y^2}\right] + \tilde{g}_T(x) - \int_x^1 dy \frac{\tilde{g}_T(x)}{y} \quad (22)$$

And hence under similar assumptions as above

$$g_T(x,\mu^2) = g_T^{reg}(x,\mu^2) - \frac{m_q}{M}\left[h_1(0+,\mu^2) - h_1(0-,\mu^2)\right]\delta(x) \quad (23)$$

To summarize these results, although the OPE ensures that the integrals of h_L and g_T are the same as those of h_1 and g_1 respectively, this statement is strictly true only if the origin is included in the integration. By analytic continuation of the moments, we find that the behavior at the origin of h_L and g_T might be singular enough and the above statement about equality of the lowest moments seems to be violated if the origin is excluded from the integrals.

4 A one-Loop Model for $h_L(x)$

In order to illustrate the results from the previous section in a concrete example, we consider $h_L(x)$ for a massive[b] quark in a one-loop model.

In such a model one finds for example

$$h_L(x)S^+ \propto \quad (24)$$
$$\bar{u}(P,S)\int \frac{d^2k_\perp dk^-}{(2\pi)^3}\gamma^\mu \frac{i}{\not{k} - m + i\varepsilon}\sigma^{+-}\gamma_5 \frac{i}{\not{k} - m + i\varepsilon}\gamma^\nu u(P,S) D_{\mu\nu}(P-k)$$

where $k^+ = xP^+$ and $D_{\mu\nu}(P-k)$ is the gluon propagator. Here we have suppressed wave function renormalization terms $[\propto \delta(x-1)]$ because they are not relevant for the behaviour near $x = 0$.

In Eq. (24) the $\delta(x)$ terms can arise from terms with k^- in numerator, because for those terms integrals of the form

$$\int dk^- \frac{k^-}{\left[k^2 - m^2 + i\varepsilon\right]^2} \frac{1}{(p-k)^2 + i\varepsilon} \quad (25)$$

[b] In the previous section, we found that the $\delta(x)$ term is explicitly proportional to the quark mass quark!

diverge linearly when $k^+ = 0$. In order to see this, we rewrite

$$k^- = P^- - \frac{\left(\vec{P}_\perp - \vec{k}_\perp\right)^2}{2(P^+ - k^+)} - \frac{(P-k)^2}{2(P^+ - k^+)} \tag{26}$$

and note that the first two terms on the r.h.s. of Eq. (26) give "regular" expression when inserted into Eq. (25). However, the third term on the r.h.s. of Eq. (26) cancels one of the denominators in Eq. (25), yielding

$$\int \frac{dk^-}{2\pi} \frac{1}{\left[k^2 - m^2 + i\varepsilon\right]^2} = \frac{i}{2} \frac{\delta(k^+)}{\vec{k}_\perp^2 + m^2}. \tag{27}$$

A detailed analysis [3] shows that in h_L, the $\delta(k^+)$-terms survive, yielding a nonzero $\delta(x)$ term that is proportional to $g_1(0+)$, [c] which confirms Eq. (21). A similar 1-loop analysis also shows that for $g_\perp$, the $\delta(k^+)$ term (at 1 loop) is multiplied by $k^+ \Rightarrow$ no $\delta(x)$. This is consistent with the result that $h_1^{1-loop}(0) = 0$ at one loop.

5 Summary and Discussion

At least in principle, one cannot exclude subtractions in the dispersion relation between real & imaginary part of Compton amplitude. Whenever such a subtraction is present, this also implies a $\delta(x)$-term in the corresponding parton distribution (defined as light-cone correlation).

A decomposition of h_L and g_T into twist-2 and twist-3 pieces suggests that $\lim_{\beta\to 0} \int_0^1 dx x^\beta g_\perp(x) \neq \int_0^1 dx g_\perp(x)$ and therefore $g_\perp(x) = g_\perp(x)^{reg} + c\delta(x)$. A similar result is derived for h_L).

The nonzero coefficient of $\delta(x)$ in $h_L(x)$ was confirmed by explicit one-loop calculations for $h_L(x)$ as well as $g_1(0+) - g_1(0-)$. At one loop, no $\delta(x)$ term was found in $g_\perp(x)$. This is consistent with the fact that $h_1^{1-loop}(0) = 0$.

The prediction of these $\delta(x)$ terms from the moment relations [Eqs.(21),(23], together with their one loop verification, are the main result of this note.

Although the one-loop calculation yields $h_1(0) = 0$, this result changes at next to leading order (NLO)[8], i.e. $h_1^{NLO}(0) \neq 0$. Therefore, using Eq. (23), we expect that the BC sum-rule[4] is also violated, but only at NLO.

In summary, both the h_L sum rule as well as the g_T are expected to be violated if the point $x = 0$ is not included in the integrals. For h_L the violation appears already at leading order, while the violation for g_T does not appear until NLO. Both for h_L and g_T, the violation is proportional to $\frac{m_q}{M}$, i.e. for

[c] At one loop, $g_1(x) = 0$ for $x < 0$ and thus $g_1(0-) = 0$.

$q = u, d$ the effect is expected to be very small. However, the contribution from strange quarks might be a significant.

Recently, it has also been suggested that $\delta(x)$ terms might be present in the twist-two distribution $g_1(x)$ [9]. Although there is not enough space here to discuss the results of Ref. [9] in detail, it should be emphasized that the general features that have been discussed in this note also apply in that case.

One of the main differences between DIS and deeply virtual Compton Scattering (DVCS) is that DIS measures only the imaginary part of the forward Compton amplitude, while DVCS should allow measurements of the full (real & imaginary part[d]) Compton amplitude — including non-forward matrix elements thereof. The virtue of DVCS in this context is twofold. First of all, since also the real part of the Compton amplitude is being measured, one could directly test whether or not a subtraction appears in the dispersion relation. Furthermore, any δ-functions on forward parton distribution functions get smeared out in non-forward distributions, i.e. the measurement of the non-forward Compton amplitude acts as some kind of regularization. Because of these features, studying near forward parton distributions using DVCS may also be of help in clarifying the issues discussed in this paper.

Acknowledgements: It is a pleasure to thank X, Ji, Y. Koike and L. Mankiewicz for helpful discussions. This work was supported in part by a grant from DOE (DE-FG03-95ER40965), by a fellowship from JSPS and in part by TJNAF.

References

1. R.L. Jaffe and X. Ji, *Nucl. Phys.* B **375**, 527 (1992).
2. D. Adams et al. (SMC), *Phys. Lett.* B **336**, 125 (1994); K. Abe et al. (E143), *Phys. Rev. Lett.* **76**, 587 (1996).
3. M. Burkardt, *Phys. Rev.* D **52**, 3841 (1995).
4. H. Burkhardt and W.N. Cottingham, *Ann. Phys.* **56**, 453 (1970).
5. M. Burkardt, *Nucl. Phys.* A **373**, 613 (1992); *Phys. Lett.* B **268**, 419 (1991);
6. Y. Koike and K. Tanaka, *Prog. Theor. Phys. Suppl.* **120**, 247 (1995).
7. R.D. Tangerman and P.J. Mulders, hep-ph/9408305.
8. A. Hayashigaki, Y. Kanazawa and Y. Koike, *Phys. Rev.* D **56**, 7350 (1997); W. Vogelsang, *Phys. Rev.* D **57**, 1886 (1998).
9. S.D. Bass, *Eur. Phys. J.*A **5**, 17 (1999).

[d]The phase information can be obtained from interference with the Bethe-Heitler process.

NUCLEON SPIN CONTENT FROM ELASTIC FORM FACTOR DATA[1]

Andrei V. Afanasev[2]
North Carolina Central University, Durham, NC 27707, USA
and
Jefferson Lab, Newport News, VA 23606, USA
E-mail: afanas@jlab.org

I use the formalism of skewed parton distributions (SPD) to describe elastic form factors of the nucleon. The results are consistent with approximate dipole behavior of $G_{Mp}(Q^2)$ and recent JLab data for the ratio of the proton form factors G_{Ep}/G_{Mp} at $Q^2 \leq 3.5$ GeV^2. Using the Angular Momentum Sum Rule, I obtain a numerical estimate for the valence quark contributions to the nucleon spin at the level of 50%, with their orbital angular momentum *included*. When combined with polarized DIS measurements, this result indicated that the orbital angular momentum of quarks contributes about 25% to the nucleon spin.

1 Introduction

The formalism of Nonforward Parton Densities [1] and Off-Forward Parton Distributions [2], recently consolidated under a general name of Skewed Parton Distributions (SPD), provide a connection between exclusive and inclusive reactions. SPD in general describe the nonperturbatibe process of quark/parton emission and subsequent absorption by a hadron in an exclusive reaction. Since SPD assume that states of the initial and final hadrons involved in the reaction are different, they appear to be sensitive to the orbital angular momentum of quarks, and it was the original motivation of Ji to introduce this new formalism [2]. This issue was also discussed earlier in Ref.[3]. An essential part of this formalism are helicity–flip SPD studied in Ref.[4].

In application to specific processes, the power of SPD formalism is that it allows to describe universally the bag-type diagrams involving current quarks. In particular, the same SPD of a nucleon would describe nucleon elastic form factors, real and virtual Compton scattering and hard electro- and photoproduction of mesons on the nucleon.

---

[1] Invited talk at the INT/JLab Workshop 'Exclusive and Semiexclusive Processes at High Momentum Transfer', Newport News, VA, May 20-22, 1999
[2] On leave from Kharkov Institute of Physics and Technology, Kharkov, Ukraine

Here I introduce a model of helicity-flip SPD which reasonably describes available data on elastic nucleon form factors and, using Angular Momentum Sum Rule[2], obtain an estimate of the valence quark contribution to the nucleon spin.

2 A Model for SPD

Angular Momentum Rum Rule derived by Ji[2] relates the first moment of SPD in the forward limit to the total angular momentum of quarks and gluons. For the case of quarks it reads

$$J_q = \frac{1}{2}[A_q(0) + B_q(0)], \tag{1}$$

where

$$A_q(t) = \int_0^1 \mathcal{F}^q(x,t)x dx B_q(t) = \int_0^1 \mathcal{K}^q(x,t)x dx \tag{2}$$

are SPD using definitions of Ref.[1]. The model for helicity–nonflip SPD was proposed in Ref.[5], assuming a Gaussian dependence of the proton light-cone wave function on the quark transverse momentum. [a] Complementing this picture with a model for helicity–flip SPD $\mathcal{K}^q(x,t)$, we earlier obtained a good agreement with available data on proton and neutron elastic form factors[6].

Key assumptions of the model are: a) helicity–flip SPD have similar functional dependence on the transverse momentum as helicity–nonflip SPD and b) contribution from sea quarks to the anomalous magnetic moment of the nucleon is neglected. The forward limit of helicity–flip SPD, $k_a(x)$ then enters full SPD as

$$\mathcal{K}^a(x,t) = k_a(x)e^{\frac{(1-x)t}{4x\lambda_k^2}}, \tag{3}$$

where λ_k is an adjustable mass parameter. The distribution $k_a(x)$ cannot be observed in deep–inelastic scattering, but may be accessed in exclusive reactions, for instance, in Deeply Virtual Compton scattering[2,1].

Here I used reasonable assumptions about the function $k_a(x)$ vs. standard parton densities (i.e., forward limit of helicity–nonflip SPD). Namely, as $x \to 1$, both pQCD and QCD Sum Rules (Ioffe current) require an extra factor of $(1-x)$ for $k_a(x)$. For the Regge limit $x \to 0$, I explored three possibilies a) $k_a(x) = (1-x)f_a(x)$ (Model A), b) $k_a(x) = x(1-x)f_a(x)$ and c) $k_a(x) =$

[a] A similar approach was also developed in Ref.[7]

$x^{\frac{1}{2}}(1-x)f_a(x)$ (Model C). The Pauli form factor of the proton is then given by the formula in terms of only valence quark distributions (contributions from sea quarks cancel):

$$F_2(t) = \int_0^1 [e_u k_u^{val} + e_d k_d^{val}] e^{\frac{(1-x)t}{4x\lambda_h^2}} dx. \tag{4}$$

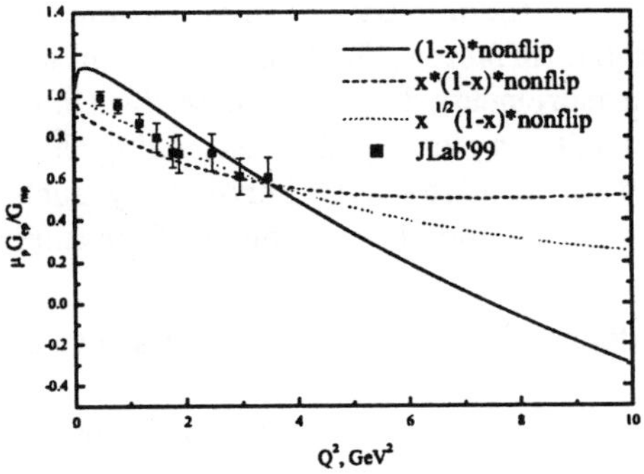

Figure 1: Ratio of electric to magnetic proton form factors for different models of helicity–flip SPD (see text). The data are from Ref.[8]

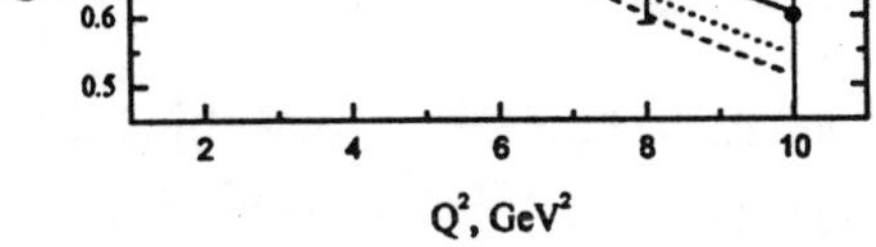

Figure 2: Neutron magnetic form factor from SPD. Notation is the same as for Fig.1. Data are from Ref.[11].

The best fit to the recent JLab data[8] on the ratio of proton form factors was obtained for $\lambda_k^2 = 0.43$ GeV^2 (Fig.1). For the Model A and B, this value was taken 0.64 and 0.34 GeV^2. These different values of $lambda_k$ correspond to approximately the same value of about 220-250 MeV for the transverse correlation momentum defined for valence quarks as

$$< k_{cor}^2 >^a = \lambda_{k;a}{}^2 \int_0^1 x\bar{x}k_a^{val}(x)dx, \tag{5}$$

where $N_u = 2$, N_d=1 are the numbers of valence a quarks in the proton. Our results for the neutron form factor (Fig.2) are consistent with data at higher Q^2, but the errors in measurements do not allow to discriminate between models for NFPD. We can also see a clear need for extending the measurements of the proton form factor ratio to higher Q^2. Fortunately, such a measurement for $Q^2 = 5.6$ GeV^2 is approved and the possibility of measurements at 10GeV^2 is being studied.[9]

Having the model for SPD, it is straightforward to evaluate the Angular Momentum Sum Rule Eq.(1). The results for all three choices of helicity–flip SPD are presented in Table 1. Total contribution of quarks to the nucleon spin appears to be close to 50%, with little dependence on the used model of helicity–flip SPD. This result is numerically close to the QCD Sum Rules calculation of Balitsky and Ji[10]. In our model functions $k_a(x)$ are positively defined, leading to cancellation between up- and down-quarks for $B(0)$ in Eq.(1). In the meantime $A(0)$ is close to 0.5 – the result known as the Momentum Sum Rule. Contributions of individual quark flavors are, however, quite sensitive to the chosen model of helicity–flip SPD. The quark total angular momentum density, $J_q(x) = x(f_q(x) + k_q(x))$ is plotted in Fig.3.

We can now combine available data from polarized deep–inelastic scattering, e.g., semi–exclusive data from SMC [12] ($\Delta u_v = 0.77 \pm 0.10 \pm 0.08$ and $\Delta d_v = -0.52 \pm 0.14 \pm 0.09$) with the values from Table 1 to obtain estimates for orbital angular momentum contribution to the proton spin from individual quark flavors. The best fit to JLab data[8] gives for the central values

$$\begin{aligned} \Delta L_u &= (\Delta_u + L_u)^{JLab} - (\Delta u)^{SMC} = -0.04 \\ \Delta L_d &= (\Delta_d + L_d)^{JLab} - (\Delta d)^{SMC} = 0.27. \end{aligned} \tag{6}$$

Earlier SLAC data[11] on the proton form factor yield different numbers, $\Delta L_u =$ 0.05 and $\Delta L_d = 0.15$.

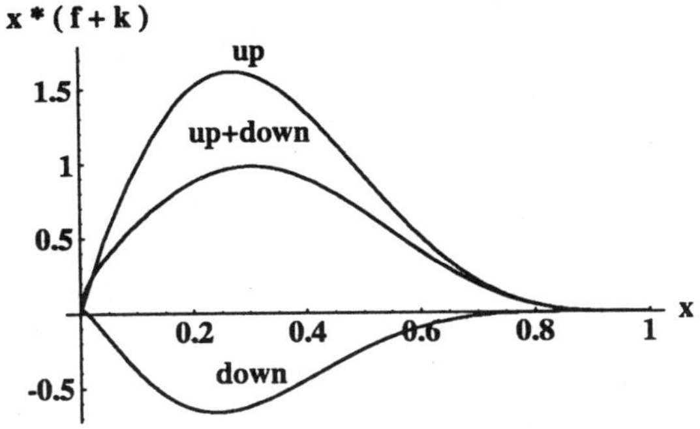

Figure 3: Total angular momentum distribution for the quarks.

Table 1: Valence Quark Contributions to the Proton Spin with Different Models for Spin-Flip SPD (see text for notation).

	A	B	C
$\Delta_u + L_u$	0.6	0.82	0.73
$\Delta_d + L_d$	-0.08	-0.37	-0.25
$\Delta_u + L_u + \Delta_d + L_d$	0.52	0.45	0.48

3 Conclusion

I obtained new estimates of quark contributions to the nucleon spin based on formalism of Skewed Parton Distributions and using the data on nucleon elestic form factors. When combined with deep-inelastic scattering data, it implies that about 25% of the proton spin is due to orbital angular momentum of quarks, while other 25% come from the quark spins. It leaves the remaining 50% to the gluon contribution.

Acknowledgments

I would like to thank Carl Carlson and Anatoly Radyushkin for invitation to this interesting workshop. This work was partially supported by the U.S. Department of Energy under contract DE–AC05–84ER40150.

References

1. A. Radyushkin, *Phys. Lett.* B **380**, 417 (1996); *ibid.*, *Phys. Lett.* B **385**, 333 (1996); *Phys. Rev.* D **56**, 5524 (1997).

2. X.Ji, *Phys. Rev.* D **55**, 7114 (1997), *Phys. Rev. Lett.* **78**, 610 (1997).
3. P. Jain, J. Ralston, Proc. of the Workshop on Future Directions in Particle and Nuclear Physics at multi-GeV Hadron Beam Facilities, BNL, Upton, NY, 1993.
4. P. Hoodbhoy, X. Ji, *Phys. Rev.* D **58**, 054006 (1998).
5. A. Radyushkin, *Phys. Rev.* D **58**, 114008 (1998).
6. A. Afanasev, E-print: hep-ph/9808291. In: Proc. Workshop on Jefferson Lab Physics and Instrumentation with 6-12-GeV Beams and Beyond, JLab, Eds.: S. Dytman, H. Fenker, P. Roos, p. 263.
7. M. Diehl et al., Phys.Lett.B460:204-212,1999; P. Kroll, these Proceedings and E-print hep-ph/9908242.
8. M.K. Jones et al. (JLab Hall A Collab.), E-print nucl-ex/9910005, (*submitted to Phys. Rev. Lett.*)
9. *Measurement of G_{Ep}/G_{Mp} to $Q^2 = 5.6$ GeV^2 by the Recoil Polarization Method*, JLab experiment 99-007, Spokespersons: C. Perdrisat, E. Brash, M. Jones, V. Punjabi; C. Perdrisat et al., Letter-of-Intent to JLab PAC-16.
10. I. Balitsky, X. Ji, *Phys. Rev. Lett.* **79**, 1225 (1997).
11. L. Andivahis et al., *Phys. Rev.* D **50**, 5491 (1994); S. Rock et al., *Phys. Rev.* D **46**, 24 (1992); A. Lung et al., *Phys. Rev. Lett.* **70**, 718 (1993).
12. SMC Collab., *Phys. Lett.* B **420**, 180 (1998).

LIGHT-FRONT QUARK MODEL PREDICTIONS OF MESON ELASTIC AND TRANSITION FORM FACTORS

Chueng-Ryong Ji and Ho-Meoyng Choi
Department of Physics, North Carolina State University, Raleigh, NC 27695-8202, USA
E-mail: ji@ncsu.edu

We investigate the electroweak form factors and semileptonic decay rates of mesons using the constituent quark model based on the light-front degrees of freedom. Our results demonstrate the broader applicability of light-front approach including the timelike region of exclusive processes.

1 Introduction

One of the distinctive advantages in the light-front approach is the well-established formulation of various form factor calculations using the well-known Drell-Yan-West ($q^+ = 0$) frame [1]. In $q^+ = 0$ frame, only parton-number-conserving Fock state (valence) contribution is needed when the "good" components of the current, J^+ and $J_\perp = (J_x, J_y)$, are used [2]. For example, only the valence diagram shown in Fig. 1(a) is used in the light-front quark model (LFQM) analysis of spacelike meson form factors. Successful LFQM description of various hadron form factors can be found in the literatures [3,4,5].

However, the timelike ($q^2 > 0$) form factor analysis in the LFQM has been hindered by the fact that $q^+ = 0$ frame is defined only in the spacelike region ($q^2 = q^+q^- - q_\perp^2 < 0$). While the $q^+ \neq 0$ frame can be used in principle to compute the timelike form factors, it is inevitable (if $q^+ \neq 0$) to encounter the nonvalence diagram arising from the quark-antiquark pair creation (so called "Z-graph"). For example, the nonvalence diagram in the case of semileptonic meson decays is shown in Fig. 1(b). The main source of the difficulty, however, in calculating the nonvalence diagram (see Fig. 1(b)) is the lack of information on the black blob which should contrast with the white blob representing the usual light-front valence wave function. In fact, we noticed [2] that the omission of nonvalence contribution leads to a large deviation from the full results. The timelike form factors associated with the hadron pair productions in e^+e^- annihilations also involve the nonvalence contributions. Therefore, it would be very useful to avoid encountering the nonvalence diagram and still be able to generate the results of timelike form factors.

In this paper, we start from the heuristic model calculations to obtain the exact result of the timelike form factor and then discuss our light-front constituent quark model predictions for the meson electroweak form factors.

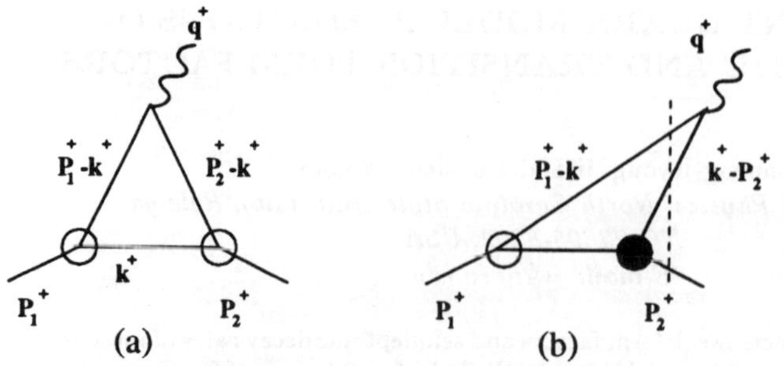

Figure 1: The LFQM description of a electroweak meson form factor: (a) the usual light-front valence diagram and (b) the nonvalence(pair-creation) diagram. The vertical dashed line in (b) indicates the energy-denominator for the nonvalence contributions. While the white blob represents the usual light-front valence wave function, the modeling of black blob has not yet been made.

Especially, we focus on the calculations of observables in timelike $|Q^2|$ region overcoming the diffculties associated with the quark-antiquark pair creation in the light-front approach. As an explicit example of application to timelike $|Q^2|$ region, we present our recent analysis of semileptonic weak decay processes. In Section 2, we present the heuristic model calculations to analytically continue the form factors in the spacelike region to those in the timelike region. In Section 3, we present our model calculations of semileptonic pseudoscalar meson decay processes as an example of application to the timelike $|Q^2|$ region. Conclusions and discussions follow in Section 4.

2 Heuristic Model Calculations

Using the solution of covariant Bethe-Salpeter equation in the ladder approximation with a relativistic version of the contact interaction[6], we can learn some lessons on the analytic continuation from the spacelike region to the timelike region [7]. The covariant model wave function is a product of two free single particle propagators, the overall momentum-conservation Dirac delta and a constant vertex function. Thus, all our form factor calculations in this section are nothing but various ways of evaluating the Feynman perturbation theory triangle diagram in scalar field theory[7].

Using this model, we show an explicit example of generating the exact result of the timelike form factor without encountering the nonvalence diagram. This can be done by the analytic continuation from the spacelike form factor calculated in the Drell-Yan-West ($q^+ = 0$) frame to the timelike region. To explicitly show it, we calculated [7]: (A) the timelike process of $\gamma^* \to M + \bar{M}$

transition in $q^+ \neq 0$ ($q^2 > 0$) frame, (B) the spacelike process of $M \to \gamma^* + M$ in $q^+ \neq 0$ ($q^2 < 0$) frame, and (C) the spacelike process of $M \to \gamma^* + M$ in $q^+ = 0$ frame. The analytic continuation from $q^2 < 0$ to $q^2 > 0$ allows us to show that the result in (C), which is obtained without encountering the nonvalence contributions at all, exactly reproduces the result in (A). In fact, all three results (A), (B), and (C) coincide with each other in the entire q^2 range. We also confirm that our results are consistent with the dispersion relations[8]. We consider here only for the equal quark/antiquark mass case such as the pion. However, the unequal mass cases such as K and D were also shown in Ref.[7].

For our numerical analysis of π meson form factor, we use the following constituent quark and antiquark masses: $m_u = m_d = 0.25$ GeV. Since our numerical results of the EM form factors obtained from (A), (B), and (C) turn out to be exactly same with each other for the entire q^2 region, each line depicted in Figs. 2 and 3 represents all three results.

As a consistency check, we also compare our numerical results of the form factor $F(q^2) = \text{Re } F(q^2) + i \text{ Im } F(q^2)$ with the dispersion relations given by

$$\text{Re } F(q^2) = \frac{1}{\pi} P \int_{-\infty}^{\infty} \frac{\text{Im } F(q'^2)}{q'^2 - q^2} dq'^2, \tag{1}$$

$$\text{Im } F(q^2) = -\frac{1}{\pi} P \int_{-\infty}^{\infty} \frac{\text{Re } F(q'^2)}{q'^2 - q^2} dq'^2, \tag{2}$$

where P indicates the Cauchy principal value.

In Fig. 2, we show the EM form factor of the pion for $-2 \text{ GeV}^2 \leq q^2 \leq 3 \text{ GeV}^2$. The imaginary part (the dotted line) of the form factor starts at $q^2_{\text{min}} = 4m^2_{u(d)} = 0.25$ GeV. It is interesting to note that the square of the total form factor $|F_\pi(q^2)|^2$ (thick solid line) produces a ρ meson-type peak near $q^2 \sim M^2_\rho$. However, it is not yet clear if this model indeed reproduces all the features of the vector meson dominance (VMD) phenomena. Even though the generated position of peak is consistent with VMD, the final state interaction is not included in this simple model calculation. We believe that much more complex mechanisms may be necessary to reproduce the realistic VMD phenomena. More detailed analysis along this line is under consideration. Nevertheless, it is remarkable that this simple model is capable of generating the peak and the position of peak is quite consistent with the VMD.

In Fig. 3, we show the timelike form factor of the pion for the entire $q^2 > 0$ region and compare the imaginary part of our direct calculations (dotted line) obtained from (A), (B), and (C) with the result (data of black dots) obtained from the dispersion relations given by Eq. (2). Our direct calculation is in an

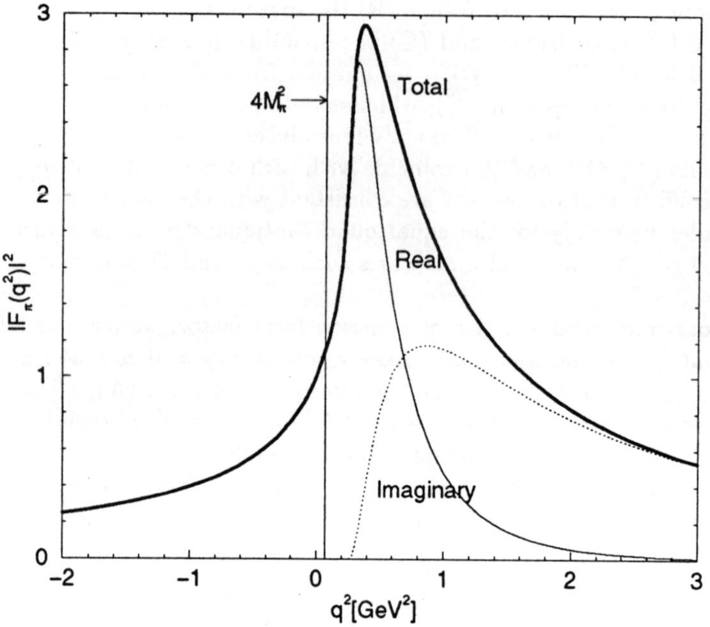

Figure 2: The electromagnetic form fator of the pion in $(3+1)$ dimensional scalar field theory for $-2 \leq q^2 \leq 3$ GeV^2. The total, real, and imaginary parts of $|F_\pi(q^2)|^2$ are represented by thick solid, solid, and dotted lines, respectively.

excellent agreement with the solution of the dispersion relations. Our result for the real part are also confirmed to be in complete agreement with the dispersion relations. For high q^2 region, the imaginary part of the form factor is dominant over the real part (thin solid line).

In Figs. 2 and 3, it is astonishing that the numerical result of (C) obtained from $q^+ = 0$ frame without encountering the nonvalence diagram coincides exactly with the numerical results of (A) and (B) obtained from $q^+ \neq 0$ frame.

3 Semileptonic Pseudoscalar Meson Decays in LFQM

The key idea in our LFQM[5] for mesons is to treat the radial wave function as a trial function for the variational principle to the QCD-motivated Hamiltonian

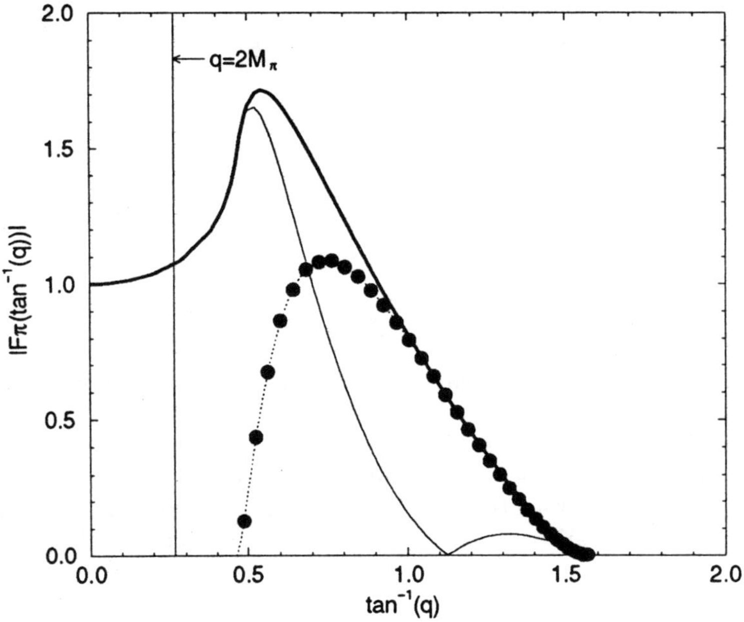

Figure 3: The electromagnetic form fator of the pion in (3+1) dimensional scalar field theory for the entire timelike region compared to the dispersion relations (data of black dots) given by Eq. (2). The same line code as in Fig. 2 is used.

saturating the Fock state expansion by the constituent quark and antiquark. The spin-orbit wave function is uniquely determined by the Melosh transformation. We take the QCD-motivated effective Hamiltonian as the well-known linear plus Coulomb interaction given by

$$H_{q\bar{q}} = H_0 + V_{q\bar{q}} = \sqrt{m_q^2 + k^2} + \sqrt{m_{\bar{q}}^2 + k^2} + V_{q\bar{q}}, \tag{3}$$

where

$$V_{q\bar{q}} = V_0 + V_{\text{hyp}} = a + br - \frac{4\kappa}{3r} + \frac{2\vec{S}_q \cdot \vec{S}_{\bar{q}}}{3m_q m_{\bar{q}}} \nabla^2 V_{\text{Coul}}. \tag{4}$$

We take the Gaussian radial wave function $\phi(k^2) = N \exp(-k^2/2\beta^2)$ as our trial wave function to minimize the central Hamiltonian [5]. Since the string tension b=0.18 GeV^2 and the constituent u and d quark masses m_u=m_d=0.22 GeV are rather well known from other quark model analyses commensurate with Regge phenomenology [9], we take them as our input parameters. The model parameters of a, κ, and $\beta_{u\bar{d}}$ are determined by the variational principle using the masses of ρ and π [5,10]. It is very important to note that all other model parameters such as m_c, m_b, β_{uc}, β_{ub}, etc. are then uniquely determined by our variational principle as shown in Ref.[10].

The semileptonic decays of a pseudoscalar meson $Q_1\bar{q}$ into another pseudoscalar meson $Q_2\bar{q}$ are governed by the weak vector current as follows

$$\langle P_2|\bar{Q}_2\gamma^\mu Q_1|P_1\rangle = f_+(q^2)(P_1+P_2)^\mu + f_-(q^2)(P_1-P_2)^\mu, \tag{5}$$

where $q^\mu = (P_1 - P_2)^\mu$ is the four-momentum transfer to the leptons. In the LFQM calculations presented in Ref.[11], the $q^+ \neq 0$ frame has been used to calculate the weak decays in the timelike region $m_l^2 \leq q^2 \leq (M_1 - M_2)^2$, with $M_{1[2]}$ and m_l being the initial[final] meson mass and the lepton(l) mass, respectively. However, when the $q^+ \neq 0$ frame is used, the inclusion of the nonvalence contributions arising from quark-antiquark pair creation("Z-graph") is inevitable and this inclusion may be very important for heavy-to-light and light-to-light decays. Nevertheless, the previous analyses [11] in $q^+ \neq 0$ frame considered only valence contributions neglecting nonvalence contributions. In this work, we circumvent this problem by calculating the processes in $q^+ = 0$ frame and analytically continuing to the timelike region. The $q^+ = 0$ frame is useful because only valence contributions are needed. However, one needs to calculate the component of the current other than J^+ to obtain the form factor $f_-(q^2)$. Since J^- is not free from the zero-mode contributions even in $q^+ = 0$ frame [12], we use $J_\perp$ instead of J^- to obtain f_-. In the $q^+ = 0$ frame, we obain the form factors $f_+(q^2)$ and $f_-(q^2)$ using the matrix element of the "+" and "$\perp$"-components of the current, J^μ, respectively, and then analytically continued to the timelike $q^2 > 0$ region by changing $q_\perp$ to $iq_\perp$ in the form facors. Our numerical results of the decay rates for $D\to\pi(K)$, $D_s\to\eta(\eta')$, and $B \to \pi(D)$ processes are consistent with the experimental data as summarized in Table 1. It is interesting to note that our value of η-η' mixing angle, $\theta_{SU(3)}$=$-19°$ presented in Ref.[5], are also in agreement with the data [13] for $D_s\to\eta(\eta')$ decays. More detailed results are given in Ref.[10].

4 Conclusions and Discussions

Using the heuristic model presented in Section 2, we discussed the important issue in the light-front approach, i.e., the calculation of the timelike form factors. Even though the model was simple, it played the role of quidance for the

Table 1: Form factors $f_+(0)$ and branching ratios (Br.) for various heavy meson semileptonic decays for $0^- \to 0^-$. We use $\theta^{\eta-\eta'}_{SU(3)}=-19°$ for $D_s \to \eta(\eta')$ decays and the following CKM matrix element: $|V_{cs}|=1.04\pm0.16$, $|V_{cd}|=0.224\pm0.016$, $|V_{ub}|=(3.3\pm0.4\pm0.7)\times10^{-3}$, and $|V_{bc}|=0.0395\pm0.003$ [13].

Processes	$f_+(0)$	Br.	Expt.[13]
$D \to K$	0.736	$(3.75 \pm 1.16)\%$	$(3.66 \pm 0.18)\%$
$D \to \pi$	0.618	$(2.36 \pm 0.34) \times 10^{-3}$	$(3.9^{+2.3}_{-1.1} \pm 0.4) \times 10^{-3}$
$D_s \to \eta$	0.421	$(1.8 \pm 0.6)\%$	$(2.5 \pm 0.7)\%$
$D_s \to \eta'$	0.585	$(9.3 \pm 2.9) \times 10^{-3}$	$(8.8 \pm 3.4) \times 10^{-3}$
$B \to \pi$	0.273	$(1.40 \pm 0.34) \times 10^{-4}$	$(1.8 \pm 0.6) \times 10^{-4}$
$B \to D$	0.709	$(2.28 \pm 0.20)\%$	$(2.00 \pm 0.25)\%$

more realistic model calculation. Since this solvable model provides the full information for the nonvalence diagram (black blob in Fig. 1(b)), we were able to check if the analytic continuation of the result in the Drell-Yan-West ($q^+ = 0$) frame (without the black blob) indeed reproduces the exact result generating the peaks analogous to the VMD phenomena. For more realistic model calculation, we came up with the LFQM [5] that described the spacelike form factors quite well. In this work, we demonstrated that our model can also be applied to the timelike exclusive processes, analyzing the semileptonic decays of a pseudoscalar meson into another pseudoscalar meson with the same model in Ref.[5]. The form factors $f_\pm$ are obtained in $q^+ = 0$ frame and then analytically continued to the timelike region by changing $q_\perp$ to $iq_\perp$ in the form factors. The matrix element of the "$\perp$" component of the current J^μ is used to obtain the form factor f_-. Our numerical results are in a good agreement with the available experimental data. This work broadens the standard light-front frame à la Drell-Yan-West to the timelike form factor calculation. The estimation of the zero-mode contribution has also been presented in Ref.[12]. To the extent that the zero-modes have a significant contribution to some physical observables [2], it seems conceivable that the condensation of zero-modes could lead to the nontrivial realization of chiral symmetry breaking in the light-front quantization approach. The success of our model calculations seems to reveal the effectiveness of light-front degrees of freedom in exclusive processes.

Acknowledgments

This work was supported by the U.S. Department of Energy(DE-FG-02-96ER4-0947). The North Carolina Supercomputing Center and the National Energy Research Scientific Computing Center are also acknowledged for the grant of computing time allocation. We would like to thank Carl Carlson and Anatoly

Radyushkin for organizing this interesting workshop.

References

1. S. D. Drell and T. M. Yan, *Phys. Rev. Lett.* **24**, 181 (1970); G. West, *Phys. Rev. Lett.* **24**, 1206 (1970); G. P. Lepage and S. J. Brodsky, *Phys. Rev.* D **22**, 2157 (1980).
2. H.-M. Choi and C.-R. Ji, *Phys. Rev.* D **59**, 034001 (1999).
3. P. L. Chung, F. Coester, and W. N. Polyzou, *Phys. Lett.* B **205**, 545 (1988); W. Jaus, *Phys. Rev.* D **44**, 2851 (1991); F. Cardarelli et al., *Phys. Lett.* B **332**, 1 (1994); *Phys. Rev.* D **53**, 6682 (1996).
4. H.-M. Choi and C.-R. Ji, *Nucl. Phys.* A **618**, 291 (1997); *ibid.* **56**, 6010 (1997).
5. H.-M. Choi and C.-R. Ji, *Phys. Rev.* D **59**, 074015 (1999).
6. M. Sawicki and L. Mankiewicz, *Phys. Rev.* D **37**, 421 (1988); L. Mankiewicz and M. Sawicki, *ibid.* **40**, 3415 (1989).
7. H.-M. Choi and C.-R. Ji, "Exploring the timelike region for the elastic form factor in a scalar field theory", hep-ph/9906225.
8. J. D. Bjorken and S. D. Drell, *Relativistic Quantum Fields* (McGraw-Hill, New York, 1965), pp. 209-282; S. Gasiorowicz, *Elementary Particle Physics* (Wiley, New York, 1966), pp. 348-362.
9. S. Godfrey and N. Isgur, *Phys. Rev.* D **32**, 189 (1985); N. Isgur, D. Scora, B. Grinstein, and M. B. Wise, *Phys. Rev.* D **39**, 799 (1989); D. Scora and N. Isgur, *Phys. Rev.* D **52**, 2783 (1992).
10. H.-M. Choi and C.-R. Ji, "Light-Front Quark Model Analysis of Exclusive $0^- \to 0^-$ Semileptonic Heavy Meosn Decays", to be published in *Phys. Lett. B* [hep-ph/9903496].
11. N.B. Demchuk, I. L. Grach, I. M. Narodetskii, and S. Simula, *Phys. At. Nuclei* **59**, 2152 (1996); H.-Y. Cheng, C.-Y. Cheng, and C.-W. Hwang, *Phys. Rev.* D **55**, 1559 (1997).
12. H.-M. Choi and C.-R. Ji, *Phys. Rev.* D **58**, 071901 (1998).
13. Particle Data Group, C. Caso et al., *Eur. Phys. J.* C **3**, 1 (1998).

KAONS AND HYPERONS AT HIGH MOMENTUM TRANSFER

O.K. BAKER[a]
Department of Physics, Hampton University, Hampton, VA 23668, and Physics Division, Jefferson Lab, Newport News, VA 23606
E-mail: baker@jlab.org

The new data from Jefferson Lab Experiment E93-018 provide better constraints on existing models of the strangeness electroproduction process than were available previously. The $^1H(e, e'K^+)\Lambda$ reaction was studied as a function of squared four-momentum transfer, Q^2, between 0.52 and 2.00 $(GeV/c)^2$, and of the virtual photon polarization parameter, ϵ. The first accurate separation of the longitudinal and transverse cross sections in this reaction is discussed in this manuscript, along with a comparison to representative model calculations. Some suggestions for future measurements are also presented.

1 Physics Motivation and Definitions

1.1 Physics Motivation

Strangeness degrees of freedom in nuclear physics provide new tests and deeper understanding of meson, baryon, and nuclear structure. The electromagnetic production of strangeness via the (e, e'K) reaction provides new and complementary information to that from hadronic reactions[1 2] such as (K, π), (π, K) and (p, p'K). This information is useful in such studies as hypernuclear production and spectroscopy, some aspects of QCD model building[3] such as the basic coupling constants needed in nucleon-meson and quark models, and conjectured strange star[5] behavior, to name only a few. The electroproduction process (using virtual photons) offers access to a much richer body of information than the photoproduction process (real photons) since both the mass and polarization of the virtual photon in the former case can can be varied independently[4].

1.2 Kinematics

The incident $(e = (E, \vec{p}_e))$ and scattered $(e' = (E_{e'}, \vec{p}_{e'}))$ electrons define the scattering plane. The recoiling K^+ meson $(k = (E_K, \vec{p}_K))$ and the associated hyperon Y $(Y = E_Y, \vec{p}_Y))$ define the production plane. The target proton is characterized by $p = (m_p, \vec{0})$ (that is, there is no momentum for the target proton in the the laboratory frame). The virtual photon lies in both planes.

[a]for the E93018 Collaboration.

The angle between the electron scattering plane and the production plane is the out-of-plane angle denoted by ϕ. The polar angle between the virtual photon and the kaon is denoted by $\theta_{\gamma K}$ while the angle between the incident and scattered electron is θ_e. The four-momentum transfer from the electron to the proton is q (= e − e′). Its components are $E - E_{e'} = \nu$ (the energy loss in the laboratory system) and $\vec{q} = \vec{p}_e - \vec{p}_{e'}$. The square of the four momentum transfer $q^2 = -Q^2$ is sometimes referred to as the mass of the virtual photon. Neglecting the mass of the electron (valid at these high energies), one can write

$$q^2 = -4EE' sin^2(\theta_e/2). \tag{1}$$

Other quantities of interest are $W^2 = m_p^2 + 2m_p\nu - Q^2$, the mass squared of the system recoiling against the electron, and the Mandelstamm variables $t = (q-k)^2 = q^2 + m_K^2 - 2qk$, $s = (e+p)^2 = m_p^2 + 2m_pE$ and $u = (k-p)^2$.

1.3 The Kaon Electroproduction Cross Section

In the one-photon-exchange approximation (valid for this measurement), the differential cross section for an unpolarized electron scattering from an unpolarized target where the kaon is detected in coincidence with the scattered electron and no final state polarizations are detected, may be written as[6]

$$\frac{d^5\sigma}{dE' d\Omega_{e'} d\Omega_{K^+}} = \Gamma(\sigma_T + \epsilon\sigma_L + \epsilon cos(2\phi)\sigma_{TT} + [\frac{\epsilon(\epsilon+1)}{2}]^{1/2} cos\phi\sigma_{LT}) \tag{2}$$

where

$$\Gamma = \frac{\alpha}{4\pi^2}\frac{E_{e'}(W^2 - m_p^2)}{Em_pQ^2(1-\epsilon)} \tag{3}$$

is the virtual photon flux. σ_T is the unpolarized (transverse) cross section, σ_L is the longitudinal cross section, σ_{TT} and σ_{LT} are the transverse-transverse and longitudinal-transverse interference cross sections. $d\Omega_{e'}$ and $d\Omega_{K^+}$ are the solid angles for electron and kaon detection. $E_{e'}$ is the scattered electron energy. The cross sections completely characterize the hadronic system. In general, these cross sections are functions of Q^2, W, and t. The last two terms in (2) vanish in parallel kinematics. Then by measuring the cross section at several values of ϵ while holding Q^2, W, and t constant, the longitudinal and transverse cross sections may be determined separately. Jefferson Lab Experiment E93-018 was the very first experiment to accurately determine these cross sections. The central kinematics can be found in Niculescu (1998)[7] while Figure (1) shows the nominal kinematics pictorially.

2 Experimental Discussion

2.1 *Overview*

The use of the detector packages in both the electron and hadron arms yielded a clean separation of the electron kaon coincidence events. The true coincidence kaon peak was well defined, as were the random coincidence peaks corresponding to the 499 MHz microstructure of the CEBAF electron beam. The ratio of real to random coincidences was better than 10 to 1 typically in the offline analysis for all kinematic settings [7]. The information from the real coincident kaons was used to calculate the missing mass spectrum shown in Figure (2). The two peaks corresponding to the masses of the $\Lambda(1115)$ and the $\Sigma^0(1192)$ are clearly identified. The production of higher mass hyperons were energetically forbidden in this measurement. In this study, all events with missing mass values between 1.100 and 1.175 GeV were accepted as valid events for the reaction $e+p \rightarrow e'+K^++\Lambda$. These were then used in subsequent calculations.

2.2 *Uncertainty Assigment*

The systematic uncertainty in the measurement is typically 5% while the statistical uncertainty in all cases is about 2% or less. The systematic uncertainty is dominated by the uncertainties in spectrometer acceptances and kaon survival probability. The point-to-point random uncertainties which are more relevant for the longitudinal-transverse cross section separation are smaller since some of the larger contributors to the systematic uncertainty are strongly correlated between the various ϵ settings. In general, the point-to-point random uncertainties are at or below 3%. Radiative corrections of up to 30% are applied to the final results [7]. Using the resulting formula for the cross section in parallel kinematics, all of the ϵ points for a given Q^2 were then fit to a straight line using linear least-squared fitting. The slope of the line yields the value of σ_L while the intercept is the resulting transverse cross section σ_T. The results of these fits are shown in Figure (3).

3 Results and Discussion

3.1 *Comparisons with Model Calculations*

The separated longitudinal cross section σ_L is shown in the top part of Figure (4) while the transverse cross section σ_T is shown in the bottom part of the figure, both as functions of Q^2. Both cross sections show a decrease with increasing Q^2. While it is expected that the longitudinal cross section approaches one in the limit of very high momentum transfers, for the kinematics in E93018, the longitudinal cross section is a decreasing function of momentum transfer. The transverse cross section is expected to be small at very high momentum transfers; this behavior is seen even at the E93-018 Q^2 range. The transverse cross section is expected to approach the photoproduction point as Q^2 goes to zero [7]. In the figures, several calculations are shown for comparison with the data. The thin curves correspond to the calculations by the Saclay-Lyon (SL) group [8] using an isoabaric model for the transition form factors. The medium thickness curves correspond to calculations by Williams, Ji, and Cotanch (WJC) [9], also using an isobaric model, while the thicker weight curves correspond to a Regge model calculation of Vanderhaeghen, Guidal, and Laget (VGL) [10].

As can be seen, the SL model shows a flat σ_L vs Q^2 dependence (after the initial rise from Q^2=0 predicted by all models), in contrast with the behavior exhibited by the data which falls with increasing Q^2. The WJC calculation shows a pronounced rise of the longitudinal cross sections initially, and then a gradual decrease with increasing Q^2 , in good agreement with the data. The VGL model shows a similar squared four-momentum transfer dependence as the WJC model for the longitudinal cross section, although the maximum of the cross section is not as pronounced in their model calculation.

In the case of the transverse cross section, the SL and the VGL model calculations correctly reproduce the Q^2 behavior of the data (although the SL curve is about 10-15% lower than the data but still consistent with it within the experimental uncertainties of the measurement), whereas the WJC model does not provide as accurate a description of the data as compared to the SL or VGL calculations.

The main differences between the two isobaric model calculations (SL [8] and WJC [9]) is that the former includes resonances up to spin 5/2 whereas the latter includes only up to spin 3/2 hadronic resonances. This difference would show up in a more striking manner in the transverse cross section since it is expected that higher mass and higher spin resonances couple more strongly to the transverse virtual photons. In the longitudinal channel, the cross section is expected to be dominated by the t-channel kaon exchange, and not as sensitive (by comparison with the transverse channel) to s and u channel exchanges. Then it is not unexpected that the WJC model can provide a good description of the trend of the σ_L data. The VGL Regge model calculation includes all resonances lying on the appropriate hadron trajectory [10]. This particular model provides a rather good description of the Q^2 dependence of both σ_L and σ_T, showing the utility of the Regge description of kaon electroproduction processes. It is also intriguing to speculate that the inclusion of even higher spin/mass resonances, or even hidden resonances (which would decay strongly into a kaon and lower mass hyperon) would provide an even better description of the Q^2 dependences of these separated cross sections [7].

3.2 Future Work

The analysis of the E93-018 data to extract the cross sections for Σ hyperon production is now ongoing. This should provide more insights into the topic of how to include higher spin and higher mass resonances since they should more strongly couple to these isovector hyperons. Polarization degrees of freedom should also give more insights into the elementary production channel; this is one of the aims of several Jefferson Lab experiments in Halls B and C. Finally, there are plans to make measurements of the elementary transverse and longitudinal cross sections up to even higher Q^2 in Hall A and to measure these separated cross sections in the deuterium nucleus in the near future.

Acknowledgments

This work is supported by the Department of Energy Award DE-FG05-91ER40620.

References

1. C. Bennhold, *Nucl. Phys.* A **547**, 79c (1992); Z. Li, *Phys. Rev.* C **52**, 1648 (1995).
2. C. Bennhold and L.E. Wright, *Phys. Rev.* C **36**, 438 (1987); S.R. Cotanch and S.S. Hsio, *Nucl. Phys.* A **450**, 419c (1986).
3. R.A. Schumacher, *Nucl. Phys.* A **585**, 63c (1995).
4. R.A. Williams, C.R. Ji, and S.R. Cotanch, *Phys. Rev.* C **46**, 1617 (1992).
5. X.-D. Li *et. al.*, *Phys. Rev. Lett.* **83**, 3776 (1999).
6. R.C.E. Devenish and D.H. Lyth, *Phys. Rev.* D **5**, 47 (1972).
7. G. Niculescu, Ph.D. Dissertation, Hampton University (unpublished), (1998).
8. J.C. David *et. al.*, *Phys. Rev.* C **53**, 2613 (1996).
9. R.A. Williams, private communications (1998).
10. M. Guidal, J.M. Laget, and M. Vanderhaeghen, *Phys. Lett.* B **400**, 6 (1997);*Nucl. Phys.* A **627**, 645 (1997); M. Vanderhaeghen, M. Guidal, and J.M. Laget,*Phys. Rev.* C **57**, 1454 (1998) .

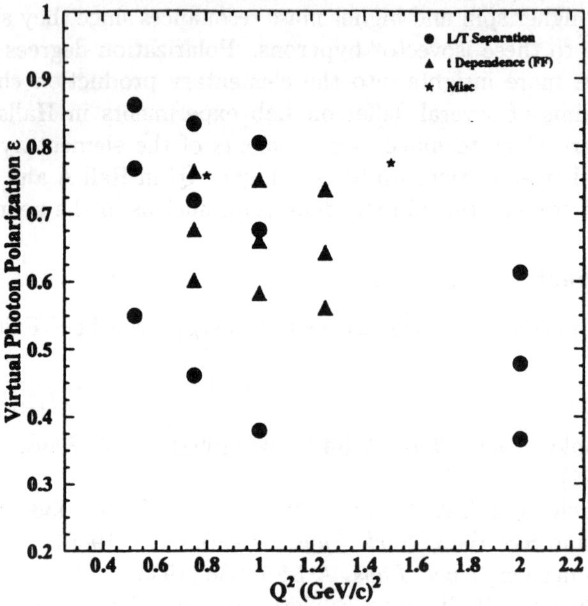

Figure 1: The nominal E93-018 kinematic settings for the separation of longitudinal and transverse cross sections in kaon electroproduction.

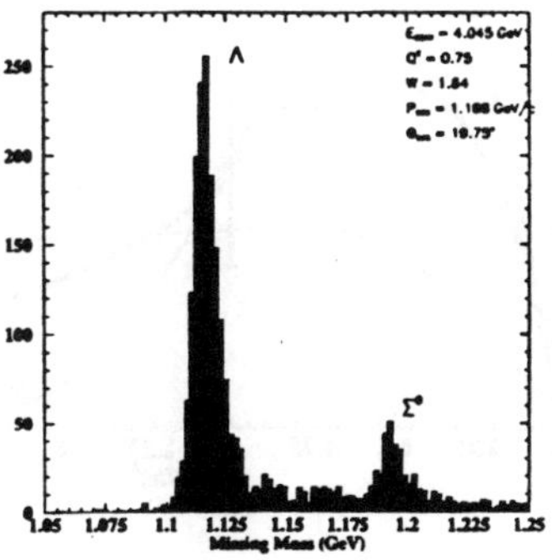

Figure 2: The missing mass calculated from the electron and kaon information. The Λ and Σ^0 peaks are clearly identified.

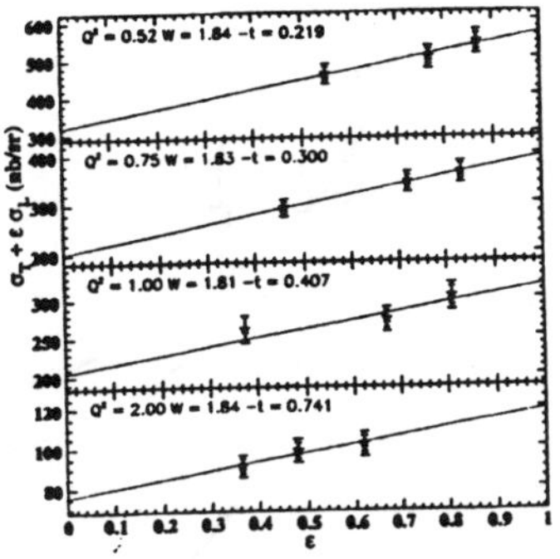

Figure 3: The linear least squares fit to the ϵ dependence of the total cross section. The slope of the line give σ_L while the intercept gives σ_T.

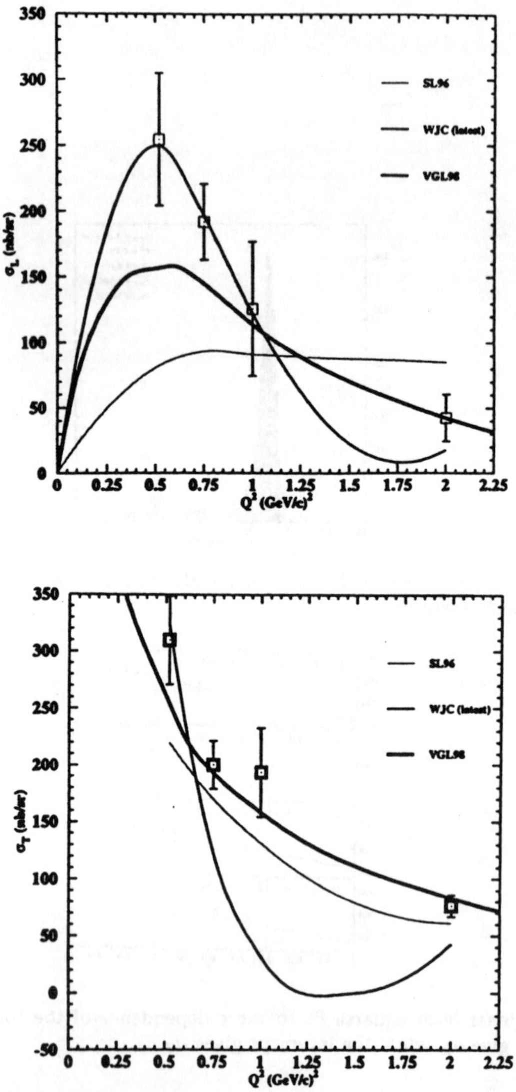

Figure 4: The longitudinal cross section for kaon electroproduction is shown in the top part of the figure while the transverse cross section is shown in the lower part. The thin curves correspond to calculations of the Saclay-Lyon group, the medium thickness curves correspond to the calculations by Williams, Ji, and Cotanch, and the thicker lines show the calculations of Vandergaegen, Guidal, and Laget as discussed in the text.

PION FORM FACTORS FROM QCD LIGHT-CONE SUM RULES

A. KHODJAMIRIAN [a,b]
Institut für Theoretische Physik, Universität Würzburg,
D-97074 Würzburg, Germany
E-mail : ak@physik.uni-wuerzburg.de

Light-cone sum rules have proved to be very useful in calculating hadronic matrix elements for exclusive processes. I present recent applications of this method to the pion electromagnetic form factor and to the form factors of $\gamma^* \rho \to \pi$ and $\gamma^* \gamma \to \pi^0$ transitions.

1 Introduction

Current and future experimental studies of exclusive and semi-exclusive processes require accurate calculations of relevant hadronic matrix elements in QCD. This task is, however, still far from being fulfilled. Of particular interest are quantitative QCD predictions for various form factors determining matrix elements of electroweak quark currents between one-hadron states. At sufficiently large values of the momentum transfer the form factors are determined by the perturbative QCD factorization. In particular, the pion electromagnetic form factor at $Q^2 \to \infty$ is given by the well known expression [1,2,3]

$$F_\pi(Q^2) = \frac{8\pi\alpha_s f_\pi^2}{9Q^2} \left| \int_0^1 du \frac{\varphi_\pi(u)}{\bar{u}} \right|^2 , \tag{1}$$

($\bar{u} \equiv 1 - u$) obtained by the convolution of twist-2 distribution amplitudes $\varphi_\pi(u)$ of the initial and final pion with the $O(\alpha_s)$ quark hard-scattering kernel. The major unsolved problem is to estimate the so-called soft, or end-point contributions to this form factor. These contributions are expected [4] to be important at intermediate momentum transfers $Q^2 \sim 1 \div 10$ GeV^2. Obviously, a consistent solution of this problem requires a calculation of both soft and hard contributions in *one and the same* framework beyond perturbation theory.

A promising and largely universal method of calculating hadronic matrix elements is provided by QCD light-cone sum rules (LCSR) [5]. This approach combines the light-cone operator-product expansion (OPE) with the conventional SVZ sum rule technique [6]. Proceeding from the basis of QCD perturbation theory, LCSR incorporate elements of nonperturbative long-distance

[a] *on leave from Yerevan Physics Institute, 375036 Yerevan, Armenia*
[b] after Sept. 1, 1999 at The Niels Bohr Institute, DK-2100, Copenhagen, Denmark

dynamics parametrized in terms of hadronic distribution amplitudes with different twist and multiplicity.

In this talk, I describe recent applications of LCSR to various pion form factors, including the pion electromagnetic form factor [7,8] and the form factors of $\gamma^*\rho(\omega) \to \pi$ and $\gamma^*\gamma \to \pi^0$ transitions [9].

2 Pion electromagnetic form factor

I begin with explaining the general idea of the LCSR approach using an important example of the pion electromagnetic form factor. More detailed derivation can be found in [8]. The starting object is the vacuum-pion correlation function

$$T_{\mu\nu}(p,q) = i \int d^4x\, e^{iqx} \langle 0|T\{j_\mu^5(0) j_\nu^{em}(x)\}|\pi^+(p)\rangle \,, \tag{2}$$

where $j_\nu^{em} = e_u \bar{u}\gamma_\nu u + e_d \bar{d}\gamma_\nu d$ is the quark electromagnetic current. One of the pions is put on-shell and the second one is replaced by the generating current $j_\mu^5 = \bar{d}\gamma_\mu\gamma_5 u$. For the on-shell pions, $p^2 = m_\pi^2$ vanishes in the chiral limit adopted here.

At fixed large $Q^2 = -q^2 >> \Lambda_{QCD}^2$ the correlation function (2) is a function of a single invariant $(p-q)^2$. Depending on the value of this variable the amplitude $T_{\mu\nu}$ corresponds to different physical pictures. At large spacelike $(p-q)^2$, $|(p-q)^2| >> \Lambda_{QCD}^2$, the interval $x^2 \to 0$ and one is able to employ the light-cone operator expansion for the product of two currents in (2). The amplitude (2) can then be factorized in a convolution of the pion distribution amplitude and the hard scattering amplitude yielding

$$T_{\mu\nu} = 2if_\pi p_\mu p_\nu \int_0^1 du \frac{u\varphi_\pi(u)}{\bar{u}Q^2 - u(p-q)^2} + \dots \,. \tag{3}$$

In the above, only the leading order, twist 2 contribution is shown, the ellipses denoting $O(\alpha_s)$ corrections and higher twist terms.

Decreasing $|(p-q)^2|$, one gradually approaches the physical region in the channel of the pion current. The interval x^2 deviates from zero, the light-cone OPE diverges and factorization is lost. Finally, at $(p-q)^2 = m_\pi^2$ one deals with the emission and propagation of an on-shell pion which is then elastically scattered by the electromagnetic current. In this region, the amplitude $T_{\mu\nu}$ is dominated by the one-pion contribution to its hadronic representation (dispersion relation)

$$T_{\mu\nu}(p,q) = 2if_\pi (p-q)_\mu p_\nu F_\pi(Q^2) \frac{1}{m_\pi^2 - (p-q)^2} + \int ds\, \frac{\rho_{\mu\nu}(s)}{s-(p-q)^2} \,. \tag{4}$$

The one-pion term is proportional to the pion decay constant f_π and to the desired form factor $F_\pi(Q^2)$. The integral over $\rho_{\mu\nu}$ in (4) contains contributions of higher mass intermediate states with the pion quantum numbers.

Equating (3) and (4) one extracts the form factor $F_\pi(Q^2)$ applying the standard elements of the QCD sum rule technique. Specifically, the quark-hadron duality is used: $\rho_{\mu\nu}(s) = 1/\pi \mathrm{Im} T_{\mu\nu}(s)\Theta(s - s_0^\pi)$, where $\mathrm{Im}T_{\mu\nu}(s)$ is calculated from (3) and $s_0^\pi \simeq 0.7$ GeV^2 is the effective threshold for the pion channel determined from the two-point QCD sum rule [6]. Furthermore, the Borel transformation in the variable $(p-q)^2$ is performed. The resulting LCSR, in the zeroth order in α_s and in the twist 2 approximation, reads [7]

$$F_\pi(Q^2) = \int\limits_{u_0}^{1} du\, \varphi_\pi(u,\mu_u) \exp\left(-\frac{\bar{u}Q^2}{uM^2}\right) \stackrel{Q^2\to\infty}{\longrightarrow} \varphi_\pi'(0) \int\limits_0^{s_0} \frac{ds\, s\, e^{-s/M^2}}{Q^4}\,, \qquad (5)$$

where $\varphi_\pi'(0) = -\varphi_\pi'(1)$, M^2 is the Borel parameter, $u_0 = Q^2/(s_0^\pi + Q^2)$. The factorization scale $\mu_u^2 = \bar{u}Q^2 + uM^2$ corresponds to the quark virtuality in the correlation function. This sum rule perfectly behaves at $Q^2 \to \infty$, in contrast to the conventional QCD sum rule [10] for $F_\pi(Q^2)$ based on the local OPE. The $1/Q^4$ behaviour of (5) corresponds to the soft end-point mechanism, provided that in $Q^2 \to \infty$ limit the integration region in (5) shrinks to a point $u = 1$.

Perturbative $O(\alpha_s)$ corrections to the correlation function (2) considerably improve the accuracy of the soft contribution. More importantly, in $O(\alpha_s)$ one recoveres the $\sim 1/Q^2$ asymptotic behaviour corresponding to the hard perturbative mechanism. The $O(\alpha_s)$ twist 2 term was calculated in [8]. Including this contribution in the LCSR and retaining the first two terms of the sum rule expansion in powers of $1/Q^2$ one obtains:

$$F_\pi(Q^2) = \frac{\alpha_s}{2\pi} C_F \int\limits_0^{s_0} \frac{ds\, e^{-s/M^2}}{Q^2} \int\limits_0^1 du \frac{\varphi_\pi(u)}{\bar{u}} + \varphi_\pi'(0) \int\limits_0^{s_0} \frac{ds\, s\, e^{-s/M^2}}{Q^4}$$

$$+ \frac{\alpha_s}{4\pi} C_F \int\limits_0^{s_0} \frac{ds\, s\, e^{-s/M^2}}{Q^4} \left\{ \varphi_\pi'(0) \left[-9 + \frac{1}{3}\pi^2 + \ln\frac{s}{\mu^2} - \ln^2\frac{s}{Q^2} \right] \right.$$

$$\left. + (2\ln\frac{s}{\mu^2} - 3) \int\limits_0^1 du \left[\frac{\varphi_\pi(u) - \bar{u}\,\varphi'(0)}{\bar{u}^2} \right] + (2\ln\frac{s}{\mu^2} - 8) \int\limits_0^1 du \frac{\varphi_\pi(u)}{\bar{u}} \right\}. \qquad (6)$$

It is remarkable that the leading asymptotic $O(1/Q^2)$ term in (6) coinsides with (1) provided that the two-point sum rule [6] for f_π yields $\int_0^{s_0} ds\, e^{-s/M^2} = 4\pi^2 f_\pi^2$,

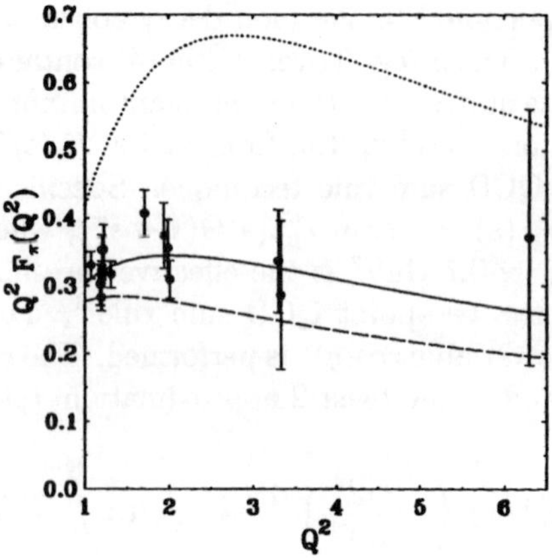

Figure 1: The light-cone sum rule predictions for the pion electromagnetic form factor [8] using asymptotic distribution amplitude (dashed), CZ distribution (dotted) and fit to the data [11] (solid).

and $\int_0^1 du\varphi_\pi^{\rm as}(u)/\bar{u} = 3$ for the asymptotic $\varphi_\pi^{\rm as}(u) = 6u\bar{u}$. Furthermore, it is instructive to split the $0 < u < 1$ integration region in LCSR into "hard" ($u < u_0$) and "soft" ($u > u_0$) parts. This separation reveals [8] that the $O(\alpha_s/Q^4)$ hard contribution is large and negative while the soft $O(1/Q^4)$ and $O(\alpha_s/Q^4)$ contributions are positive yielding considerable cancellations in the sum rule.

To further improve the accuracy of the LCSR, one also has to include the higher twist corrections. Physically, these corrections take into account both the transverse momentum of the quark-antiquark state and the contributions of higher Fock states in the pion wave function. In addition to the twist 4 terms, in [8] the factorizable twist 6 contributions proportional to the square of the quark condensate have been calculated. The latter corrections turn out to be comfortably small. Adding twist 4,6 terms to the twist 2 (leading and $O(\alpha_s)$) parts yields the LCSR prediction for $F_\pi(Q^2)$ shown in Fig. 1 for the pion distribution amplitude $\varphi_\pi(u,\mu) = 6u\bar{u}\left[1 + a_2(\mu)C_2^{3/2}(u-\bar{u})\right]$ with two choices: $a_2 = 0$ (asymptotic) and $a_2(1\,\text{GeV}) = 2/3$ (CZ). The available experimental data [11] seem to rule out the CZ option. The fit of the LCSR to these data yields $a_2(1\,\text{GeV}) = 0.12 \pm 0.07^{+0.05}_{-0.07}$ where the first (second) uncertainty is experimental (theoretical).

Another way to employ the LCSR prediction [8] is to subtract the leading asymptotic $1/Q^2$ term (the first term in (6)) from the sum rule. The remaining power suppressed part can be called "nonperturbative" contribution $F_{nonp}(Q^2)$. One then obtains a quantitative prediction for the full pion form factor by adding $F_{nonp}(Q^2)$ to the perturbative QCD result which is currently

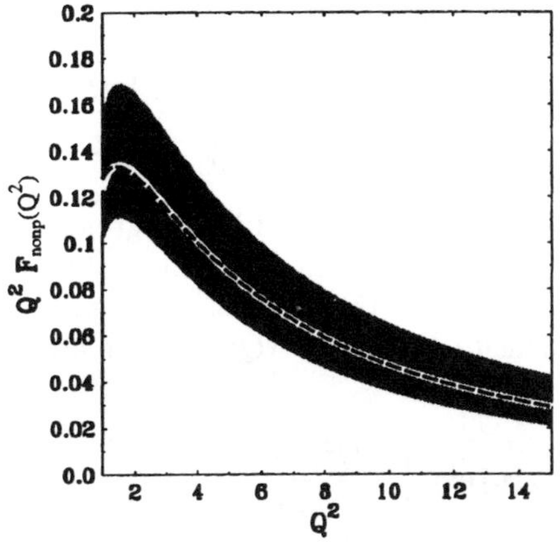

Figure 2: The light-cone sum rule prediction[8] for the nonperturbative correction to the pion form factor with the asymptotic pion distribution amplitude. The grey band shows the sensitivity of the result to variation of the Borel parameter within 0.8 GeV^2 $< M^2 <$ 1.5 GeV^2. The dashed curve is the calculation for the nominal value $M^2 = 1\text{GeV}^2$.

known with the NLO accuracy. The LCSR prediction for $F_{nonp}(Q^2)$ is plotted in Fig. 2. It turns out to be numerically moderate due to abovementioned cancellations between soft and hard $O(1/Q^4)$ contributions in the twist 2 part.

3 $\gamma^*\rho \to \pi$ and $\gamma^*\gamma \to \pi$ transition form factors

Similar to the pion electromagnetic form factor, the $\gamma^*\rho \to \pi$ or $\gamma^*\omega \to \pi$ transition form factors can be measured by extracting the one-pion exchange in the electroproduction of ρ or ω. The form factors are defined as $\frac{1}{3}\langle\pi^0(p) \mid j_\mu^{em} \mid \omega(p-q)\rangle \simeq \langle\pi^0(p) \mid j_\mu^{em} \mid \rho^0(p-q)\rangle = F^{\rho\pi}(Q^2)m_\rho^{-1}\epsilon_{\mu\nu\alpha\beta}e^\nu q^\alpha p^\beta$, e_ν being the polarization vector of ρ or ω. The derivation of LCSR [9] for $F^{\rho\pi}(Q^2)$ basically follows the procedure described in the previous section. The underlying correlation function is

$$\int d^4x e^{-iqx}\langle\pi^0(p)|\, T\{j_\mu^{em}(x)j_\nu^{em}(0)\}\,|0\rangle = i\epsilon_{\mu\nu\alpha\beta}q^\alpha p^\beta F^{\gamma^*\pi}(Q^2,(p-q)^2). \quad (7)$$

If both Q^2 and $|(p-q)^2|$ are sufficiently large, the light-cone OPE is valid for the amplitude $F^{\gamma^*\pi}(Q^2,(p-q)^2)$ starting from the leading twist 2 term [3]:

$$F^{\gamma^*\pi}(Q^2,(p-q)^2) = \frac{\sqrt{2}f_\pi}{3}\int_0^1 \frac{du\, \varphi_\pi(u)}{\bar{u}Q^2 - u(p-q)^2} + \ldots \quad (8)$$

Physical states in the $(p-q)^2$–channel include vector mesons $\rho, \omega, \rho', \omega', \ldots$ and a continuum of hadronic states with the same quantum numbers. The hadronic dispersion relation can be written as

$$F^{\gamma^*\pi}(Q^2,(p-q)^2) = \frac{\sqrt{2}f_\rho F^{\rho\pi}(Q^2)}{m_\rho^2-(p-q)^2} + \frac{1}{\pi}\int\limits_{s_0^\rho}^{\infty} ds\, \frac{\mathrm{Im}F^{\gamma^*\pi}(Q^2,s)}{s-(p-q)^2}. \qquad (9)$$

Here, the ρ and ω contributions are combined in one ground-state resonance term using $m_\rho \simeq m_\omega$ and $3\langle\omega \mid j_\nu^{em} \mid 0\rangle \simeq \langle\rho^0 \mid j_\nu^{em} \mid 0\rangle = (f_\rho/\sqrt{2})m_\rho e_\nu^*$. For the higher states above the effective threshold s_0^ρ, the quark-hadron duality is used with the spectral density $1/\pi \mathrm{Im}F^{\gamma^*\pi}(Q^2,s)$ calculated from (8). The resulting LCSR

$$F^{\rho\pi}(Q^2) = \frac{f_\pi}{3f_\rho}\int_{u_0}^1 \frac{du}{u}\varphi_\pi(u)\exp\left(-\frac{Q^2(1-u)}{uM^2}+\frac{m_\rho^2}{M^2}\right) + O(\text{twist } 4) \qquad (10)$$

was obtained in [9] where one can find additional details and numerical predictions. Note that the form factor (10) corresponds to the soft mechanism with $1/Q^4$ behaviour at large Q^2. Due to helicity suppression, the perturbative $O(\alpha_s)$ correction to $F^{\rho\pi}(Q^2)$ should have the same $1/Q^4$ behaviour. This and twist 6 corrections still have to be included in (10).

Finally, let me focus on the $\gamma^*\gamma \to \pi$ transition. Since the real photon is a large-distance object, the transition form factor $F^{\gamma\pi}(Q^2)$ contains nonperturbative contributions beyound the light-cone expansion of two electromagnetic quark currents. The leading $1/Q^2$ asymptotics of $F^{\gamma\pi}(Q^2)$ is well known [3] and given by (8) at $(p-q)^2 \to 0$. The calculation of the contributions suppressed by powers of $1/Q^2$ is a nontrivial task. Within LCSR approach, one has to invoke the photon distribution amplitudes. A reliable estimate is nevertheless possible [9] using the hadronic dispersion relation (9) where the resonance term is determined by the LCSR (10) and the integral over higher states is estimated using duality. One can analytically continue this relation to $(p-q)^2 \to 0$ provided that it does not contain subtraction terms (a similar approach was used in [12] for the structure function of the real photon). The result shown in Fig. 3 again has a better agreement with the experimental data [13] in the case of the asymptotic pion distribution amplitude. The recent update [14] of $F^{\gamma\pi}(Q^2)$ includes the $O(\alpha_s)$ correction decreasing the leading order results in Fig. 3 by 15-20%.

4 Conclusions

Light-cone sum rules provide a powerful tool for calculating various form factors at intermediate momentum transfers. In this framework, both soft end-point and hard perturbative mechanisms can be incorporated. The LCSR

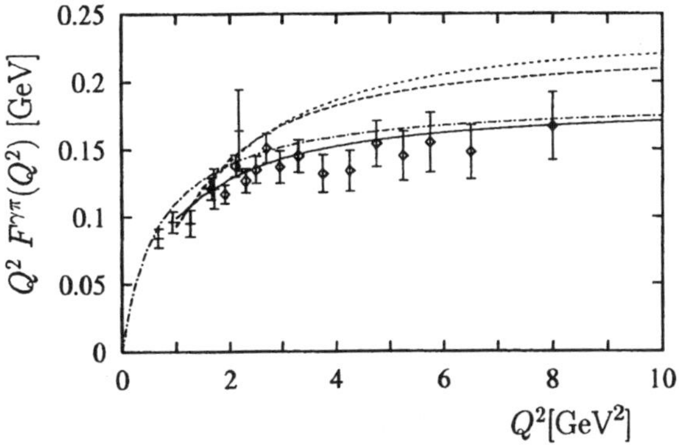

Figure 3: Form factor of the $\gamma^*\gamma \to \pi^0$ transition calculated [9] with the asymptotic (solid), CZ (long dashed) and BF (short-dashed) wave function of the pion in comparison with the experimental data points [13] and with the interpolation formula from [3] (dash-dotted).

results for the pion electromagnetic form factor indicate that the soft contribution is indeed large. However, an essential part of it is cancelled by the $O(\alpha_s/Q^4)$ corrections originating from the hard region. As a result, the overall power suppressed correction to the perturbative QCD factorization turns out to be relatively moderate. Comparison of LCSR predictions for the pion form factors with the available data indicates an almost asymptotic pion distribution amplitude $\varphi_\pi(u)$. More precise measurements of the form factors are needed to tightly constrain the nonasymptotic parts of $\varphi_\pi(u)$.

The LCSR approach is extremely useful also for heavy flavour exclusive decays, in particular, for the form factors of $B \to \pi, K, \rho, K^*$ transitions [15]. The sum rules for heavy-to-light and light-to-light transition form factors share a common nonperturbative input, that is the set of light hadron distribution amplitudes. Moreover, the structure of perturbative corrections to the pion form factor [8] turns out to be very similar to the heavy quark limit [16] of the LCSR for $B \to \pi$ in $O(\alpha_s)$. One may conclude that within a universal approach of QCD light-cone sum rules, both fields, the hard exclusive processes and the heavy flavour decays can mutually benefit.

Acknowledgments

I am grateful to C. Carlson and A. Radyushkin for the invitation to this fruitful workshop. Collaboration with V. Braun and M. Maul on the subject of this talk is acknowledged. This work was supported by the German Federal Ministry for Education and Research (BMBF) under contract number 05 7WZ91P (0).

References

1. V.L. Chernyak and A.R. Zhitnitsky, JETP Lett. **25** (1977) 510; Yad. Fiz. **31** (1980) 1053; Phys. Rept. **112** (1984) 173.
2. A.V. Efremov and A.V. Radyushkin, Phys. Lett. B **94** (1980) 245; Theor. Math. Phys. **42** (1980) 97.
3. G.P. Lepage and S.J. Brodsky, Phys. Lett. B **87** (1979) 359; Phys. Rev. D **22** (1980) 2157, Phys. Rev. **D24** (1981) 1808.
4. N. Isgur and C.H. Llewellyn Smith, Phys. Lett. **B217** (1989) 535; A.V. Radyushkin, Nucl. Phys. **A527** (1991) 153C; *ibid.* **A532** (1991) 141.
5. I.I. Balitsky, V.M. Braun and A.V. Kolesnichenko, Nucl. Phys. **B312** (1989) 509; V.M. Braun and I.E. Filyanov, Z. Phys. **C 44** (1989) 157; V.L. Chernyak and I.R. Zhitnitskii, Nucl. Phys. **B345** (1990) 137.
6. M.A. Shifman, A.I. Vainshtein and V.I. Zakharov, Nucl. Phys. **B147** (1979) 385.
7. V.M. Braun and I. Halperin, Phys. Lett. **B328** 457 (1994) 457 .
8. V.M. Braun, A. Khodjamirian and M. Maul, hep-ph/9907495.
9. A. Khodjamirian, Eur. Phys. J. **C6** (1999) 477.
10. B.L. Ioffe and A.V. Smilga, Phys. Lett. **114B** (1982) 353; Nucl. Phys. **B216** (1983) 373;
    V.A. Nesterenko and A.V. Radyushkin, Phys. Lett. **115B** (1982) 410.
11. C.J. Bebek *et al.*, Phys. Rev. **D17** (1978) 1693; S.R. Amendolia *et al.* [NA7 Collaboration], Nucl. Phys. **B277** (1986) 168.
12. A.S. Gorsky, B.L. Ioffe, A.Yu. Khodjamirian and A.G. Oganesian, Z. Phys. **C44** (1989) 523.
13. CELLO Collaboration (H.-J. Behrend et. al), Z. Phys. **C49** (1991) 401; CLEO Collaboration (J. Gronberg et al.), Phys. Rev. **D57** (1998) 33.
14. A. Schmedding and O. Yakovlev, hep-ph/9905392.
15. V.M. Belyaev, A. Khodjamirian and R. Rückl, Z. Phys. **C 60** (1993) 349; V.M. Belyaev, V.M. Braun, A. Khodjamirian and R. Rückl, Phys. Rev. **D51** (1995) 6177;
    P. Ball and V.M. Braun, Phys. Rev. **D55** (1997) 5561; Phys. Rev. **D58** (1998) 094016;
    A. Khodjamirian and R. Rückl, hep-ph/9801443, in Heavy Flavours II, eds. A.J. Buras and M. Lindner (World Scientific, Singapore,1998), p.345;
    P. Ball, JHEP **09** (1998) 005.
16. E. Bagan, P. Ball and V.M. Braun, Phys. Lett. **B417** (1998) 154.

ON THE PQCD PREDICTION FOR THE PION FORM FACTOR

B. MELIĆ, B. NIŽIĆ, K. PASSEK
Theoretical Physics Division, Rudjer Bošković Institute, P.O. Box 1016, HR-10001 Zagreb, Croatia
E-mail: melic@thphys.irb.hr, nizic@thphys.irb.hr, passek@thphys.irb.hr

We comment on the results of a complete leading-twist next-to-leading order QCD analysis of the spacelike pion electromagnetic form factor at large-momentum transfer Q. For the asymptotic distribution amplitude, we have examined the sensitivity of the predictions to the choice of the renormalization scale. The results show that, regarding the size of the radiative corrections, reliable perturbative predictions for the pion electromagnetic form factor can already be made at a momentum transfer Q of the order of 5 to 10 GeV.

1 Introduction

The study of exclusive processes at large-momentum transfer represents a challenging area for application of perturbative QCD (PQCD). Although the PQCD approach [1] undoubtedly represents an adequate and efficient tool for analyzing exclusive processes at very large momentum transfer, its applicability to these processes at experimentally accessible momentum transfer has been long debated and attracted much attention. In a moderate energy region (a few GeV) soft contributions could still be substantial.[2] Further on, the self-consistency of the PQCD approach was questioned regarding the non-factorizing end-point contributions.[3] It has been shown, that the incorporation of the Sudakov suppression in the so-called modified hard-scattering approach (mHSA) [4,5] effectively eliminates soft contributions from the end-point regions and that the PQCD approach to the pion form factor begins to be self-consistent for a momentum transfer of about $Q^2 > 4$ GeV^2. However, in the PQCD approach to exclusive processes one still has to check its self-consistency by studying radiative corrections. It is well known that, unlike in QED, the leading-order (LO) predictions in PQCD do not have much predictive power, and that higher-order corrections are important. They have a stabilizing effect reducing the dependence of the predictions on the schemes and scales.

In our recent papers [6] we have clarified some discrepancies between previous calculations,[7] and by including the complete closed form for the NLO evolution of the pion distribution amplitude (DA) derived recently,[8] we have obtained the complete NLO PQCD prediction for the pion electromagnetic form factor (within the so-called standard hard-scattering approach (sHSA) and using different candidate DAs). The size of the NLO correction as well as the size of the expansion parameter, i.e. QCD running coupling constant, serve as sensible measures of the self-consistency of PQCD prediction. But, as the truncation of the perturbative expansion for the pion form factor at finite order causes the residual dependence of the prediction on the choice of the renormalization and factorization scales (as well as on the renormalization scheme), the choices for these scales represent the major ambiguity in the interpretation of the results.

In this paper we would like to outline our calculation [6] and to address the scale ambiguity problem.

2 Pion electromagnetic form factor in the sHSA

In leading twist, the pion electromagnetic form factor can be expressed by a convolution formula

$$F_\pi(Q^2) = \int_0^1 dx \int_0^1 dy \, \Phi^*(y, \mu_F^2) \, T_H(x, y, Q^2, \mu_R^2, \mu_F^2) \, \Phi(x, \mu_F^2) \,. \tag{1}$$

Here $Q^2 = -q^2$ is the momentum transfer in the process and is supposed to be large, μ_R is the renormalization scale, and μ_F is the factorization scale at which soft and hard physics factorize; x and y ($\overline{x} = 1 - x$ and $\overline{y} = 1 - y$) denote incoming and outgoing quark (antiquark) momentum fractions.

The (process dependent) hard-scattering amplitude $T_H(x, y, Q^2, \mu_R^2, \mu_F^2)$ is calculated in perturbation theory and represented as a series in the QCD running coupling constant $\alpha_S(\mu_R^2)$. We have used the dimensional regularization method and the $\overline{MS}$ renormalization scheme in our calculation.[6]

The (process independent) pion DA Φ is intrinsically nonperturbative quantity, whose evolution can be calculated perturbatively and represented as a series in $\alpha_S(\mu_F^2)$. There are compelling theoretical results [2,9] which disfavor the end-point concentrated distributions, and one expects that the pion DA does not differ much from the asymptotic form. In this work we comment only on the results obtained with the asymptotic distribution $\phi_{as}(x, \mu_F^2)$.

Generally, one can express the NLO form factor as

$$F_\pi(Q^2, \mu_R^2, \mu_F^2) = F_\pi^{(0)}(Q^2, \mu_R^2, \mu_F^2) + F_\pi^{(1)}(Q^2, \mu_R^2, \mu_F^2) \,. \tag{2}$$

The first term in (2) is the LO contribution, while the second term is the NLO contribution coming from the NLO correction to the hard-scattering amplitude as well as arising from the inclusion of the NLO evolution of the DA. For the results obtained using $\phi_{as}(x, \mu_F^2)$ distribution, the effect of the NLO evolution of the DA is negligible ($\approx 1\%$).

Truncation of the perturbative series of $F_\pi(Q^2)$ at any finite order causes a residual dependence on the scheme as well as on the scales (which is explicitly denoted in Eq. (2)). As we approximate $F_\pi(Q^2)$ only by two terms of the perturbative series, we hope that we can minimize higher-order corrections by a suitable choice of μ_R and μ_F, so that the LO term $F_\pi^{(0)}(Q^2, \mu_R^2, \mu_F^2)$ gives a good approximation to the complete sum $F_\pi(Q^2)$.

We take that a PQCD prediction for pion form factor can be considered reliable provided the corrections to the LO prediction are reasonably small ($< 30\%$) and the expansion parameter (effective QCD coupling constant) is acceptably small ($\alpha_S(\mu_R^2) < 0.3$ or 0.5). The consistency with the experimental data is not of much use here since reliable experimental data[10] for the pion form factor exist for $Q^2 \leq 4$ GeV^2 i.e., outside the region in which the perturbative treatment based on Eq. (1) is justified. It should also be mentioned that there are controversial arguments[11] regarding the reliability of existing experimental data.[10] The new data in this energy region are expected from the CEBAF experiment E-93-021.

3 Examining the scale dependence of the NLO prediction

The simplest and widely used choice for the μ_R and μ_F scales is

$$\mu_R^2 = \mu_F^2 = Q^2 \,, \tag{3}$$

the justification for the use of which is mainly pragmatic. The prediction for the pion form factor depends very weakly on the choice of the factorization scale μ_F. Actually, taking μ_F^2 to be an effective constant, i.e., $\mu_F^2 = \langle \mu_F^2 \rangle$, the only μ_F^2 dependence of the results obtained using $\phi_{as}(x, \mu_F^2)$ distribution comes from the NLO evolution of the DA which is negligible.

Physically, a more appropriate choice for μ_R^2 would be that corresponding to the characteristic virtualities of the particles in the parton subprocess (which are considerably lower than the overall momentum transfer Q^2 i.e., virtuality of the probing photon)

$$\mu_R^2 = a(x, y)\, Q^2 \,. \tag{4}$$

For example, some of the physically motivated choices[6] are

$$a(x, y) \in \{\bar{x}\bar{y}\,, \sqrt{\bar{x}\bar{y}\bar{y}}\,, e^{-5/3}\bar{x}\bar{y}\} \,. \tag{5}$$

These correspond, respectively, to the (LO) gluon virtuality, geometrical mean of the gluon and quark virtualities (an attempt to take into account that the QCD coupling is renormalized not only by the vector particle propagator, but also by the quark-gluon vertex and the quark-propagator), and to the choice of

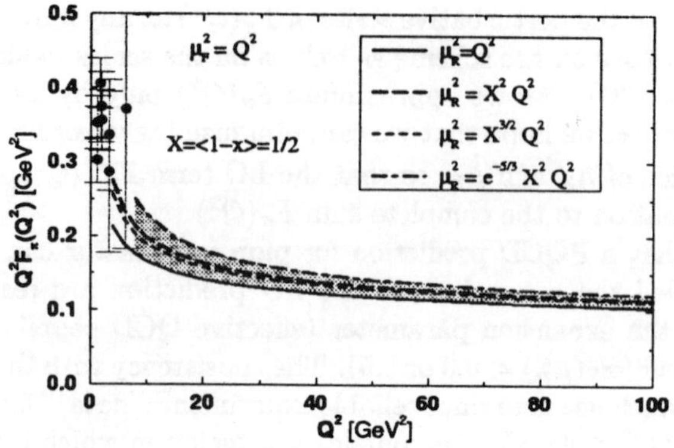

Figure 1: NLO prediction for $Q^2F_\pi(Q^2)$ obtained using the $\phi_{as}(x,\mu_F^2)$ distribution amplitude and the choices of μ_R^2 given by Eqs. (3), (6), and (7), while $\mu_F^2 = Q^2$ and $\langle \overline{x} \rangle = 0.5$. The shaded area denotes the theoretical uncertainty introduced by the renormalization scale ambiguity.

the renormalization scale according to the Brodsky-Lepage-Mackenzie (BLM) procedure [12] the essence of which is that all vacuum-polarization effects from the QCD β function should be resummed into the running coupling constant (the NLO coefficient of T_H becomes n_f (i.e. β_0) independent).

A glance at Eq. (1), where the coupling constant $\alpha_S(\mu_R^2)$ appears under the integral sign, reveals that the choice (4) leads immediately to the problem if the usual one-loop formula for the effective QCD running coupling constant is employed. Namely, regardless of how large Q^2 is, the integrations over x and y allow $\alpha_S(\mu_R^2)$ to be evaluated at low momenta i.e., in the region where usual one-loop formula is not a good representation of the effective QCD coupling. There are number of proposals [13,14] for the form of the coupling constant $\alpha_S(\mu_R^2)$ for small μ_R^2, but its implementation in this calculation demands the more refined treatment.[6] Alternatively, one can choose μ_R^2 to be an effective constant

$$\mu_R^2 = \langle \mu_R^2 \rangle = \langle a(x,y) \rangle \, Q^2 = a\, Q^2 \,. \tag{6}$$

Hence, the expressions (5) get replaced by their respective averages

$$a \in \{\langle \overline{x} \rangle^2, \langle \overline{x} \rangle^{3/2}, e^{-5/3} \langle \overline{x} \rangle^2\} \,. \tag{7}$$

Now, the key quantity in the above expressions is $\langle \overline{x} \rangle$, the average value of the momentum fraction. Owing to the fact that $\phi_{as}(x,\mu_F^2)$ is centered around the value $x = 0.5$, the simplest choice is $\langle \overline{x} \rangle = 0.5$.

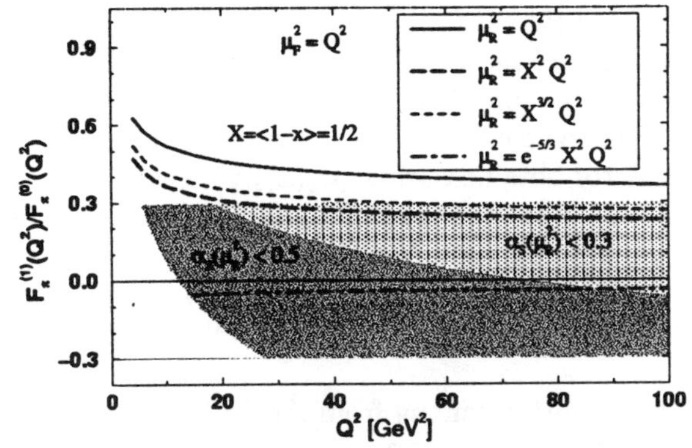

Figure 2: The ratio $F_\pi^{(1)}(Q^2)/F_\pi^{(0)}(Q^2)$ obtained using the same DA and the same choices for μ_R^2 and μ_F^2 scales as in Fig. 1 The shaded area denotes the region of predictions which corresponds to $|F_\pi^{(1)}(Q^2)/F_\pi^{(0)}(Q^2)| < 30\%$, and $\alpha_S(\mu_R^2)<0.5$ ($\alpha_S(\mu_R^2)<0.3$).

Numerical results of our complete NLO QCD calculation for $Q^2F_\pi(Q^2)$, obtained using the $\phi_{as}(x,\mu_F^2)$ distribution, with $\mu_F^2 = Q^2$ and different choices for the renormalization scale μ_R^2 given by (3) (6), (7), and $\langle \overline{x} \rangle = 1/2$, are displayed in Fig. 1 (in our calculation we take $\Lambda_{\overline{MS}} = 0.2$ and $f_\pi = 0.131$ GeV). The ratio of the NLO to the LO contribution to $F_\pi(Q^2)$, i.e., $F_\pi^{(1)}(Q^2)/F_\pi^{(0)}(Q^2)$, as a useful measure of the importance of the NLO corrections, is plotted as a function of Q^2 in Fig. 2.

One notices that the total NLO perturbative prediction for $Q^2F_\pi(Q^2)$ is somewhat below the trend indicated by the presently available experimental data, but what alarms us and seems to question the self-consistency of the PQCD approach is the fact that the ratio $F_\pi^{(1)}/F_\pi^{(0)}$ corresponding to the often encountered choice $\mu_R^2 = Q^2$ (solid line) is rather high: $F_\pi^{(1)}/F_\pi^{(0)} \leq 30\%$ is not reached until $Q^2 \approx 500$ GeV^2. The answer to this problem lies in the previously stated inappropriateness of the choice $\mu_R^2 = Q^2$. Namely, owing to the partitioning of the overall momentum transfer Q^2 among the particles in the parton subprocess, the essential virtualities of the particles are smaller than Q^2, so that the "physical" renormalization scale, better suited for the process of interest, is inevitably lower than Q^2. It follows from Fig. 2 that by choosing the renormalization scale determined by the dynamics of the pion rescattering process, the size of the NLO corrections is significantly reduced and reliable predictions are obtained at considerably lower values of Q^2, namely, for $Q^2 < 100$ GeV^2.

The total NLO prediction for $Q^2F_\pi(Q^2)$ depends weakly on the choice of μ_R^2. This is a reflection of the stabilizing effect that the inclusion of the NLO corrections has on the LO predictions. Let us explore this point more closely. By taking μ_R^2 and μ_F^2 to be effective constants, the LO and NLO contributions to the pion form factor obtained using $\phi_{as}(x,\mu_F^2)$ distribution amount to

$$Q^2F_\pi^{(0)}(Q^2,\mu_R^2,\mu_F^2) = 8\pi f_\pi^2 \alpha_S(\mu_R^2), \tag{8}$$

$$Q^2F_\pi^{(1)}(Q^2,\mu_R^2,\mu_F^2) = 8 f_\pi^2 \alpha_S^2(\mu_R^2) \left[\frac{\beta_0}{4}(\ln\frac{\mu_R^2}{Q^2}+\frac{14}{3}) - 3.92\right], \tag{9}$$

where the NLO contribution coming from the NLO evolution of the DA is neglected. As it is seen from (8), all of the μ_R dependence of the LO result $Q^2F_\pi^{(0)}$ is contained in the strong coupling constant $\alpha_S(\mu_R^2)$. Thus, as μ_R decreases the LO result increases, and it increases without bound. In contrast to the LO, the NLO contribution $Q^2F_\pi^{(1)}$, as evident from the explicit expression given in (9), decreases (becomes more negative) with decreasing μ_R. Upon adding up the LO and NLO contributions, we find that the full NLO result, as a function of μ_R stabilizes and reaches a maximum value for $\mu_R^2 = \mu_{extreme}^2 \approx Q^2/18$. If we take that the renormalization scale continuously changes in the interval defined by $a \in [1/18, 1]$ the curves representing the NLO predictions for $Q^2F_\pi(Q^2)$ fill out the shaded region in Fig. 1, and this shaded region essentially determines the scale ambiguity related theoretical uncertainty of the NLO calculation.

4 Towards resolving the renormalization scale ambiguity problem

The optimization of the scale and scheme choice according to some sensible criteria remains an important task for the application of PQCD. Several scale-setting procedures were proposed in the literature: the principle of fastest apparent convergence (FAC) [15] ($F_\pi^{(1)}(Q^2,\mu_{FAC}^2)=0$), the principle of minimal sensitivity (PMS) [16] ($\mu_{PMS}^2=\mu_{extreme}^2$), and the BLM method. The application of those methods can give strikingly different results in some calculations.[17]

As it is known, the relations between physical observables must be independent of renormalization scale and scheme conventions to any fixed order of perturbation theory. It was argued [18] that applying the BLM scale-fixing to perturbative predictions of two observables in, for example, $\overline{MS}$ scheme and then algebraically eliminating $\alpha_{\overline{MS}}$ one can relate any perturbatively calculable observables without scale and scheme ambiguity, where the choice of BLM scale ensures that the resulting "commensurate scale relation" (CSR) is independent of the choice of the intermediate renormalization scheme. Following this approach, in paper by Brodsky *et al.*[19] the exclusive hadronic amplitudes

Table 1: NLO PQCD results for the pion form factor, $Q^2F_\pi(Q^2)$ in the α_V scheme.

Q^2 [GeV^2]	$\alpha_V(\mu_V^2)$	$F_\pi^{(1)}(Q^2)/F_\pi^{(0)}(Q^2)$ [%]	$Q^2F_\pi(Q^2)$ [GeV^2]
20	0.434	-26.5	0.138
100	0.289	-17.7	0.103

were calculated in α_V scheme, in which the effective coupling $\alpha_V(\mu^2)$ is defined from the heavy-quark potential $V(\mu^2)$.

It follows from (9) that the β_0 dependent term vanishes for $\mu_R^2 = \langle\mu_{BLM}^2\rangle = e^{-14/3}Q^2$. Using the scale-fixed relation [19] between $\alpha_V(\mu_V^2)$ and $\alpha_{\overline{MS}}(\mu_{\overline{MS}}^2) \equiv \alpha_S(\mu_R^2)$, and Eqs. (8-9) one obtains the NLO prediction for the pion form factor in α_V scheme [19]

$$Q^2F_\pi(Q^2,\mu_V^2) = 8\pi f_\pi^2\,\alpha_V(\mu_V^2)\left(1 - 1.92\frac{\alpha_V(\mu_V^2)}{\pi}\right), \qquad (10)$$

where $\mu_V^2 = e^{5/3}\mu_R^2 = e^{-3}Q^2 \approx Q^2/20$. Considering the energy region we are interested in, for the purpose of this calculation we approximate α_V with usual one-loop expression. Thus obtained numerical predictions are given in Table 1. Considering the size of the $F_\pi^{(1)}/F_\pi^{(0)}$ ratio ($\approx -20\%$) and the size of the effective coupling, we find that NLO PQCD predictions for the pion form factor obtained in α_V scheme can be considered reliable for Q^2 below 100 GeV^2 i.e., already for momentum transfer Q of the order of $5-9$ GeV.

5 Conclusions

We have shown that, regarding the size of the radiative corrections, the sHSA can be consistently applied to the calculation of the pion electromagnetic form factor already for momentum transfer Q of the order of $5-10$ GeV. In order to improve this PQCD based prediction it is necessary to obtain and apply the proper form of the QCD coupling in the low-momentum regime as well as to investigate the corrections (present in the few GeV region) introduced by the mHSA.[20] Reliable experimental data and inclusion of the soft contributions should then enable complete confrontation between theory and experiment.

Acknowledgments

This work was supported by the Ministry of Science and Technology of the Republic of Croatia under Contract No. 00980102. One of us (K.P.) would like to thank Anatoly Radyushkin and Carl Carlson for the opportunity to participate in this stimulating workshop.

References

1. S.J. Brodsky and G.P. Lepage in *Perturbative QCD*, ed. A.H. Mueller (World Scientific, Singapore, 1989), and references therein.
2. V. Braun and I. Halperin, *Phys. Lett.* B **328**, 457 (1994); P. Kroll and M. Raulfs, *Phys. Lett.* B **387**, 848 (1996).
3. N. Isgur and C.H. Llewellyn Smith, *Nucl. Phys.* B **317**, 526 (1989); A.V. Radyushkin, *Nucl. Phys.* A **532**, 141 (1991).
4. H.-N. Li and G. Sterman, *Nucl. Phys.* B **381**, 129 (1992); D. Tung and H.-N. Li, *Chin. J. Phys.* **35**, 651 (1997).
5. R. Jakob and P. Kroll, *Phys. Lett.* B **315**, 463 (545 (E)) (1993).
6. B. Melić, B. Nižić, and K. Passek, *Phys. Rev.* D **60**, 074004 (1999); B. Melić, B. Nižić, and K. Passek, *Fizika* B **8**, 327 (1999), hep-ph/9903426.
7. R.D. Field, R. Gupta, S. Otto, and L. Chang, *Nucl. Phys.* B **186**, 429 (1981); F.-M. Dittes and A.V. Radyushkin, *Sov. J. Nucl. Phys.* **34**, 293 (1981); M.H. Sarmadi, Ph. D. thesis, University of Pittsburgh, 1982; R.S. Khalmuradov and A.V. Radyushkin, *Sov. J. Nucl. Phys.* **42**, 289 (1985); E.P. Kadantseva, S.V. Mikhailov, and A.V. Radyushkin, *Sov. J. Nucl. Phys.* **44**, 326 (1986); E. Braaten and S.-M. Tse, *Phys. Rev.* D **35**, 2255 (1987).
8. D. Müller, *Phys. Rev.* D **49**, 2525 (1994); *Phys. Rev.* D **51**, 3855 (1995).
9. A.V. Radyushkin, *Acta Phys. Polon.* B **26**, 2067 (1995); S. Ong, *Phys. Rev.* D **52**, 3111 (1995); P. Kroll and M. Raulfs, *Phys. Lett.* B **387**, 848 (1996); A.V. Radyushkin, *Few Body Syst.Suppl* **11**, 57 (1999).
10. J. Bebek *et al*, *Phys. Rev.* D **17**, 1693 (1978).
11. C.E. Carlson and J. Milana, *Phys. Rev. Lett.* **65**, 1717 (1990). S. Dubnička and L. Martinovič, *J. Phys.* G **15**, 1349 (1989).
12. S.J. Brodsky, G.P. Lepage, and P.B. Mackenzie, *Phys. Rev.* D **28**, 228 (1983).
13. J.M. Cornwall, *Phys. Rev.* D **26**, 1453 (1982); A. Donnachie and P.V. Landshoff, *Nucl. Phys.* B **311**, 509 (1989).
14. D.V.Shirkov and I.L.Solovtsov, *Phys. Rev. Lett.* **79**, 1209 (1997).
15. G. Grunberg, *Phys. Rev.* D **29**, 2315 (1984).
16. P.M. Stevenson, *Nucl. Phys.* B **231**, 65 (1984).
17. G. Kramer and B. Lampe, *Z. Phys.* A **339**, 189 (1991).
18. S.J. Brodsky and H.J. Lu, *Phys. Rev.* D **51**, 3652 (1995);
19. S.J. Brodsky, C.-R. Ji, A. Pang and D.G.Robertson, *Phys. Rev.* D **57**, 245 (1998).
20. N.G. Stefanis, W. Schroers and H.-Ch. Kim, hep-ph/9812280.

QCD CONSTRAINTS ON FORM FACTOR SHAPES*

Richard F. Lebed
Jefferson Lab, Newport News, Virginia 23606, USA

ABSTRACT

This talk presents an introduction to the use of dispersion relations to constrain the shapes of hadronic form factors consistent with QCD. The applications described include methods for studying $|V_{cb}|$ and $|V_{ub}|$, the strange quark mass, and the pion charge radius.

1. Introduction and History

Between the mid 1950s and late 1960s, a great deal of theoretical activity focused on attempting to solve (or at least severely constrain) problems of strong interaction physics using dispersion theory. An extensive and elegant body of work was developed to study the analyticity properties of form factors and scattering amplitudes. Eventually, however, when theorists believed they had reached the the limits of what could be gleaned from dispersive techniques, their attentions were drawn elsewhere: to the quark-parton model, to current algebra, and eventually to gauge theories, especially QCD.

The appeal of dispersion theory lies in its ability to incorporate in a completely rigorous and model-independent fashion those features shared by all well-defined field theories, namely, causality, unitarity, and crossing symmetry. Moreover, it works equally well in perturbative and nonperturbative regimes of the underlying dynamical theory. However, no specific Lagrangian is demanded by this scheme, and this lack of specificity acts as a double-edged sword: Without dynamical input, one can only deduce those consequences common to all possible dynamics. On the other hand, we now possess QCD, which is the fundamental, albeit unsolved, theory of strong interactions. A combination of the two, in which QCD inputs are inserted directly into dispersion relations, should yield a rich harvest of rigorous and model-independent bounds on hadronic quantities. In this talk we explore the implementation of this idea to the specific cases of weak and electromagnetic hadronic form factors.

Dispersion theory has been with us in particle physics for quite some time. The origin of the name traces directly back to the famous Kramers-Kronig relation[1] in electromagnetism, used to describe the dispersion of light in an arbitrary medium. The standard formula reads

$$\mathrm{Re}\, f(\omega) = \mathrm{Re}\, f(0) + \frac{\omega^2}{2\pi^2} P \int_0^\infty d\omega' \frac{\sigma_{\rm tot}(\omega')}{\omega'^2 - \omega^2}, \tag{1}$$

*Work supported by the Department of Energy under contract No. DE-AC05-84ER40150.

where f, ω, $\sigma_{\rm tot}$, and P denote, respectively, the forward scattering amplitude, frequency, total cross section, and principal value prescription to remove the denominator singularity. It follows that the dispersive ($\mathrm{Re}\, f$) and absorptive ($\mathrm{Im}\, f \sim \sigma_{\rm tot}$) amplitudes are intimately connected. This relation is derived by using causality and unitarity, which lead to restrictions on the analyticity properties of f in the space of complex ω. From there, the elegant theorems of complex analysis, especially Cauchy's theorem, provide the identities known as dispersion relations.

Since quantum mechanics and quantum field theory are expressed over the field of complex numbers, it is natural to expect that some variant of the dispersive approach should also exist in particle theory. Indeed, as early as 1951, Gell-Mann, Goldberger, and Thirring[2] described how causality and unitarity lead to a dispersion relation for the vacuum polarization two-point function of QED. A flurry of other dispersion relations followed in the literature, each presented with more or less rigor, depending upon assumptions about the analytic structure of the quantity under scrutiny; however, the particular dispersion relation used below is nothing more than the QCD version of the one first studied in 1951.

2. Formalism

We begin by defining the vacuum polarization tensor as the two-point current correlator in momentum space:

$$\Pi^{\mu\nu}(q) \equiv i \int d^4x\, e^{iqx} \langle 0 | \mathrm{T} J^\mu(x) J^{\dagger\nu}(0) | 0 \rangle, \tag{2}$$

Here J is some chosen current; since we are working with QCD, we choose it to be a quark bilinear. Moreover, we choose J to be a weak or electromagnetic, rather than gluonic, current, so that the individual (perturbative) current insertions are easier to identify. Suppressing for now the Lorentz indices $\mu\nu$, we would like to use Cauchy's theorem to write an expression

$$\Pi(q^2) = \frac{1}{2\pi i} \int_C dt\, \frac{\Pi(t)}{(t-q^2)}, \tag{3}$$

which relates Π at two different momentum arguments, q^2 and t. However, in order to do this, we must identify a closed contour C, inside of which Π is analytic in t. In the present case, causality implies that $\Pi(t)$ is analytic in t except on parts of the positive t axis, where $J^\dagger$ can create on-shell hadrons, which generates a discontinuity only in the imaginary (absorptive) part of Π. We choose C to consist of the lower and upper sides of this branch cut, together with the circle with $|t| \to \infty$; the latter contribution vanishes as long as $\Pi \to 0$ for large $|t|$. Then

$$\Pi(q^2) = \frac{1}{2\pi} \int_0^\infty dt\, \frac{\mathrm{Im}\,\Pi(t+i\epsilon)}{(t-q^2)} + \frac{1}{2\pi} \int_\infty^0 dt\, \frac{\mathrm{Im}\,\Pi(t-i\epsilon)}{(t-q^2)}. \tag{4}$$

Using the Schwarz reflection principle ($\Pi(z^*) = \Pi^*(z)$ if Π is real on some segment of the real axis, which is true for $t<0$ since there are no on-shell thresholds and hence is no imaginary part there), the two terms in (4) are equal:

$$\Pi(q^2) = \frac{1}{\pi}\int_0^\infty dt\,\frac{\operatorname{Im}\Pi(t+i\epsilon)}{(t-q^2)}. \tag{5}$$

If $\Pi(q^2)$ diverges, or the contribution from the circle $|t| \to \infty$ does not vanish, such terms may be removed through the process called "subtraction": Since the offending terms appear as coefficients of a polynomial in q^2, taking a sufficient number n of q^2 derivatives yields a finite result,

$$\frac{\partial^n \Pi(q^2)}{(\partial q^2)^n} = \frac{\partial^n \Pi_{\text{finite}}(q^2)}{(\partial q^2)^n}. \tag{6}$$

Then the expression for the dispersion relation reads

$$\Pi^{(n)}(q^2) \equiv \frac{1}{n!}\frac{\partial^n \Pi(q^2)}{(\partial q^2)^n} = \frac{1}{\pi}\int_0^\infty dt\,\frac{\operatorname{Im}\Pi(t+i\epsilon)}{(t-q^2)^{n+1}}. \tag{7}$$

Restoring the Lorentz indices and inserting a complete set of states between J and $J^\dagger$ (unitarity) yields

$$\operatorname{Im}\Pi^{\mu\nu}(t+i\epsilon) = \frac{1}{2}\sum_\Gamma \int d\Phi(\Gamma)(2\pi)^4\delta^4\left(t - \sum_\Gamma p\right)\langle 0|J^\mu|\Gamma\rangle\langle\Gamma|J^{\dagger\nu}|0\rangle, \tag{8}$$

where only on-shell states Γ with phase space Φ are included in the sum (a consequence of reducing the step functions in the time ordering). The matrix elements $\langle 0|J^\mu|\Gamma\rangle$ are nothing more than decay constants and form factors—pure hadronic quantities—while q^2 can be chosen so that $\Pi(q^2)$ can be evaluated directly in the fundamental theory of QCD. In particular, one chooses q^2 to be far from the hadronic (strong coupling) region, and then $\Pi(q^2)$ may be computed using an operator product expansion. A very useful observation due to Meiman[3] in 1963 is that the $\mu = \nu$ components of Eq. (8) are positive definite, meaning that each hadronic contribution serves only to saturate further the partonic (perturbative QCD) side of the dispersion relation. In this way one obtains a rigorous inequality between partonic and hadronic physics.

One path from Eqs. (7)–(8) leads to the famous QCD sum rules,[4] which study the saturation of the equality between the partonic and hadronic sides. We focus also on what this equality tells us about the behavior of matrix elements $\langle 0|J^\mu|\Gamma\rangle$. The first work to use the Meiman inequality with QCD inputs was by Bourrely, Machet, and de Rafael[5] in 1981.

As an explicit example, consider the pion electromagnetic form factor:

$$\langle \pi^+(p')|J^\mu_{\text{EM}}|\pi^+(p)\rangle = f(q^2)(p+p')^\mu, \tag{9}$$

where $q = p - p'$. Then Eq. (8) becomes

$$\operatorname{Im}\Pi^{ii}(t) \geq \frac{1}{48\pi}\left(t - 4m_\pi^2\right)^{3/2} t^{-1/2}|f(t)|^2\theta(t-4m_\pi^2), \tag{10}$$

while the partonic side, finite after two subtractions, is computed to be

$$\Pi^{ii\,(2)}(q^2) = \frac{1}{8\pi(-q^2)}\left\{1 + \frac{\alpha_s(q^2)}{\pi} + O\left[\left(\frac{\alpha_s^2(q^2)}{\pi}\right)^2\right] + n.p.\right\}, \tag{11}$$

where *n.p.* stands for nonperturbative corrections such as vacuum condensates. The combined inequality reads

$$\frac{1}{8\pi(-q^2)}\left\{1 + \frac{\alpha_s(q^2)}{\pi} + O\left[\left(\frac{\alpha_s^2(q^2)}{\pi}\right)^2\right] + n.p.\right\} \geq \frac{1}{48\pi^2}\int_{4m_\pi^2}^{\infty} dt\, \frac{(t-4m_\pi^2)^{3/2}}{t^{1/2}(t-q^2)^3}|f(t)|^2. \tag{12}$$

In general, one obtains an inequality of the form

$$\frac{1}{\pi}\int_{t_+}^{\infty} dt\, \frac{W_F(t)|F(t)|^2}{(t-q^2)^{n+1}} \leq \Pi^{(n)}(q^2), \tag{13}$$

where t_+ is the lowest threshold and W_F is a positive weighting factor arising from phase space and the quantum numbers of the form factor $F(t)$. As discussed by Okubo and Fushih[6] in 1971, it is very convenient to map the complex t plane with a cut for $t_+ \leq t < +\infty$ to the unit disc using a complex kinematic variable z:

$$z(t;t_s) \equiv \frac{\sqrt{t_+ - t} - \sqrt{t_+ - t_s}}{\sqrt{t_+ - t} + \sqrt{t_+ - t_s}}, \tag{14}$$

which maps the upper (lower) side of the cut to the lower (upper) half of the unit circle. The parameter[7] $t_s < t_+$ is chosen later for convenience. One then defines a weighting function

$$\phi_F(t;t_s) = \sqrt{\frac{W_F(t)}{\Pi^{(n)}(q^2)(t-q^2)^{n+1}|dz(t;t_s)/dt|}}, \tag{15}$$

which is analytic inside the unit circle, and in terms of which the dispersive bound reads

$$\frac{1}{2\pi i}\oint \frac{dz}{z}\,|\phi_F(z)F(z)|^2 \leq 1. \tag{16}$$

If any subthreshold poles remain inside the unit circle at points $z = z_p$, they may be removed by means of so-called Blaschke factors,

$$P_F(z) = \prod_p \frac{z - z_p}{1 - z_p^* z}. \tag{17}$$

$P_F(z)$ has the feature that for $|z| = 1$, $|P_F(z)| = 1$, so that the dispersive bound is unchanged,

$$\frac{1}{2\pi i}\oint \frac{dz}{z}\,|\phi_F(z)P_F(z)F(z)|^2 \leq 1, \tag{18}$$

and $\phi_F(z)P_F(z)F(z)$ is analytic on the whole unit disc. Crossing symmetry relates the form factor in all kinematic regimes by analytic continuation. We have thus isolated the analytic structure of the form factor,

$$F(t) = \frac{1}{P_F(t)\phi_F(t;t_s)} \sum_{n=0}^{\infty} a_n z^n(t;t_s), \tag{19}$$

where the coefficients a_n are unknown; however, inserting Eq. (19) back into (18) gives

$$\sum_{n=0}^{\infty} |a_n|^2 \leq 1. \tag{20}$$

Equations (19) and (20) first appeared[8] in 1995, and re-express in a very compact and explicit notation all of the analyticity, unitarity, and explicit QCD information implicit in Eqs. (7) and (8). Since ϕ_F and P_F are known functions, the form factor is known except for a set of parameters a_n, each of which must be less than unity in magnitude. A randomly chosen shape for a form factor would almost inevitably have some $|a_n| > 1$, and thus would be disallowed by the dispersive bound (20).

One more point that makes the model-independent parameterization (19) useful is that for spacelike and semileptonic processes, the allowed kinematic range for z tends to have $|z| \ll 1$. Indeed, the parameter t_s is chosen to enhance this effect. For example, for $\bar{B} \to D\ell\bar{\nu}$, $|z| < 0.03$. This means that the convergence of Eq. (19) is geometrically fast, and only the first few a_n's are relevant to the shape of the form factor. The theoretical uncertainty incurred by ignoring the other, infinite set of a_n's is called "truncation error,"[8] and falls off geometrically fast with the number of a_n's used to parameterize the form factor.

3. A Gallery of Results

1) *$|V_{ub}|$ and $|V_{cb}|$*. The need for a parameterization describing all solutions of Eqs. (7)–(8) was recognized[9] in studies of the $\bar{B} \to \pi\ell\bar{\nu}$ form factor, useful for the extraction of $|V_{ub}|$. There it was seen that the inclusion of each (at that time, hypothetical) form factor data point served to decrease the region allowed by the dispersion relation geometrically fast (Fig. 3). Similar comments apply to using points from a lattice simulation.[11]

The model-independent form factor parameterization Eq. (19) was first used[8] to extrapolate measured $\bar{B} \to D^{(*)}\ell\bar{\nu}$ form factor data to a point where phase space vanishes. In order to extract $|V_{cb}|$ from the form factor* in the differential width

$$\frac{d\Gamma}{dq^2}(\bar{B} \to D^*\ell\bar{\nu}) \propto |V_{cb}|^2 \left|F(q^2)\right|^2 \sqrt{(M_B + M_{D^*})^2 - q^2}, \tag{21}$$

one must separate $|V_{cb}|$ from $|F(q^2)|$. The normalization of F, namely, $F(q^2 = (M_B - M_{D^*})^2) = 1$ (up to small corrections), is determined by the heavy quark limit.[10] However, Eq. (21) shows that phase space vanishes at exactly this q^2; therefore, an extrapolation is needed. Previously, experimental measurements of the form factor used an

*Strictly speaking, 4 form factors contribute; however, in the limit of heavy quarks, each one either vanishes or is proportional to a single "Isgur-Wise" form factor.[10]

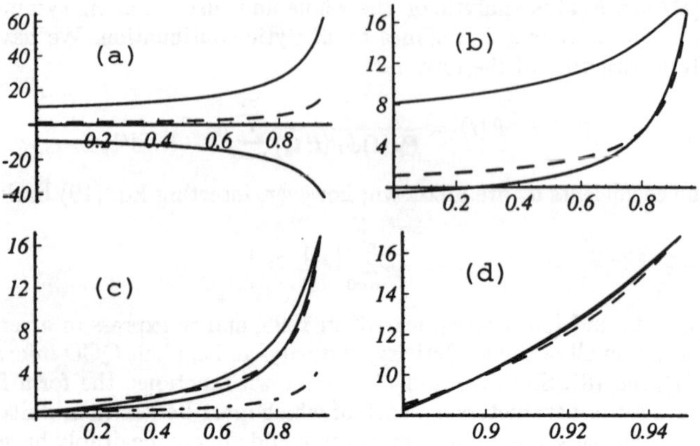

Fig. 1. Bounds on the $\bar{B} \to \pi \ell \bar{\nu}$ form factor f using the dispersive bound and fixing zero, one, and two points in (a), (b), and (c), respectively. Dashed lines indicate pole dominance models. (d) shows how certain choices of B^* pole parameters can violate the dispersive bounds.

Experiment	Process	$\lvert V_{cb}\rvert \cdot 10^3$
CLEO[13]	$\bar{B} \to D^* \ell \bar{\nu}$	$36.9^{+2.0}_{-2.1}$
CLEO[16]	$\bar{B} \to D \ell \bar{\nu}$	44.8 ± 6.1
ALEPH[13]	$\bar{B} \to D^* \ell \bar{\nu}$	31.9 ± 2.4
ALEPH[13]	$\bar{B} \to D \ell \bar{\nu}$	29.2 ± 7.3
DELPHI[17]	$\bar{B} \to D^* \ell \bar{\nu}$	38.0 ± 1.3

Table 1. Determinations of $|V_{cb}|$ using Eqs. (19) and (20). Footnotes reference the source of the fit. Uncertainties are statistical plus theoretical.

ad hoc linear or quadratic extrapolation, which implies a theoretical uncertainty of unknown size (see Fig. 2a). Using (19) and (20) removes this uncertainty, and subsequent work[12,13] refined the analysis to the point that it is used by both theorists and the experimental groups themselves (Table 1).

2) *Strange quark mass.* $K_{\ell 3}$ decays possess two form factors, one of which appears with the coefficient m_ℓ^2 in the rate and is called the scalar form factor. The corresponding $\Pi^{(n)}(q^2)$, evaluated deep in the Euclidean region, is proportional to $(m_s - m_u)^2$, *i.e.*, is sensitive to m_s. One can invert the program of 1), so that a large amount of form factor data, thus delineating its shape, is used[14] to constrain the function $\Pi^{(n)}(q^2)$ and hence m_s. Indeed, one finds that $F(t) \propto m_s a_n$, so that (20) implies a rigorous lower bound on m_s.

Currently, not enough data exists in the world sample for such a determination, although DAΦNE expects to increase the available pool many times over. Until such data exists, one may apply the results of a model, or better, a chiral perturbation theory

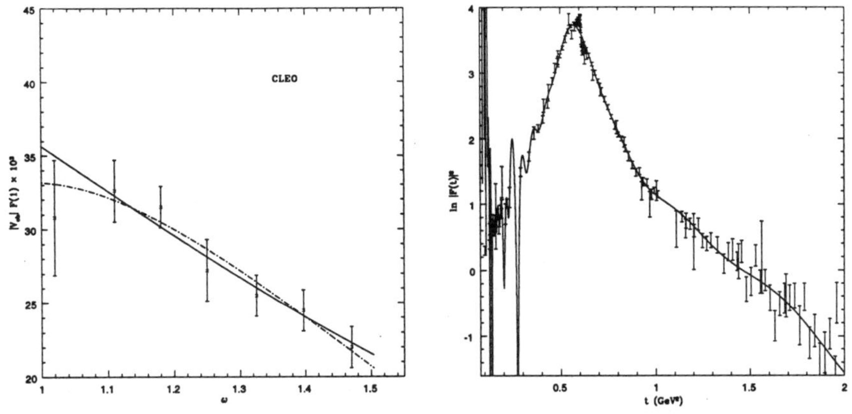

Fig. 2. (*a*). Fit to CLEO data for the $\bar{B} \to D^* \ell \bar{\nu}$ form factor using the model-independent parameterization (19) with (solid) and without (dashed) the dispersive constraint Eq. (20). The eye prefers the dashed curve, but it is forbidden by QCD. (*b*). Fit to timelike pion electromagnetic form factor data. Contrast the smoothness of the fit above $t = 0.4$ GeV^2 to the oscillations below.

(χPT) calculation to compute the required shape parameters a_n. Believing χPT to a level of 5%, 1%, or 1/20% leads to lower bounds on $m_S^{\overline{\rm MS}}(1 \text{ GeV})$ of 40, 90, and 140 MeV, respectively.

3) *Pion form factor.* The parameterization (19) exhibits geometric convergence for $t < t_+$ ($|z| < 1$). On the other hand, one often possesses data directly on the cut $t > t_+$ ($|z| = 1$), the timelike region. Does (19) have anything to say about this region? Although one must be much more careful about convergence, the answer[15] appears to be yes. Theorems of complex convergence plus knowledge of asymptotic ($t \to \infty$) properties of form factors allow one to use (19) even in the timelike region.

For example, for the pion electromagnetic form factor, one obtains the fit of Fig. 2*b*. The presence of the ρ peak is not put in by hand, but simply emerges from fitting to (19). Note, however, the wild oscillations for $t < 0.4$ GeV^2; one can show that these occur due to large gaps in the data for $\theta > \pi/2$ on the unit circle $|z| = 1$. These large oscillations persist when one analytically continues into the spacelike region, where one obtains

$$|F(t=0)| = 2.56 \pm 2.00, \quad \langle r^2 \rangle = 2.66 \pm 3.44 \, \text{fm}^2, \tag{22}$$

which are rather loose bounds, considering that, *e.g.*, $|F(t=0)| = 1$ by charge conservation. This points to the well-known problem of the instability of analytic continuation of discrete timelike data to the spacelike region; now, however one can quantify exactly how unstable this continuation is.

One can also proceed directly with spacelike pion form factor data alone, where the geometric convergence of (19) is restored. Then one finds,[15] using this model-independent parameterization, $\langle r^2 \rangle = 0.480 \pm 0.020$ fm^2, a few σ larger than the usual numbers quoted in the literature ($\simeq 0.42$ fm^2), which use *ad hoc* parameterizations.

4. Conclusions

Dispersive techniques provide an elegant, rigorous bridge between hadronic and elementary quantities. For semileptonic or electromagnetic decays, they provide a model-independent, rapidly convergent parameterization of form factors. We have seen that a number of different problems have already been studied using this method. Nucleon form factors, the K charge radius, and improvement of timelike form factors are obvious future directions. The reader can doubtless imagine many others.

References

1. R. Kronig, J. Op. Soc. Am., **12**, 547 (1926); H. A. Kramers, Atti. Congr. Intern. Fisici, **2**, 545 Como (1927).
2. M. Gell-Mann, M. L. Goldberger, and W. Thirring, Phys. Rev. **95**, 1612 (1951).
3. N. N. Meiman, Sov. Phys. JETP **17**, 830 (1963).
4. For a collection of seminal works, see M. A. Shifman (editor), *Vacuum Structure and QCD Sum Rules*, North-Holland, New York, 1992.
5. C. Bourrely, B. Machet, and E. de Rafael, Nucl. Phys. B **189**, 157 (1981).
6. S. Okubo and I. Fushih, Phys. Rev. D **4**, 2020 (1971).
7. C. G. Boyd and R. F. Lebed, Nucl. Phys. B **485**, 275 (1997).
8. C. G. Boyd, B. Grinstein, and R. F. Lebed, Phys. Lett. B **353**, 306 (1995).
9. C. G. Boyd, B. Grinstein, and R. F. Lebed, Phys. Rev. Lett. **74**, 4603 (1995).
10. N. Isgur and M. B. Wise, Phys. Lett. B **232**, 113 (1989); B **237**, 527 (1990).
11. L. Lellouch, Nucl. Phys. B **479**, 353 (1996).
12. C. G. Boyd, B. Grinstein, and R. F. Lebed, Nucl. Phys. B **461**, 493 (1996); I. Caprini, L. Lellouch, and M. Neubert, Nucl. Phys. B **530**, 153 (1998).
13. C. G. Boyd, B. Grinstein, and R. F. Lebed, Phys. Rev. D **56**, 6895 (1997).
14. R. F. Lebed and K. Schilcher, Phys. Lett. B **430**, 341 (1998).
15. W. W. Buck and R. F. Lebed, Phys. Rev. D **58**, 056001 (1998).
16. CLEO Collaboration (J. Bartelt *et al.*), Phys. Rev. Lett. **82**, 3746 (1999).
17. DELPHI Collaboration (M. Margoni and F. Simonetto, spokespersons), Report No. DELPHI 99-107 CONF 294, submitted to the HEP '99 Conference, Tampere, Finland, 15–21 July, 1999.

HARD EXCLUSIVE PHOTOPRODUCTION OF Φ AND J/Ψ MESONS [a]

C. F. BERGER
Department of Physics and Astronomy, SUNY Stony Brook
Stony Brook, NY 11794-3800, USA
E-mail: carola.berger@sunysb.edu

W. SCHWEIGER
Institut für Theoretische Physik, Universität Graz, Universitätsplatz 5
A-8010 Graz, Austria
E-mail: wolfgang.schweiger@kfunigraz.ac.at

We consider the reaction $\gamma p \to M p$, with M being either a Φ or a J/Ψ meson, within perturbative QCD treating the proton as a quark-diquark system. The phenomenological couplings of gauge bosons to (spatially extended) diquarks and the quark-diquark distribution amplitude of the proton are adopted from previous investigations of baryon form factors and two-photon processes. Going beyond leading order, we take into account hadron-mass effects by a systematic expansion in the small parameter (hadron mass/ photon energy). With the meson distribution amplitudes taken from the literature our predictions for the differential cross section at $|t| \gtrsim 3\,\mathrm{GeV}^2$ seem to provide a reasonable extrapolation of the low-t data in case of the Φ. They fail, however, completely for the J/Ψ. A closer inspection reveals that "hadron-mass corrections" become dominant for t values of only a few GeV in case of the J/Ψ. This indicates that one has to go to much larger energies and momentum transfers in order to achieve a reliable description of J/Ψ production within the hard-scattering approach.

1 Introduction

The forthcoming data from the 93-031 experiment at CEBAF will provide precise information on exclusive photoproduction of vector mesons (ρ, ω, Φ) in the momentum-transfer range $1 \lesssim |t| \lesssim 5$ GeV^2. For this kinematics non-perturbative (vector-meson dominance) and perturbative (quark and gluon exchange) production mechanisms are supposed to compete. From the perturbation-theoretical viewpoint the Φ channel is certainly the cleanest and simplest of these three production channels. The valence Fock state of the Φ, i.e. the s-$\bar{s}$ state, can only be produced via two-gluon exchange. Quark exchange is OZI suppressed if the strangeness content of the nucleon is small. Both quark and gluon exchanges are possible, however, in the case of the ρ and the ω. In the present contribution we are interested in describing Φ production in the

[a] Talk given by W. Schweiger

perturbative regime. We employ a modified version of the hard-scattering approach (HSA) [1] in which the proton is considered as a quark-diquark system. This kind of perturbative model has already been applied successfully to a number of exclusive photon-induced reactions [2-5]. A consistent description of baryon electromagnetic form factors, Compton scattering off baryons, etc. has been achieved in the sense that the corresponding large momentum-transfer data ($p_\perp^2 \gtrsim 3$ GeV^2) are reproduced with the same set of model parameters. In addition to Φ production we investigate whether the same model could be applied to $\gamma p \to J/\Psi\, p$ in the few-GeV momentum-transfer region, where experimental measurements seem to be feasible in the future. In the following we start with a short outline of the diquark model, say a few words about helicity amplitudes for photoproduction of vector mesons and their scaling behavior within our model. Finally, we compare our differential cross-section results for photoproduction of Φs and J/Ψs with the existing (low-t) data and with predictions of other models and draw our conclusions.

2 The Hard-Scattering Approach with Diquarks

Within the hard-scattering approach a helicity amplitude $M_{\{\lambda\}}$ for the reaction $\gamma p \to M\, p$ is (to leading order in $1/p_\perp$) given by the convolution integral [1]

$$M_{\{\lambda\}}(\hat{s},\hat{t}) = \int_0^1 dx_1 dy_1 dz_1 \phi^{M\dagger}(z_1,\tilde{p}_\perp)\phi^{p\dagger}(y_1,\tilde{p}_\perp)\widehat{T}_{\{\lambda\}}(x_1,y_1,z_1;\hat{s},\hat{t})\phi^{p}(x_1,\tilde{p}_\perp)\,. \tag{1}$$

The distribution amplitudes (DAs) ϕ^H are probability amplitudes for finding the valence Fock state in the hadron H with the constituents carrying certain fractions of the momentum of their parent hadron and being collinear up to a factorization scale $\tilde{p}_\perp$. In our model the valence Fock state of an ordinary baryon is assumed to consist of a quark (q) and a diquark (D). We fix our notation in such a way that the momentum fraction appearing in the argument of ϕ^H is carried by the quark – with the momentum fraction of the other constituent (either diquark or antiquark) it sums up to 1 (cf. Fig. 1). For our actual calculations the (logarithmic) $\tilde{p}_\perp$ dependence of the DAs is neglected throughout since it is of minor importance in the restricted energy and momentum-transfer range we are interested in. The hard scattering amplitude $\widehat{T}_{\{\lambda\}}$ is calculated perturbatively in collinear approximation and consists in our particular case of all possible tree diagrams contributing to the elementary scattering process $\gamma\, q\, D \to Q\,\bar{Q}\, q\, D$. Eight of the, altogether, sixteen diagrams which contribute to photoproduction of (heavy) quarkonia are depicted in Fig. 2. The subscript $\{\lambda\}$ of $\widehat{T}$ represents a set of possible photon, proton and vector-meson helicities. The Mandelstam variables s and t are written

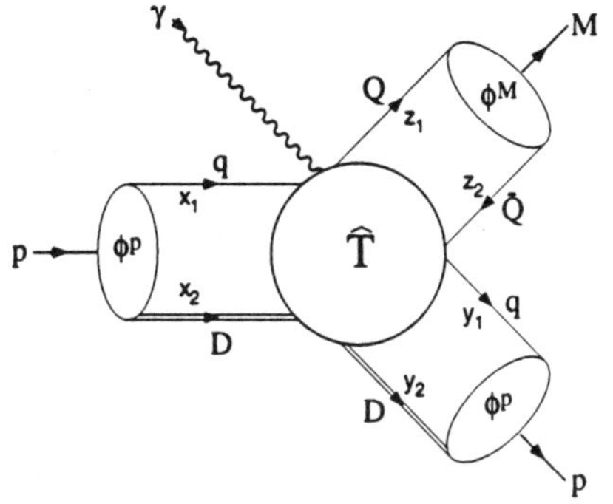

Figure 1: Graphical representation of the hard-scattering formula, Eq. (1), for $\gamma p \to M p$. x_i, y_i, and z_i denote longitudinal momentum fractions of the constituents.

with a hat to indicate that they are defined for vanishing hadron masses. Our calculation of the hard-scattering amplitude involves an expansion in powers of $(m_H/\sqrt{\hat{s}})$ which is performed at fixed scattering angle. We keep only the leading order and next-to-leading order terms in this expansion. Hadron masses, however, are fully taken into account in flux and phase-space factors.

The model, as applied in Refs. 2-5, comprises scalar (S) as well as axial-vector (V) diquarks. V-diquarks become relevant if one wants to describe spin observables which require the flip of baryonic helicities. For the Feynman rules of electromagnetically and strongly interacting diquarks, as well as for the choice of the quark-diquark DAs of octet baryons we refer to Ref. 2. Here it is only important to mention that the composite nature of diquarks is taken into account by multiplying each of the Feynman diagrams entering the hard-scattering amplitude with diquark form factors. These are parameterized by multipole functions with the power chosen in such a way that in the limit $p_\perp \to \infty$ the scaling behavior of the pure quark HSA is recovered. We restrict the (running) strong coupling constant α_s to be smaller than 0.5. For further details of the diquark model we refer to the publication of R. Jakob et al. [2]. The numerical values of the model parameters for the present study are also taken from this paper.

The only new ingredient which appears in the photoproduction of mesons is the meson DA. A model for the Φ-meson DA which fulfills QCD sum-rule constraints has been proposed by Benayoun and Chernyak [6]. In order to treat Φ and J/Ψ production on the same footing we adapt this DA by attaching an exponential factor which provides a flavor dependence. The effect of this factor is negligible for Φs, but it becomes crucial for J/Ψs. The form of the exponen-

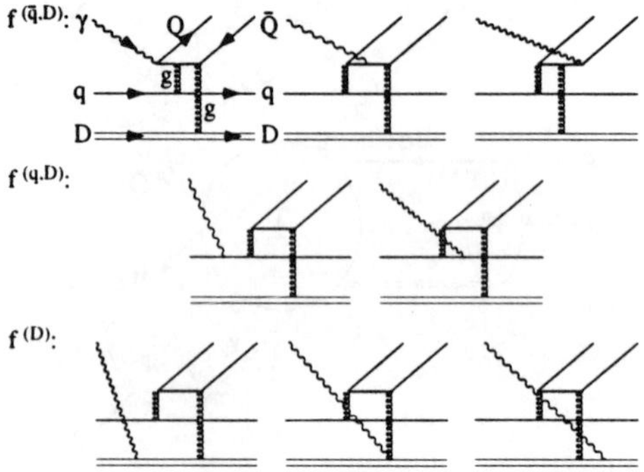

Figure 2: Eight of the, altogether, sixteen tree-level diagrams which contribute to the elementary process $\gamma\, q\, D \to Q\, \bar{Q}\, q\, D$. The remaining diagrams are obtained by interchanging the two gluons. We have also indicated how the diagrams are grouped to gauge invariant expressions.

tial factor is inspired by a simple relativistic treatment of heavy mesons [7] and provides a flavor dependence in accordance with heavy-quark effective theory. For longitudinally polarized Φs and J/Ψs the DA reads

$$\phi^M(z_1, \lambda_\Phi = 0) \propto z_1(1 - z_1) \exp\left[\tilde{b}^2 m_M^2 (z_1 - 0.5)^2\right] . \tag{2}$$

The DA for transversially polarized Φs and J/Ψs contains an additional polynomial in z_1 which enhances the maximum at $z_1 = 0.5$ and makes it narrower (cf. Ref. 6). For the oscillator parameter $\tilde{b}$ we take a value of 0.97 in accordance with estimates for the radius of the J/Ψ meson (see, e.g., the discussion in Ref. 8). The absolute "normalization" of the DA, Eq. (2), is related to the experimentally determinable decay constant of the Φ and J/Ψ meson, respectively.

At the end of this section we want to make two technical remarks. The first remark concerns the numerical treatment of the convolution integral, Eq. (1), which is not straightforward since propagator singularities are encountered in the range of integration. We have carefully separated the propagator singularities and integrated analytically so that the remaining integrals could be performed by a rather fast fixed-point Gaussian quadrature. For details of this integration procedure we refer to Ref. 5. The second remark concerns the treatment of hadron masses. These are taken into account in the hard-scattering amplitudes by assigning to every constituent of the hadron H an effective mass $x m_H$, where x is the fraction of the four-momentum of the hadron H carried by the constituent. Keeping in mind that the momentum fractions are weighted by the hadron DA in the convolution integral, Eq. (1), this means that, e.g., the quarks acquire (on the average) a mass which is rather the mass of a constituent quark than a current quark. This may be interpreted in such a way that at intemediate momentum transfers a quark surrounded by a cloud of $q\bar{q}$ pairs and gluons is not resolved but rather acts like a single particle with a corresponding mass. For a more detailed explanation of our treatment of mass-effects which, by the way, also guarantees the gauge invariance of the mass-correction terms we refer to Ref. 9.

3 Helicity Amplitudes

For exclusive photoproduction of vector mesons $\gamma p \to M p$ one finds, altogether, 24 complex helicity amplitudes. By virtue of parity invariance only 12 of these helicity amplitudes are independent. Following the notation of Ref. 10 we denote them by

$$H_{1,\lambda_\phi} = M_{\lambda_\phi,\lambda_f=+1/2,\lambda_\gamma=1,\lambda_i=-1/2}\,, \quad H_{2,\lambda_\phi} = M_{\lambda_\phi,\lambda_f=+1/2,\lambda_\gamma=1,\lambda_i=+1/2}\,,$$
$$H_{3,\lambda_\phi} = M_{\lambda_\phi,\lambda_f=-1/2,\lambda_\gamma=1,\lambda_i=-1/2}\,, \quad H_{4,\lambda_\phi} = M_{\lambda_\phi,\lambda_f=-1/2,\lambda_\gamma=1,\lambda_i=+1/2}\,, \tag{3}$$

with $\lambda_\phi = 0, \pm 1$. For our normalization of the helicity amplitudes the unpolarized differential cross section takes on the form

$$\frac{d\sigma}{dt} = \frac{1}{32\pi(s-m_p^2)^2} \sum_{\lambda_\phi=0,\pm1} \sum_{i=1}^{4} |H_{i,\lambda_\phi}|^2 \,. \tag{4}$$

Within the HSA the energy dependence of the helicity amplitudes at fixed cm angle and large s is roughly

$$\begin{aligned}
&H_{2,0}\;,\; H_{3,0} \propto s^{-5/2}\,,\\
&H_{1,0}\;,\; H_{4,0}\,,\, H_{2,1}\,,\, H_{3,1},\, H_{2,-1}\,,\, H_{3,-1} \propto s^{-3}\,,\\
&H_{1,1}\;,\; H_{4,1},\, H_{1,-1}\,,\, H_{4,-1} \propto s^{-7/2}\,,
\end{aligned} \tag{5}$$

depending on whether the helicity of the hadronic constituents is conserved or flipped by one or two units. This scaling behavior is modified by logarithms due to the running coupling α_s and, eventually, the evolution of the hadron distribution amplitudes. Further deviations are due to the diquark form factors. For small momentum tranfers diquarks appear nearly point-like and the decay behavior of the helicity amplitudes is weakened. The energy dependence of Eqs. (5) is only approached at large enough momentum transfer, where the diquark form factors become fully operational.

If one neglects mass effects altogether, the helicity-flip amplitudes vanish and only the hadron-helicity conserving amplitudes $H_{2,0}$ and $H_{3,0}$ are non-zero. Due to (anti)symmetry properties of the Feynman graphs under interchange of the momentum fractions z_1 and z_2, only diagrams in which the photon couples to the heavy quark contribute in the massless case. Mass corrections suppressed by the order $(m_H/\sqrt{\hat{s}})$ do not contribute to the non-flip amplitudes, but provide for non-vanishing single-flip amplitudes. In order to get a non-vanishing result for the double-flip amplitudes it would even be necessary to consider mass corrections suppressed by order $(m_H/\sqrt{\hat{s}})^2$. Since such corrections would also show up in the helicity-conserving amplitudes and we do not want to make things too complicated, we neglect them completely.

4 Results and Conclusions

Fig. 3 shows (unpolarized) differential cross sections for the photoproduction of Φ (left) and J/Ψ (right) mesons. In the absence of large-t data we are not able to directly confront the diquark-model predictions (solid line) with experiment. However, comparing with the experimental points which are only available up to $t \lesssim 1.5$ GeV^2, one can recognize that our results provide a reasonable extrapolation in case of the Φ, but seem to fail completely for the J/Ψ. This failure can be traced back to the "hadron-mass corrections" which enter the cross section only via the hadron-helicity-flip amplitudes (apart from flux and phase-space factors). Whereas the two hadron-helicity conserving amplitudes provide already about 60% of the cross section in case of the Φ, they are negligible in case of the J/Ψ (solid line vs. long-dashed line). This means for the J/Ψ that hadron-mass effects become dominant for the kinematics considered in the plot. Thus they are not reliably estimated by keeping only the lowest order terms in a mass expansion. By comparing the kinematics of the two plots in Fig. 3 one observes that the relevant parameter of our mass expansion, namely $(m_M/\sqrt{\hat{s}})$, is in both cases small. Both reactions are considered within the same momentum-transfer range, but within very different angular ranges. $|t|$ values between 3 and 6 GeV^2 correspond to a scattering angle between 70 and 110 degrees for a photon lab-energy of 6 GeV, but to nearly-forward scattering (between 10 and 20 degrees) for a photon lab-energy of 150 GeV. Keeping in mind that our mass expansion is an expansion in the parameter $(m_H/\sqrt{\hat{s}})$ at **fixed** angle, the different magnitude of mass corrections in Φ and J/Ψ production becomes clearer. The expansion coefficient of the first mass-correction term is angular dependent and large for small scattering angles. This indicates that perturbation theory for J/Ψ production is only trustworthy at large enough energies and scattering angles.

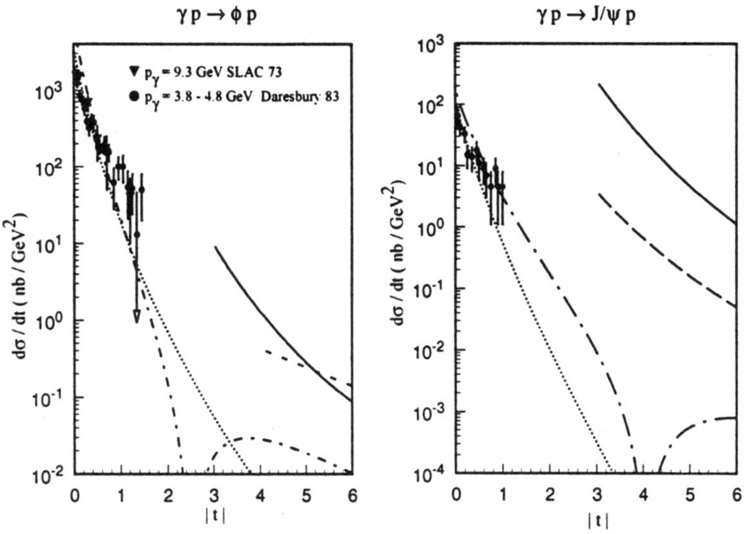

Figure 3: Differential cross section for the photoproduction of the ϕ-p (*left figure*) and the J/Ψ-p (*right figure*) final states for photon lab-energies of 6 and 150 GeV, respectively. The solid line corresponds to the diquark-model prediction. Cross sections resulting from the Pomeron-exchange mechanism [11] (dotted line) and a two-gluon-exchange model [12] (dash-dotted line) are plotted for comparison. The short-dashed line in the left figure represents an attempt to include correlations into the two-gluon-exchange model [13]. The long-dashed line in the right figure has been obtained within the diquark model by neglecting mass corrections, i.e. by taking into account only the hadron-helicity conserving amplitudes $H_{2,0}$ and $H_{3,0}$. ϕ-data are taken from Ballam et al. [14] and Barber et al. [15]. The J/Ψ data points have been obtained by Binkley et al. [16] by averaging the measured cross sections over photon lab-energies between 60 and 300 GeV.

The problem that Φ production is well reproduced, whereas J/Ψ production is overestimated does not only occur within our model for the hard-scattering region, but is also encountered if one describes the diffractive scattering by means of a pomeron-exchange mechanism [11] (in which the pomeron directly couples to the quarks) or the two-gluon-exchange model proposed by Laget and Mendez-Galain [12]. The latter model represents an attempt to describe the transition from diffractive to hard scattering. Within this model the two gluons may be considered as a remnant of the pomeron. For both models the situation in J/Ψ production is improved by arguing that the coupling of the pomeron to a charmed quark is weaker than the coupling to a strange quark. Predictions of these models are also depicted in Fig 3. The two-gluon-exchange model exhibits a characteristic node which, however, is completely washed out if correlations between the protonic constituents are taken into account, i.e. if the two gluons are allowed to couple to different constituents of the proton. For Φ production and $|t| \gtrsim 4\ \mathrm{GeV}^2$ our results and the results of the two-gluon-exchange model **with** correlations become comparable. This is not surprising, since correlations are automatically contained in our HSA-based model.

To conclude, within our perturbative diquark model we are not able to make reliable predictions for exclusive photoproduction of J/Ψ mesons in the few-GeV momentum-transfer range where experimental data can be expected. The reason is that effects due to the charm-quark mass are only reliably estimated (within our approach) at very large energies and momentum tranfers. For exclusive photoproduction of Φ mesons, however, our results provide a reasonable extrapolation of the low-t data and are also comparable with a two-gluon exchange model which continues the pomeron exchange for diffractive scattering to higher momentum transfers. CEBAF, or even better, an upgrade of CEBAF, could help to figure out how different mechanisms (pomeron exhange $\rightarrow$ two-gluon exchange $\rightarrow$ hard-scattering mechanism) in high-energy photoproduction evolve when going from low to high t. Our model will meet the first severe test in the near future when the data analysis of the CEBAF 93-031 experiment for $\gamma p \rightarrow \Phi p$ will be completed.

References

1. See, e.g., S. J. Brodsky and G. P. Lepage in *Perturbative Quantum Chromodynamics*, ed. A. H. Mueller (World Scientific, Singapore, 1989).
2. R. Jakob et al., *Z. Phys.* A **347**, 109 (1993).
3. P. Kroll et al., *Phys. Lett.* B **316**, 546 (1993).
4. P. Kroll et al., *Nucl. Phys.* A **598**, 435 (1996).
5. P. Kroll et al., *Phys. Rev.* D **55**, 4315 (1997).
6. M. Benayoun and V.L. Chernyak, *Nucl. Phys.* B **329**, 285 (1990).
7. M. Wirbel, B. Stech, and M. Bauer, *Z. Phys.* C **29**, 637 (1985).
8. T. Feldmann and P. Kroll, *Phys. Lett.* B **413**, 410 (1997).
9. C.F. Berger and W. Schweiger, paper in preparation.
10. M. Pichowsky, Ç. Şavkli, F. Tabakin, *Phys. Rev.* C **53**, 593 (1996).
11. A. Donnachie and P.V. Landshoff, *Phys. Lett.* B **185**, 403 (1987).
12. J.-M. Laget and R. Mendez-Galain, *Nucl. Phys.* A **581**, 397 (1995).
13. J.-M. Laget, preprint DAPNIA/SPhN-98-54 (1998).
14. J. Ballam et al., *Phys. Rev.* D **5**, 3150 (1973).
15. D.P. Barber et al., *Z. Phys.* C **12**, 1 (1983).
16. M. Binkley et al., *Phys. Rev. Lett.* **48**, 73 (1982).

THE $\gamma^*\gamma \to \pi^0$ TRANSITION IN A BOUND-STATE APPROACH

DUBRAVKO KLABUČAR
Physics Department, Faculty of Science, POB 162, University of Zagreb, Croatia

DALIBOR KEKEZ
Rudjer Bošković Institute, POB 1016, 10001 Zagreb, Croatia

We present a recent treatment of the $\pi^0\gamma^*\gamma$ transition form factor in the coupled Schwinger-Dyson and Bethe-Salpeter approach to stress the desirability of measuring it at CEBAF.

The form factor $T_{\pi^0}(-Q^2, 0)$ for the transition $\gamma^*(k)\gamma(k') \to \pi^0(p)$ (where $k^2 = -Q^2 \neq 0$ is the momentum-squared of the spacelike off-shell photon γ^*), is usually thought to be almost completely understood on the basis of the perturbative QCD (pQCD) over the whole range of the momenta-squared covered by the new CLEO data [1] (1.5 GeV$^2 \lesssim Q^2 \lesssim 9$ GeV^2). However, Radyushkin and Ruskov's [2] QCD sum rule analysis indicates the presence of large nonperturbative contributions even for the largest of these presently accessible momenta. This motivated us to address in Ref. [3] the transition form factor in the coupled Schwinger–Dyson (SD) and Bethe–Salpeter (BS) approach in all spacelike-momentum regimes, from very large Q^2 all the way to $Q^2 = 0$. The $Q^2 = 0$ limit corresponds to the π^0 decay into two real photons ($k^2 = k'^2 = 0$) explained by the Abelian axial *alias* Adler-Bell-Jackiw (ABJ) anomaly, which is usually difficult to incorporate into a bound-state approach [4]. Fortunately, as shown already in Refs. [5,6] and used in the first analysis of $T_{\pi^0}(-Q^2, 0)$ in a SD approach [7], such approaches which rely on the chirally well–behaved SD (and BS) equations for the light pseudoscalar mesons ($\pi, K, \eta, ...$) while respecting the Ward-Takahashi identities (WTI) of QED, reproduce (in the chiral and soft limit) the fundamental anomaly result

$$T_{\pi^0}(0,0) = \frac{1}{4\pi^2 f_\pi} \tag{1}$$

naturally and without any fine tuning. This is an advantage with respect to pQCD approaches, which have problems at low Q^2. (*E.g.*, see Ref. [8].)

In the present coupled SD-BS approach [4,9,10], the anomalous amplitude (1) is also obtained model-independently and without imposing any requirements on the solutions for the π^0 wave function or the quark propagator solutions (*e.g.*, see [5,6]). This is exactly as it should be, since the anomaly – and its result (1) – must not depend on the internal structure of the pion. Subsequently, in Ref. [3], we showed that the SD-BS approach to modeling QCD provides a good description for *both* low and high values of Q^2. We thereby obtained the behavior similar to the Brodsky–Lepage interpolation formula $T_{\pi^0}(-Q^2,0) = (1/4\pi^2 f_\pi) / (1+Q^2/8\pi^2 f_\pi^2)$ proposed [11] as a desirable behavior for $T_{\pi^0}(-Q^2,0)$ as it reduces to the ABJ anomaly amplitude (1) at $Q^2=0$, while agreeing with the following type of leading behavior for large Q^2:

$$T_{\pi^0}(-Q^2,0) = \mathcal{J}\,\frac{f_\pi}{Q^2} \qquad (\mathcal{J} = \textit{constant} \text{ for large } Q^2), \tag{2}$$

which is favored both experimentally [1] and theoretically (*e.g.*, [2,8,11,12]).

In the coupled SD-BS approach, the BS equation for the pion bound-state $q\bar{q}$ vertex $\Gamma_{\pi^0}(q,p)$ employs the dynamically dressed quark propagator $S(k) = [A(k^2)\not{k} - B(k^2)]^{-1}$, obtained by solving its SD equation. Then, in the case of light pseudoscalars, the $q\bar{q}$ bound states are simultaneously also the (pseudo-)Goldstone bosons of dynamical chiral symmetry breaking (DχSB).

Following Jain and Munczek [13,14], we adopt the ladder-type approximation employing bare quark–gluon–quark vertices but dressed propagators. For the gluon propagator we use an effective, (*partially*) modeled one in Landau-gauge [13,14], given by $G(-l^2)(g^{\mu\nu} - l^\mu l^\nu/l^2)$. (This Ansatz is often called the "Abelian approximation" [15].) The effective propagator function G is the sum of the perturbative contribution G_{UV} and the nonperturbative contribution G_{IR}: $G(Q^2) = G_{UV}(Q^2) + G_{IR}(Q^2)$, $(Q^2 = -l^2)$. The perturbative part G_{UV} is required to reproduce correctly the ultraviolet (UV) asymptotic behavior that unambiguously follows from QCD in its high–energy regime. Therefore, G_{UV} must be given by $\alpha_s(Q^2)/Q^2$, where the running coupling constant $\alpha_s(Q^2)$ is well-known from pQCD, so that G_{UV} is *not* modeled. As in Refs. [4,9,10], we follow Refs. [13,14] and employ the two–loop asymptotic expression for $\alpha_s(Q^2)$. For the modeled, IR part, we adopt from Ref. [14] $G_{IR}(Q^2) = (16\pi^2/3)\, a\, Q^2 e^{-\mu Q^2}$ and $a = (0.387\,\text{GeV})^{-4}$, $\mu = (0.510\,\text{GeV})^{-2}$.

Our calculational procedures are detailed in our Refs. [4,9,10]. We reproduce the solutions of Ref. [14] for the dressed propagator $S(q)$, *i.e.*, $A(q^2)$ and $B(q^2)$, as well as the solutions for the four functions comprising the pion bound-state vertex Γ_{π^0}. Actually, Ref. [14] employs the BS amplitude $\chi_{\pi^0}(q,p) \equiv S(q+p/2)\Gamma_{\pi^0}(q,p)S(q-p/2)$, which is completely equivalent.

We assume that the $\pi^0\gamma^\star\gamma$ transition (and other pseudoscalar meson $\to$ two photon transitions) proceeds through the pseudoscalar-vector-vector triangle graph (Fig. 1), and that we calculate the pertinent amplitude $T^{\mu\nu}_{\pi^0}(k,k') = \varepsilon^{\alpha\beta\mu\nu} k_\alpha k'_\beta T_{\pi^0}(k^2,k'^2)$ as in Refs. [4,9,10], using the framework advocated by, *e.g.*, Refs. [5–7] in the context of electromagnetic interactions of BS bound

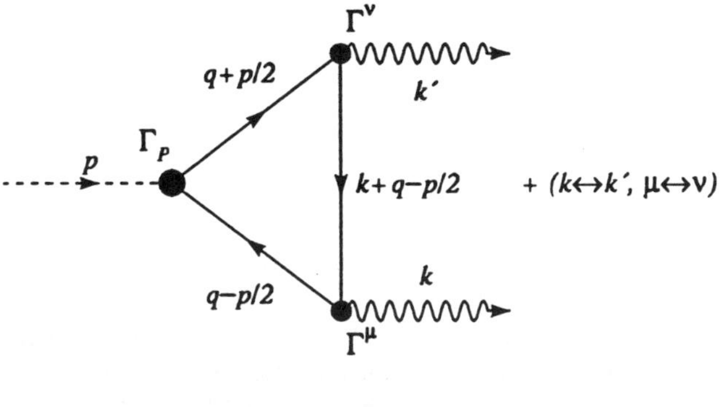

Figure 1: The GIA diagram for $P^0 \to \gamma^{(*)}\gamma^{(*)}$ processes ($P^0 = \pi^0, \eta, \eta', \eta_c, \eta_b$).

states, and often called (*e.g.*, by Ref. [7]) the generalized impulse approximation (GIA). We therefore use the *dressed* quark propagator $S(q)$ and the pseudoscalar BS bound–state vertex $\Gamma_P(q,p)$, as well as consistently dressed *electromagnetic* vertex $\Gamma^\mu(q',q)$, which satisfies the vector Ward–Takahashi identity (WTI) $(q'-q)_\mu\Gamma^\mu(q',q) = S^{-1}(q') - S^{-1}(q)$. Namely, assuming that photons couple to quarks through the bare vertex γ^μ would be inconsistent with our dressed quark propagator $S(q)$, which contains the momentum-dependent functions $A(q^2)$ and $B(q^2)$. The bare vertex γ^μ obviously violates the vector WTI, implying the nonconservation of the electromagnetic vector current and charge. Solving the pertinent SD equation for the dressed quark–photon–quark ($qq\gamma$) vertex Γ^μ is still an unsolved problem, and using the realistic Ansätze for Γ^μ still remains the only practical way to satisfy the WTI. The simplest particular solution of the vector WTI is the Ball–Chiu (BC) [16] vertex

$$\Gamma^\mu_{BC}(q',q) = A_+(q'^2,q^2)\frac{\gamma^\mu}{2} + \frac{(q'+q)^\mu}{(q'^2-q^2)}\{A_-(q'^2,q^2)\frac{(\not q'+\not q)}{2} - B_-(q'^2,q^2)\}\ , \quad (3)$$

where $H_\pm(q'^2,q^2) \equiv [H(q'^2) \pm H(q^2)]$, for $H = A$ or B. It does *not* introduce any new parameters as it is completely determined by the dressed quark propagator $S(q)$. In the SD-BS approach, this minimal WTI-satisfying Ansatz (3) is still the most widely used $qq\gamma$ vertex (*e.g.*, Refs. [6,7,9,10]). A general WTI-satisfying vertex can be written [16] as $\Gamma^\mu = \Gamma^\mu_{BC} + \Delta\Gamma^\mu$, where the addition $\Delta\Gamma^\mu$ doesn't contribute to the WTI, since it is transverse, $(q'-q)_\mu\Delta\Gamma^\mu(q',q) = 0$. That is, $\Delta\Gamma^\mu(q',q)$ lies in the hyperplane spanned by the vectors $\mathcal{T}_i^\mu(q',q)$ ($i = 1, ..., 8$) transverse to the photon momentum $k = q' - q$. Curtis and Pennington (CP) [17] advocated an Ansatz for $\Delta\Gamma^\mu(q',q)$ exclusively along the

$i = 6$ basis vector: $\Delta\Gamma^\mu(q',q) = \mathcal{T}_6^\mu(q',q)[A_-(q'^2,q^2)/2d(q',q)]$. The coefficient of $\mathcal{T}_6^\mu(q',q)$ can be chosen to ensure renormalizability [17], *e.g.*,

$$d_\pm(q',q) = \frac{1}{q'^2+q^2}\left\{(q'^2 \pm q^2)^2 + \left[M^2(q'^2)+M^2(q^2)\right]^2\right\}, \tag{4}$$

where $M(q^2) \equiv B(q^2)/A(q^2)$ is the DχSB-generated dynamical mass function, which in our case has the large-q^2 dependence [4] in agreement with pQCD.

The choice $d = d_-$ corresponds to the CP vertex Ansatz Γ^μ_{CP} [17]. We use it in analytic calculations of $T_{\pi^0}(-Q^2, 0)$, which are possible for $Q^2 = 0$ and $Q^2 \to \infty$. However, in the numerical calculations, which are necessary for finite values of $Q^2 \neq 0$, we prefer the ***modified*** CP (mCP) vertex, Γ^μ_{mCP} [3], resulting from the choice $d = d_+$, since it is easier to deal with numerically. Unlike the BC one, the mCP vertex is consistent with renormalizability just like the CP one. In the present context, the important *qualitative* difference between the BC vertex, and the CP as well as the modified, mCP vertex, is that $\Gamma^\mu_{BC}(q',q) \to \gamma^\mu$ when *both* $q'^2, q^2 \to \pm\infty$, whereas $\Gamma^\mu_{CP}(q',q) \to \gamma^\mu$ and $\Gamma^\mu_{mCP}(q',q) \to \gamma^\mu$ as soon as *one* of the squared momenta tends to infinity.

Assuming the isospin symmetry, $\chi \equiv \chi_{u\bar{u}} = \chi_{d\bar{d}}$ is the BS amplitude for both $u\bar{u}$ and $d\bar{d}$ pseudoscalar bound states, and $\chi_{\pi^0}(q,p) \equiv \chi(q,p)\,\lambda^3/\sqrt{2}$. Then, for π^0, GIA yields (see Fig. 1) the tensor amplitude

$$T^{\mu\nu}_{\pi^0}(k,k') = -N_c\frac{1}{3\sqrt{2}}\int\frac{d^4q}{(2\pi)^4}\mathrm{tr}\{\Gamma^\mu(q-\frac{p}{2},k+q-\frac{p}{2})S(k+q-\frac{p}{2})$$
$$\times\Gamma^\nu(k+q-\frac{p}{2},q+\frac{p}{2})\chi(q,p)\} + (k\leftrightarrow k',\mu\leftrightarrow\nu), \tag{5}$$

leading to the analytical reproduction of the $\pi^0 \to \gamma\gamma$ amplitude (1) with the CP and mCP vertices, in the same way as with the BC vertices.

Let us stress that in the regime of asymptotically large Q^2, we obtained [3] analytical results for the transition form factor in agreement with the behavior (2). Since the pion is light, $k\cdot k' \approx Q^2/2$ for large negative $k^2 = -Q^2$ and $k'^2 = 0$. Taking into account the behavior of the propagator functions $A(q^2)$ and $B(q^2)$ (*e.g.*, see [4]) we can then in Eq. (5) approximate those quark propagators that depend on the photon momenta k and k', by their asymptotic forms: $S(q-p/2+k) \approx S(k-p/2) \approx -(2/Q^2)(\not{k}-\not{p}/2)$ and analogously for the propagator where k is replaced by k'. Although the relative loop momentum q can be large in the course of integration, its neglecting is justified because the BS amplitude $\chi(q,p)$ decays quickly and thus strongly damps the integrand for the large q's. Ref. [3] then showed that $T^{\mu\nu}_{\pi^0}(k,k') = -(2/N_c Q^2)\,\varepsilon^{\mu\lambda\nu\sigma}(k-k')_\lambda p_\sigma f_\pi$, finally leading to Eq. (2) with the coefficient $\mathcal{J} = 4/3$. We thus found ***model independently*** that the asymptotic behavior in the present approach (with $qq\gamma$ vertices such as the bare γ^μ, mCP and CP), agrees exactly with the leading term predicted by OPE [12].

The asymptotic behavior $\propto 1/Q^2$ obtained in the present approach is especially satisfying when compared with the one known to result from the simple constituent quark model (with the constant light–quark mass parameter m_u), where $T_{\pi^0}(-Q^2, 0) \propto (m_u^2/Q^2)\ln^2(Q^2/m_u^2)$ as $Q^2 \to \infty$, which overshoots the data considerably because of the additional $\ln^2(Q^2)$-dependence.

It is important to note that this asymptotic behavior, $T_{\pi^0}(-Q^2, 0) = (4/3)f_\pi/Q^2$, was obtained for the $qq\gamma$ vertices which reduce to the bare one, γ^μ, as soon as ***one*** of the squared momenta tends to infinity, such as the renormalizable CP and mCP vertices. Subsequently, Ref. [18] showed that this asymptotic behavior must in fact hold for ***any*** renormalizable $qq\gamma$ vertex. For the simplest dressed WTI-preserving vertex, the BC one, which reduces to the bare one only when momenta are large in both of its fermion legs, and which is not consistent with multiplicative renormalizability [17], the asymptotic behavior is $(4/3)\widetilde{f}_\pi/Q^2$, where the quantity $\widetilde{f}_\pi$ is given by the same Mandelstam-formalism expression as f_π except that its integrand is modified by the factor $[1 + A(q^2)]^2/4$ [3]. For our model solutions [4], $\widetilde{f}_\pi \approx 1.334 f_\pi$, so that the usage of the BC vertices causes the increase of $\mathcal{J}$ in Eq. 2 from $\mathcal{J} = 4/3$ to $\mathcal{J} \approx 1.78$.

Of course, the model-independent asymptotic coefficient $4f_\pi/3$ is the one having the more fundamental meaning, resulting from the renormalizable $qq\gamma$ vertices such as the CP or mCP ones, which have properties closer to the true vertex solution. Also indicative is the asymptotics found [3] for such $qq\gamma$ vertices when both photons are off-shell, $k^2 = -Q^2 << 0$ and $k'^2 = -Q'^2 \leq 0$:

$$T_{\pi^0}(-Q^2, -Q'^2) = \frac{4}{3}\,\frac{f_\pi}{Q^2 + Q'^2}\,. \tag{6}$$

The usage of the BC vertex (3) would again modify this result by the substitution $f_\pi \to \widetilde{f}_\pi$. However, Eq. (6) agrees with the leading term of the OPE result of Novikov ***et al.*** [19] for the special case $Q^2 = Q'^2$. Also, the distribution-amplitude-dependence of the pQCD approach cancels out for that symmetric case, so that $T_{\pi^0}(-Q^2, -Q'^2)$ in this approach (***e.g.***, see [20]), in the limit $Q^2 = Q'^2 \to \infty$, exactly agrees with both our Eq. (6) and Ref. [19]. For that symmetric case, we should thus have even the precise agreement of the coefficients irrespective of the description of the pion internal structure encoded in the distribution amplitude. Obviously, this favors the renormalizable $qq\gamma$ vertices, such as CP and mCP ones, over the BC vertex. However, the BC vertex may anyway be the one which is more accurate not only for the presently accessible Q^2, but also for larger values before starting to fail.

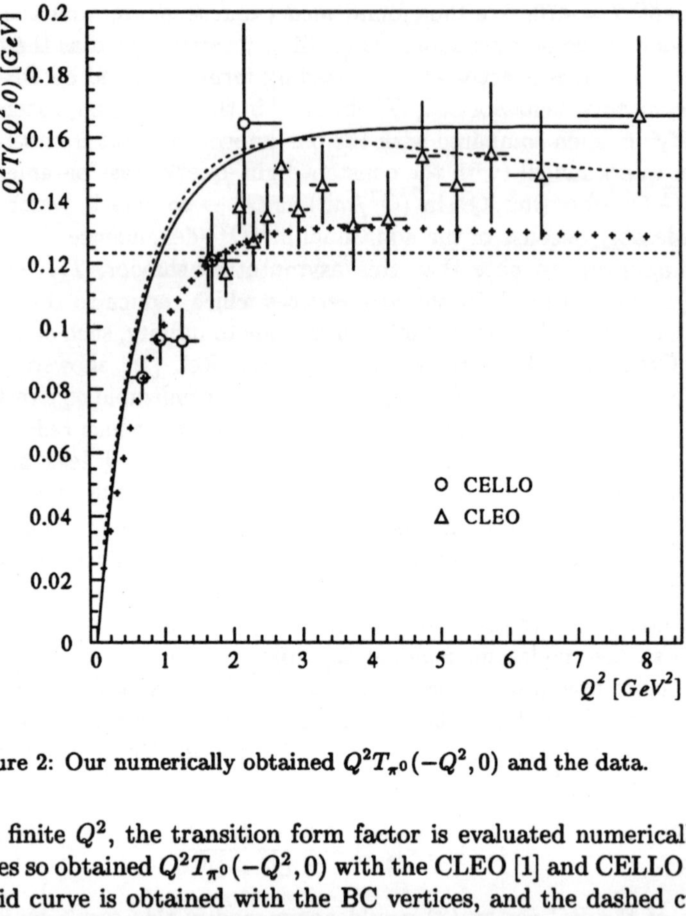

Figure 2: Our numerically obtained $Q^2T_{\pi^0}(-Q^2,0)$ and the data.

For finite Q^2, the transition form factor is evaluated numerically. Fig. 2 compares so obtained $Q^2T_{\pi^0}(-Q^2,0)$ with the CLEO [1] and CELLO [21] data. The solid curve is obtained with the BC vertices, and the dashed curve with the mCP ones. The horizontal dashed line denotes $4f_\pi/3$, the asymptotic value for renormalizable $qq\gamma$ vertices such as the mCP one. It barely touches the experimental error bars from below. Note that also Manohar [12] warned that his OPE approach (also yielding $4f_\pi/3$) indicates the possibility of large corrections to his leading term, but possibly also to the pQCD asymptotic coefficient $2f_\pi$, which is however exact in the *strict* $\ln(Q^2) \to \infty$ limit [11]. We get it lower by the factor 2/3 because our present approach does not incorporate the effects of the pQCD evolution. However, as Ref. [2] seems to indicate that effects other than the pQCD evolution (including the presently interesting dynamical dressing of quarks) may still play important (and maybe even dominant) role even at Q^2-values larger than the presently accessible ones, one should consider seriously also nonperturbative approaches such as ours. On the other hand, high-precision measurements of $T_{\pi^0}(-Q^2,0)$ can test (and give a hint on how to improve) the SD and BS model solutions which have so far been successful in fitting the low-energy hadron properties such as the meson spectrum. For example, our model choice [14] somewhat overshoots (for both BC and mCP vertices) the present $T_{\pi^0}(-Q^2,0)$ data [1,21] in the region $Q^2 \lesssim 4$ GeV^2.

Once one has a solution for $\chi(q,p)$ that leads to the correct value of f_π, the transition form factor is most sensitive on $A(q^2)$, or, more precisely, on its values at small and intermediate momenta $-q^2$, where $A(q^2)$ is still appreciably different from 1 [3,22]. To illustrate what happens when the $A(q^2)$-profile is decreased, let us enforce by hand the extreme, artificial case $A(q^2) \equiv 1$. (To avoid confusion, we stress it is for illustrative purposes only, as we cannot have such a SD-solution in the adopted approach of Refs. [13,14].) This leads to the curve traced on Fig. 2 by small crosses, pertaining to the usage of both BC and mCP vertices (as well as the CP ones), since $A(q^2) \equiv 1$ makes $\Gamma^\mu_{mCP} \to \Gamma^\mu_{BC}$. This curve reveals how the heights of the curves depicting $Q^2 T_{\pi^0}(-Q^2,0)$ depend on how much the $A(q^2)$-profile exceeds 1. Obviously, for both the solid curve and the dashed one, the agreement with experiment would be improved by lowering them somewhat (at least in the momentum region $Q^2 \lesssim 4$ GeV), which could be achieved by modifying the model [14] and/or its parameters so that such a new solution for $A(q^2)$ is somewhat lowered towards its asymptotic value $A(q^2 \to \infty) \to 1$. (Of course, in order to be significant, this must not be a specialized re-fitting aimed *only* at $A(q^2)$. Lowering of $A(q^2)$ should be a result of a broad fit to many meson properties, comparable to the original fit [14]. This, however, is beyond the scope of this paper.)

By the same token, high precision measurements of $T_{\pi^0}(-Q^2,0)$ can be especially helpful in obtaining information on which solutions for $A(q^2)$ are empirically acceptable and which are not. Of course, measurements of $T_{\pi^0}(-Q^2,0)$ give information on the integrated strength of $A(q^2)$ rather than on $A(q^2)$ itself. However, since it is known that the form of that function must be a rather smooth transition (*e.g.*, see [4]) from $A(q^2) > 1$ for q^2 near 0, to $A(q^2) \to 1$ in the q^2-domain where QCD is perturbative, such measurements [23] would give a useful hint even about $A(q^2)$ itself – namely about what solutions for $A(q^2)$ one may have in sensible descriptions of dynamically dressed quarks and their bound states. Therefore, the intermediate-momentum ($Q^2 \lesssim 4$ GeV^2) measurements of $T_{\pi^0}(-Q^2,0)$ at Jefferson Lab, such as those proposed in Ref. [23], would be very desirable. Hopefully, the study of virtual Compton scattering at CEBAF can produce the data on $T_{\pi^0}(-Q^2,0)$ as envisioned in Ref. [23].

Acknowledgments

D. Klabučar thanks the organizers C. Carlson and A. V. Radyuskin for the wonderful workshop. The authors acknowledge the support of the Croatian Ministry of Science and Technology.

References

1. J. Gronberg *et al.* (CLEO collaboration), Phys. Rev. D **57**, 33 (1998).
2. A. V. Radyushkin and R. T. Ruskov, Nucl. Phys. B **481**, 625 (1996).
3. D. Kekez and D. Klabučar, Phys. Lett. **457B**, 359 (1999).
4. D. Kekez *et al.*, Int. J. Mod. Phys. A **14**, 161 (1999).
5. M. Bando, M. Harada, and T. Kugo, Prog. Theor. Phys. **91**, 927 (1994).
6. C. D. Roberts, Nucl. Phys. A **605**, 475 (1996).
7. M.R. Frank *et al.*, Phys. Lett. B **359**, 17 (1995).
8. I. V. Musatov and A. V. Radyushkin, Phys. Rev. D **56**, 2713 (1997).
9. D. Kekez and D. Klabučar, Phys. Lett. B **387**, 14 (1996).
10. D. Klabučar and D. Kekez, Phys. Rev. D **58**, 096003 (1998).
11. G. P. Lepage and S. J. Brodsky, Phys. Rev. D **22**, 2157 (1980).
12. A. Manohar, Phys. Lett. **B244**, 101 (1990).
13. P. Jain and H. J. Munczek, Phys. Rev. D **44**, 1873 (1991); Phys. Rev. D **46**, 438 (1992).
14. P. Jain and H. J. Munczek, Phys. Rev. D **48**, 5403 (1993).
15. P. Maris and C. D. Roberts, Phys. Rev. C **56**, 3369 (1997).
16. J. S. Ball and T.-W. Chiu, Phys. Rev. D **22**, 2542 (1980).
17. D. C. Curtis and M. R. Pennington, Phys. Rev. D **42**, 4165 (1990).
18. C. D. Roberts, Fizika B (Zagreb) **8**, 285 (1999).
19. V. A. Novikov *et al.*, Nucl. Phys. B **237**, 525 (1984).
20. P. Kessler and S. Ong, Phys. Rev. D **48**, R2974 (1993).
21. H.-J. Behrend *et al.* (CELLO Collaboration), Z. Phys. C **49**, 401 (1991).
22. D. Klabučar and D. Kekez, Fizika B (Zagreb) **8**, 303 (1999).
23. CEBAF Letter of Intent # LOI–94/005. Co–Spokesmen: A. Afanasev, J. Gomez, and S. Nanda; A. Afanasev, **hep-ph/9608305**.

MEASUREMENT OF THE RATIO OF THE PROTON'S ELECTRIC TO MAGNETIC FORM FACTORS BY RECOIL POLARIZATION

M. K. Jones for the Jefferson Lab Hall A Collaboration
College of William and Mary, Williamsburg, VA, 23187, USA

The longitudinal and transverse polarizations of the outgoing proton were measured for the reaction $^1H(\vec{e},e'\vec{p})$ at four-momentum transfer squared of 0.5 to 3.5 GeV^2. The ratio of the electric to magnetic form factors of the proton is proportional to the ratio of the transverse to longitudinal polarizations.

1 Introduction

Efforts to understand the structure of the nucleon have been ongoing for the past century. Studies of the exclusive and inclusive electron-proton scattering have been an important tool in these efforts.

The nucleon can be characterized by the electric, $G_{Ep}(Q^2)$, and magnetic, $G_{Mp}(Q^2)$, Sachs form factors. In the Breit frame, $G_{Ep}(Q^2)$ ($G_{Mp}(Q^2)$) is the Fourier transform of the charge (current) distribution. In the limit $Q^2 \rightarrow 0$, $G_{Ep} = 1$ and $G_{Mp} = \mu_p$, the proton magnetic moment. The unpolarized *ep* cross section is:

$$\frac{d\sigma}{d\Omega} \propto G_{Ep}^2 + \frac{\tau}{\epsilon} G_{Mp}^2, \tag{1}$$

where $\epsilon = [1 + 2(1+\tau)\tan^2(\frac{\theta_e}{2})]^{-1}$, $\tau = Q^2/4M^2$, M is the mass of the proton and θ_e is the scattered electron angle. $G_{Ep}(Q^2)$ and $G_{Mp}(Q^2)$ can be extracted by the Rosenbluth technique[1] in which ϵ is varied at a fixed Q^2. For $Q^2 < 0.5$ GeV^2, Hofstadter[2] *et al.* found that

$$\mu_p G_{Ep}(Q^2) \approx G_{Mp}(Q^2) \approx (1 + \frac{Q^2}{0.71})^{-2}. \tag{2}$$

This means that the charge and current distributions have an exponential radial shape. Unfortunately, as one can see from Eq. 1, $G_{Mp}(Q^2)$ dominates the cross section at Q^2 above 1 GeV^2. This makes precise measurement of $G_{Ep}(Q^2)$ difficult, but, as a consequence, $G_{Mp}(Q^2)$ is well measured[3] up to $Q^2 = 31$ GeV^2.

A different approach to extracting G_{Ep} and G_{Mp} is by measuring polarization observables in elastic $\vec{e}\,p$ or $\vec{e}\,\vec{p}$ scattering. With the new generation of continuous beam electron accelerators, these types of experiments have become feasible. G_{Ep} and G_{Mp} are related to the transverse, P_t, and longitudinal, P_l, polarizations of the outgoing proton in the reaction $^1H(\vec{e},e'\vec{p})$. In the one-photon exchange approximation, the relationship[4,5] is:

$$I_0 P_t = -2\sqrt{\tau(1+\tau)} G_{Ep} G_{Mp} \tan\frac{\theta_e}{2}, \tag{3}$$

$$I_0 P_\ell = \frac{1}{M}(E_e + E_{e'})\sqrt{\tau(1+\tau)} G_{Mp}^2 \tan^2\frac{\theta_e}{2}, \tag{4}$$

where $I_0 = G_{Ep}^2 + \frac{\tau}{\epsilon} G_{Mp}^2$, E_e is the beam energy and $E_{e'}$ is the scattered electron energy. For the outgoing proton, the component of the polarization normal to the reaction plane is zero. The ratio G_{Ep}/G_{Mp} can be expressed as:

$$\frac{G_{Ep}}{G_{Mp}} = -\frac{P_t}{P_l}\frac{(E_e + E_{e'})}{2M}\tan(\frac{\theta_e}{2}) \tag{5}$$

The recoil polarization technique was used first at Bates to measure[6] G_{Ep}/G_{Mp} at $Q^2 = 0.38$ and 0.5 GeV^2.

2 Experiment

Our experiment was done in Hall A at Thomas Jefferson National Accelerator Facility (JLab). Longitudinally polarized electron beams with energies between 0.934 GeV and 4.090 GeV were scattered by a liquid hydrogen target. For the four highest Q^2 points, beams with ~0.39 polarization and currents up to ~ 115 μA were used, and the helicity was flipped at 30 Hz. For the lower Q^2 points, a strained GaAs wafer was used which limited the current to ~15 μA with typical polarizations of ~0.60 and the helicity was flipped at 1 Hz. This was the first experiment to use the strained GaAs wafer. Since that time, the polarized source at JLab, using the strained GaAs wafer, has increased both the available polarization and current.

Elastic *ep* events were selected by detecting the scattered electrons and protons in coincidence in the two identical High Resolution Spectrometers (HRS) of Hall A [7]. The spectrometer consists of two quadrupole magnets followed by a dipole magnet and then a third quadrupole magnet. The experiment must determine the polarization components of the proton at the target by measuring the polarizations in detectors which are located after the HRS. For a simple dipole magnet, P_t is unchanged after the proton travels through the dipole while P_l is rotated by a precession angle,χ , given by:

$$\chi = \frac{E_p}{M}\cdot(\mu_p - 1)\cdot(\theta_{in} + \theta_b - \theta_{out}) \tag{6}$$

where E_p is the proton's energy, θ_{in} (θ_{out}) is angle into (out of) the dipole and θ_b is bend angle of the dipole, which is 45° for the Hall A spectrometers. The focal plane polarization components are related to the target components by a spin matrix:

$$\begin{pmatrix} P_n^{fpp} \\ P_t^{fpp} \end{pmatrix} = \begin{pmatrix} S_{nt} \approx 0 \; S_{nl} \approx -\sin\chi \\ S_{tt} \approx 1 \;\; S_{tl} \approx 0 \end{pmatrix} \begin{pmatrix} P_t \\ P_l \end{pmatrix} \tag{7}$$

where the fpp superscript indicates a polarization at the focal plane. In general, the target transverse and longitudinal polarization components are mixed together as the magnetic moment of the proton precesses while traveling through the HRS. The spin matrix elements have to be calculated as a function of the target coordinates on an event by event basis.

The focal plane polarimeter (FPP) is used to determine the transverse, P_t^{fpp}, and normal, P_n^{fpp}, polarizations at the focal plane by measuring the azimuthal angular distribution of the $\vec{p}$+^{12}C reaction. The FPP consists of two front straw-tube drift chambers which determine the trajectory of the proton incident onto the carbon analyzer, and two rear straw-tube drift chambers which reconstruct the track of the scattered proton. The azimuthal (φ) angular distribution in each bin of the polar scattering angle, ϑ, has the form :

$$\begin{aligned} N^{\pm}(\vartheta,\varphi) = N_p(\vartheta)[1 &+ (hA_c(\vartheta)P_t^{fpp} + \alpha)\sin\varphi \\ &- (hA_c(\vartheta)P_n^{fpp} + \beta)\cos\varphi], \end{aligned} \tag{8}$$

where h is the electron beam polarization, $N_p(\vartheta)$ is the number of protons scattered in the polarimeter and $A_c(\vartheta)$ is the analyzing power. The angular distributions, $N^+(\varphi,\vartheta)$ and $N^-(\varphi,\vartheta)$, are measured for the plus and minus helicity states of the beam. Since the instrumental asymmetries (α and β) are helicity independent, the difference, $N^+(\varphi,\vartheta) - N^-(\varphi,\vartheta)$, depends only on the physics quantities, $hA_c(\vartheta)P_t^{fpp}$ and $hA_c(\vartheta)P_n^{fpp}$ which are extracted by Fourier analysis of the distribution. The target polarizations are extracted from the focal plane polarizations by a method explained in Ref. [8], and the ratio G_{Ep}/G_{Mp} is determined from Eq. 5.

3 Results and conclusion

The ratios, $\mu G_{Ep}/G_{Mp}$, are plotted in Fig. 1. The absolute statistical error bars on the new Jefferson Lab data range from 0.022 at Q^2= 0.5 GeV^2 to 0.047 at Q^2= 3.5 GeV^2 and are a distinct improvement on the previous measurements that used the Rosenbluth separation technique (especially for $Q^2 > 1.5$ GeV^2). The size of the systematic errors is given by the shaded area in Fig. 1. From Eq. 7, one can see that a major source of systematic error is precise knowledge of χ when it is near 180°, which occurs at $Q^2 = 2.168$ GeV^2. Another source of systematic error is the knowledge of the exact analytic form of the spin matrix. A detailed table of the measured points and errors is given in Ref.[15].

In Fig. 1, one can clearly see a downward trend in the $\mu G_{Ep}/G_{Mp}$ ratio above $Q^2 = 1$ GeV^2, where before there was a conflicting set of measurements. The striking feature of the new data is the fall-off of the $\mu G_{Ep}/G_{Mp}$ ratio from 0.966 starting at $Q^2 = 0.5$ GeV^2 to a value of 0.609 at $Q^2 = 3.5$ GeV^2. This indicates that the shapes of G_{Ep} and G_{Mp} are different. The improved quality of the data will have a significant impact on theoretical models.

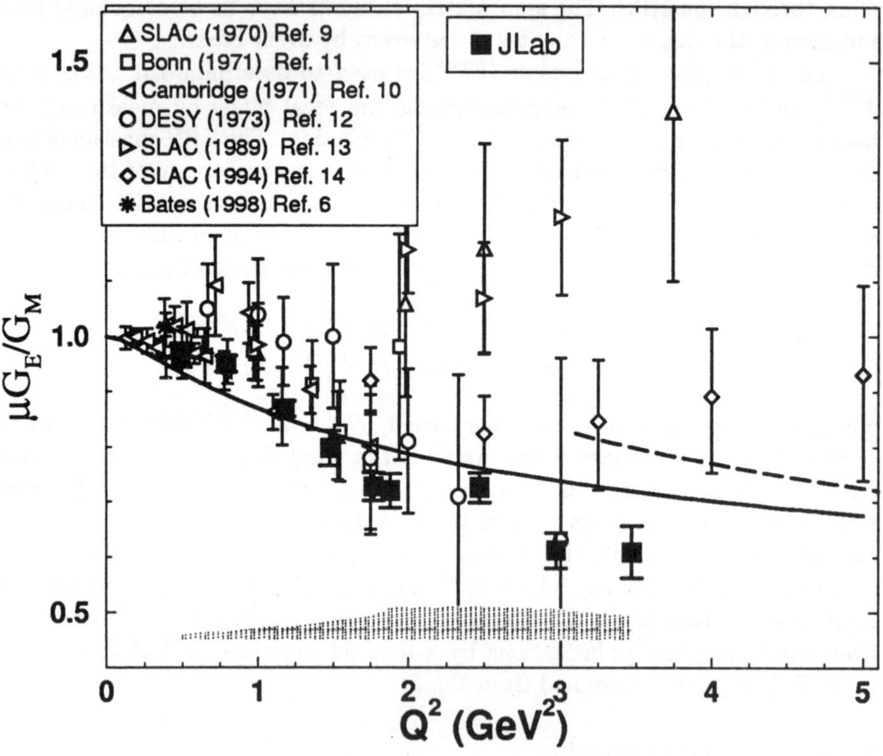

Figure 1: New $\mu G_{Ep}/G_{Mp}$ measurement at Jlab compared to previous world data. The grey area is the size of the systematic uncertainty for the JLab data.

The Q^2 region between 1 and 10 GeV^2 has been anticipated as the region where the transition from mesonic to quark-gluon degrees of freedom occurs. This transition region is a difficult area for theoretical models, and has led to numerous approaches to predicting the nucleon form factors. The models tend to have parameters which have to be constrained by the data. In Fig. 1, two model predictions are compared to the data from this experiment. The vector meson dominance model of Ref. [16], an improvement on the original work of Ref. [17], is plotted as a solid line. It was fit to both the space-like and time-like nucleon form factors (not including this data) and includes a super-convergence condition to constrain the behavior of the form factors to the perturbative QCD predicted fall-off at large Q^2. In Ref.[18] the nucleon form factors were calculated in the diquark model, and the predicted $\mu G_{Ep}/G_{Mp}$ is plotted as a dashed line. In the limit of $Q^2 \rightarrow \infty$ the di-quark framework becomes the hard scattering formalism of perturbative QCD.

It is important to compare the theoretical predictions to both the proton and neutron form factors. G_{Mp} has been measured to <5% accuracy up to $Q^2 = 15$ GeV^2. This experiment has reduced the error on G_{Ep}/G_{Mp} to a level comparable to that on G_{Mp}. With the new data tighter constraints will be placed on theory to fit both form factors of the proton. In addition, at JLab, experiments are planned to measure G_{Mn} and G_{En} which will further constrain theory. Extension of this experiment to measure G_{Ep}/G_{Mp} to $Q^2 =$ 5.6 GeV^2 is scheduled to run in Fall of 2000.

Acknowledgments

We would like to thank the JLab accelerator and Hall A staff for their support during this experiment. Results presented here are the thesis work of Gilles Quéméner (College of William and Mary). This work is supported in part by the U.S. National Science Foundation and the U.S. Department of Energy.

References

1. M.N. Rosenbluth, Phys. Rev. **79**, (1950) 615.
2. E.E. Chambers and R. Hofstadter, Phys. Rev. **103**, (1956) 1454.
3. A.F. Sill *et al.*, Phys. Rev. D **48**, (1993) 29.
4. A. I. Akhiezer and M. P. Rekalo, Sov. J. Particles Nucl. **3**, (1974) 277.
5. R. Arnold, C. Carlson and F. Gross, Phys. Rev. **C23**, (1981) 363.
6. B. Milbrath, J. McIntyre et al., Phys. Rev. Lett. **80** (1998) 452.
7. http://www.jlab.org/Hall-A/equipment/
8. G. Quéméner *et al.*, Nucl. Phys. **A654**, (1999) 543c.
9. J. Litt *et al.*, Phys. Lett. **31B**, (1970) 40.
10. L. E. Price, J. R. Dunning, M. Goitein, K. Hanson, T. Kirk, and Richard Wilson, Phys. Rev. D **4**, (1971) 45.
11. Ch. Berger, V. Burkert, G. Knop, B. Langenbeck, K. Rith, Phys. Letters **35B**, (1971) 87.

12. W. Bartel *et al.*, Nuc. Phys. **B58**, (1973) 429.
13. R. C. Walker *et al.*, Phys. Rev. D **49** , (1994) 5671. R. C. Walker et al. , Phys. Lett. B **224**, (1989) 353.
14. L. Andivahis *et al.*, Phys. Rev. D **50**, (1994) 5491. P. E. Bosted et al., Phys. Rev. Lett. **68**, (1992) 3841.
15. M. K. Jones *et al.*, to be published in Phys. Rev. Lett., LANL E-Print nucl-ex/9910005.
16. A.W. Hammer, U.G. Meissner and D. Drechsel, Phys. Lett. **B385** (1996) 343.
17. G. Hoehler *et al.*,Nucl. Phys. B **114**, (1976) 505.
18. P. Kroll, M. Schurmann and W. Schweiger, Z. Phys. A - Hadrons and Nuclei **338**, (1991) 339.

DEUTERON AND HELIUM FORM FACTOR MEASUREMENTS AT JEFFERSON LAB[a]

R. S. SULEIMAN, G. G. PETRATOS
Department of Physics, Kent State University, Kent,
OH 44242, USA

Recent JLab measurements of the deuteron elastic structure functions $A(Q^2)$ and $B(Q^2)$ up to large momentum transfers are presented. Prospects for elastic electron scattering off the deuteron and the helium isotopes with higher JLab beam energies are also presented. The future measurements will be critical in the search for a transition from the conventional meson-nucleon to the quark-gluon descriptions of the nuclear few-body systems.

Measurements of the elastic form factors of the deuteron and the helium isotopes are of crucial importance in understanding their electromagnetic structure and testing the light nuclei "standard model" of the meson-nucleon framework based on the impulse approximation (IA) and meson-exchange currents (MEC). [1] Such measurements offer unique opportunities to study the short-range nucleon-nucleon interaction, the few-body wave functions, isobar and three-body force contributions, and effects from possible quark-cluster admixtures. Large momentum transfer measurements can test "nuclear chromodynamics" predictions based on quark-dimensional scaling (QDS) and perturbative QCD (pQCD). [2]

The starting point of the conventional theoretical approach of elastic scattering from few-body systems is the impulse approximation, where the incident electron interacts with one of the nucleon constituents of deuterium or helium. The light nuclei form factors are then convolutions of the nuclear wave function with the form factors of the constituent nucleons. At large momentum transfers the effects of relativity cannot be ignored and either corrections to the IA or fully relativistic approaches, as in the deuteron case, [3,4,5] have been developed. It has long been understood and overwhelmingly supported by the available data that the few-body form factors are sensitive to the presence of meson-exchange currents and isobar configurations that augment the IA picture. [1]

It is widely recognized that at distances much less than the nucleon size, the underlying quark substructure of the nucleons cannot be ignored. This has lead to the formulation of so called hybrid quark models [6] that treat few-body nuclei as quark clusters when the internucleon separations become

[a]Work supported by the U.S. National Science Foundation

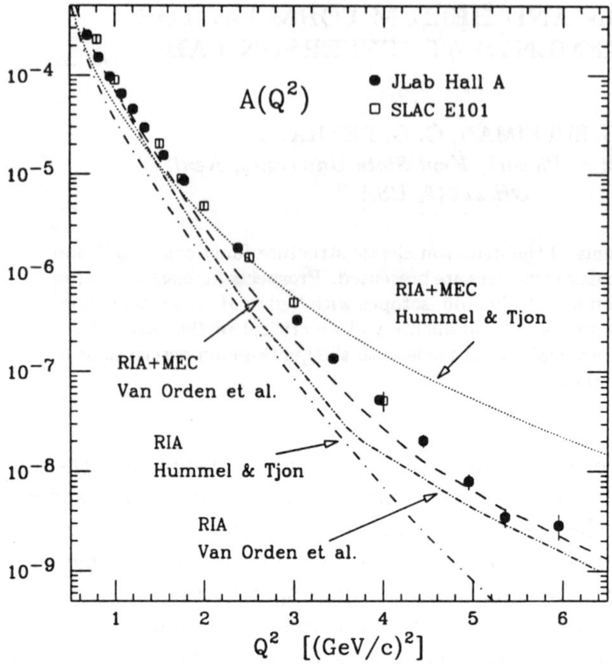

Figure 1: The deuteron electric structure function $A(Q^2)$ from JLab experiment E91-26 compared to RIA and RIA+MEC theoretical calculations (see text). Also shown are previous SLAC data.

smaller than ~1 fm. At sufficiently "large" momentum transfers, the few-body form factors are expected to be calculable in terms of only quarks and gluons within the framework of pQCD. The first attempt at a quark-gluon description of elastic form factors was based on the dimensional scaling quark model (DSQM), [7] where the underlying dynamical mechanism during elastic scattering is the hard rescattering of the constituent quarks via exchange of hard gluons. The Q^2 dependence of this process is then predicted by simply counting the number n of gluon propagators (n=5 for deuterium, 8 for ^{3}He, and 11 for ^{4}He), which implies that the elastic structure functions $A(Q^2)$ of the few-body systems should follow the power law: $\sqrt{A(Q^2)} \sim (Q^2)^{-n}$. This prediction was later substantiated, for the deuteron case, in the pQCD framework, where it was shown [8] that to leading-order: $\sqrt{A(Q^2)} = [\alpha_s(Q^2)/Q^2]^5 \sum_{m,n} d_{mn}[\ln(Q^2/\Lambda^2)]^{-\gamma_n-\gamma_m}$, where $\alpha_s(Q^2)$ and Λ are the QCD strong coupling constant and scale parameter, and $\gamma_{m,n}$ and d_{mn} are QCD anomalous dimensions and constants.

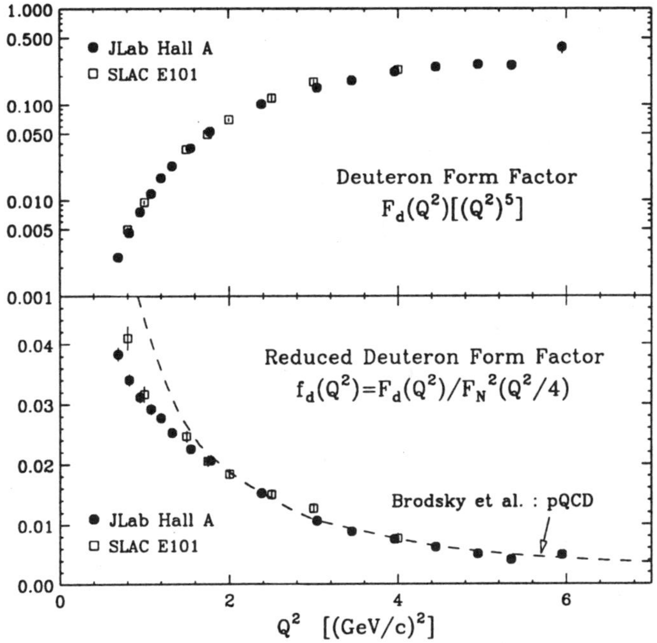

Figure 2: The deuteron form factor $F_d(Q^2)$ times $(Q^2)^5$ (top) and the reduced deuteron form factor $f_d(Q^2)$ (bottom) from JLab E91-26 and SLAC E101 measurements. The curve is the asymptotic pQCD prediction arbitrarily normalized to the data at $Q^2 = 4$ $(GeV/c)^2$.

The unique features of the Jefferson Lab (JLab) electron beam and spectrometer facilities are ideal for improving and extending the existing elastic structure function measurements of light nuclei to higher momentum transfers. The first step was taken with experiment E91-26 that measured the electric, $A(Q^2)$, and magnetic, $B(Q^2)$, elastic structure functions of the deuteron up to large momentum transfers. The experiment performed elastic electron-deuteron scattering in coincidence using the Hall A spectrometer facility. Elastically scattered electrons were detected in the electron High Resolution Spectrometer (HRS). Recoil deuterons were detected in coincidence in the hadron HRS. Coincidence events were identified using time-of-flight between an electron trigger and a recoil deuteron trigger. The $A(Q^2)$ elastic structure function was extracted in the squared four-momentum transfer range $0.7 < Q^2 < 6.0$ $(GeV/c)^2$ from cross section measurements at forward angles. Measurements of the cross section at a backward angle (144.5°) enabled separation of the $B(Q^2)$ elastic structure function by means of a Rosenbluth separation in the range $0.7 < Q^2 < 1.4$ $(GeV/c)^2$.

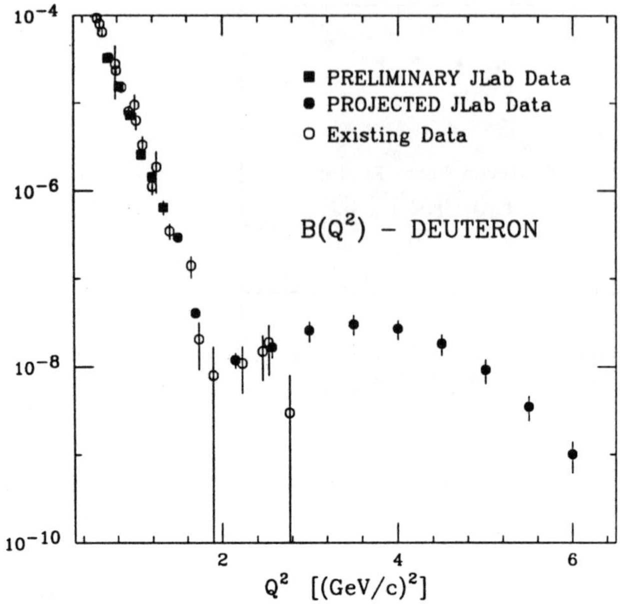

Figure 3: Preliminary and projected JLab data for the deuteron magnetic elastic structure function $B(Q^2)$. Also shown are existing data from previous Bonn, Saclay and SLAC experiments.

E91-26 has significantly extended the Q^2 range of the previous $A(Q^2)$ SLAC measurements, as shown in Figure 1, and has measured a record low cross section in electron-nucleus scattering: the average cross section for the highest Q^2 point (measured with two different beam energies) is approximately 2×10^{-41} cm^2/sr. The JLab data[9] continue the trend of the SLAC E101 data[10]: they are indicative of a smooth $A(Q^2)$ fall off with Q^2 and do not exhibit any apparent diffractive feature. The double-dot-dashed and dot-dashed curves in Fig. 1 represent the relativistic impulse approximation (RIA) calculations of Van Orden, Devine and Gross (VDG) [4] and Hummel and Tjon (HT) [3], respectively. The VDG curve is based on the Gross equation. The HT curve is based on a quasi-potential approximation of the Bethe-Salpeter equation. Both groups have augmented their models by including the $\rho\pi\gamma$ meson-exchange current contribution. The magnitude of this contribution depends on the $\rho\pi\gamma$ coupling constant and vertex form factor choices. Although our data favor the VDG RIA+MEC calculation, a complete test of the RIA+MEC framework will require improved/extended measurements of the nucleon form factors and of the deuteron $B(Q^2)$, in progress or planned at JLab.

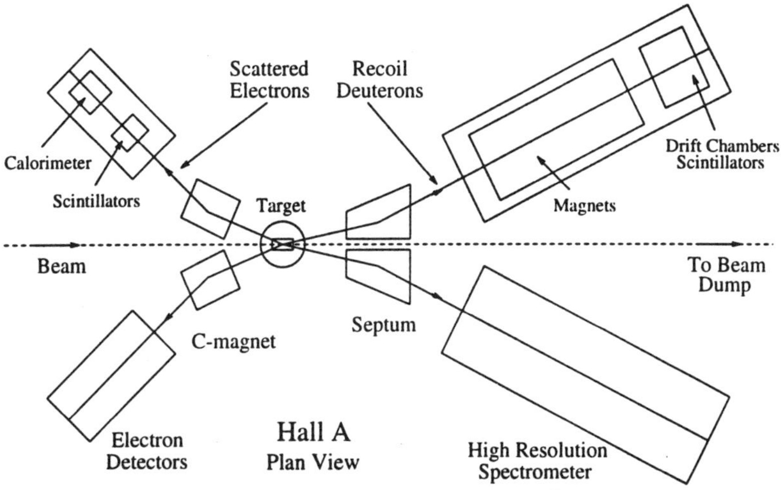

Figure 4: Schematic of a JLab Hall A dual double-arm spectrometer configuration for measuring the deuteron magnetic elastic structure function $B(Q^2)$ at large momentum transfers.

Figure 2 (top) shows the JLab and SLAC data on the "deuteron form factor", $F_d(Q^2) \equiv \sqrt{A(Q^2)}$, multiplied by $(Q^2)^5$. It is evident that the data show an approach to a scaling behavior consistent with the power law of DSQM and pQCD. Figure 2 (bottom) shows values for the "reduced" deuteron form factor [11] $f_d(Q^2) \equiv F_d(Q^2)/F_N^2(Q^2/4)$ where the two powers of the nucleon form factor $F_N(Q^2) = (1 + Q^2/0.71)^{-2}$ remove in a minimal and approximate way the effects of nucleon compositeness. The $f_d(Q^2)$ data appear to follow, for $Q^2 > 2$ (GeV/c)^2, the asymptotic Q^2 prediction of pQCD [8]: $f_d(Q^2) \sim [\alpha_s(Q^2)/Q^2][\ln(Q^2/\Lambda^2)]^{-\Gamma}$. Here $\Gamma = -(2C_F/5\beta)$, where $C_F = (n_c^2 - 1)/2n_c$, $\beta = 11 - (2/3)n_f$, with $n_c = 3$ and $n_f = 2$ being the numbers of QCD colors and effective flavors. Although several authors have questioned the validity of QDS and pQCD at the momentum transfers of this experiment, [12] similar scaling behavior has been reported in deuteron photodisintegration at moderate photon energies. [13]

The magnetic structure function $B(Q^2)$ E91-26 data analysis is still in progress. Preliminary $B(Q^2)$ data are shown in Figure 3 along with previous data. [14] E91-26 has provided precise data that will put severe constraints on the parameters of the IA calculations. It is very important to extend these precision measurements to higher momentum transfers. The high luminosity capability of JLab can provide precision measurements of $B(Q^2)$ over its apparent diffraction minimum and up to possibly $Q^2 = 5$-6 (GeV/c)^2. Due to the

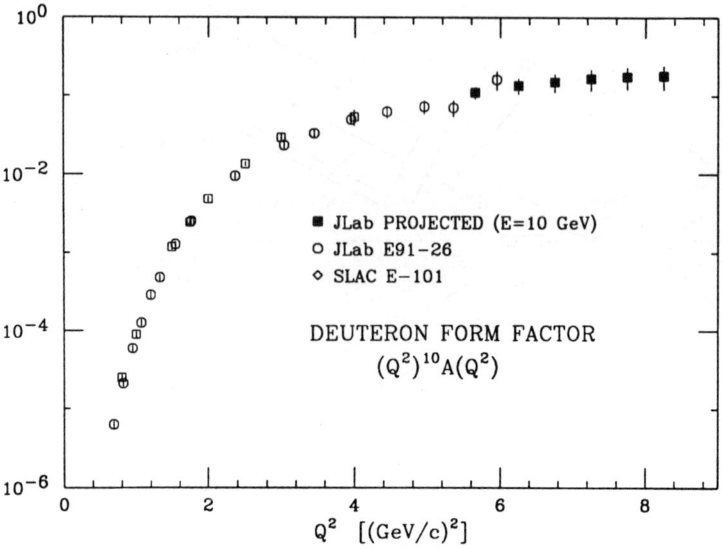

Figure 5: Existing JLab and SLAC data, and JLab projected data, using a 10 GeV upgraded electron beam, for the deuteron $A(Q^2)$ (multiplied by $(Q^2)^{10}$).

smallness of the ratio B/A, in order to separate B from A, scattered electrons must be detected at very backward angles (close to 180°), and recoil deuterons at very forward angles (close to 0°). Detection of deuterons in coincidence with electrons is necessary to separate elastic from inelastic scattering. A possible design, that meets the above criterion and provides a very large solid angle to maximize the counting rate (the expected cross section at large Q^2 is of order 10^{-42} cm^2/sr), is shown in Figure 4. Deuterons will be detected in the two Hall A HRS spectrometers. Two septa magnets, to be provided by INFN-Rome (currently under final design stage), placed between the target and the HRS's, will direct deuterons recoiling at $\sim 3°$ to the HRS's. Electrons, scattered at $\sim 170°$, will be directed, with the help of two C-type dipole magnets, to two electron detection systems, as shown in Fig. 4. Projected data from this experiment are shown in Figure 3 for a one-month running period using this apparatus in JLab Hall A. The expected data precision should provide severe constraints for the theoretical calculations of the electromagnetic structure of the two-body nuclear system. These data will offer unique grounds for testing models of the nucleon-nucleon interaction, meson-exchange currents, isobar configurations and the possible influence of quark degrees of freedom.

An upgrade of JLab to beam energies between 6-12 GeV will offer unique opportunities to advance our understanding of the structure of few-body nu-

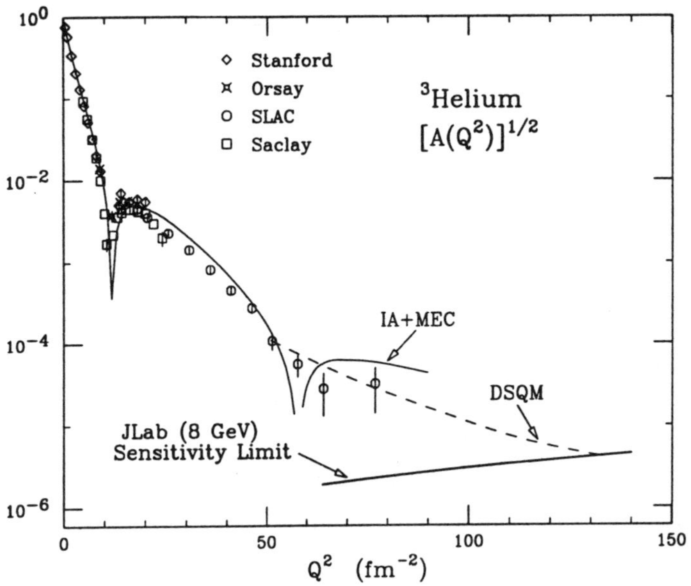

Figure 6: The sensitivity limit (±25% stat. error in about a week of beam time) for a future $A(Q^2)$ measurement of ^{3}He with an upgraded 8 GeV JLab beam. Also shown are existing data and large momentum transfer predictions of IA+MEC and the dimensional scaling quark model.

clear systems. Large momentum transfer measurements of the elastic form factors of the few-body systems, in particular, should be able to explore a possible transition from meson-nucleon to quark degrees of freedom and to test the QDS and pQCD predictions.

It is evident that higher Q^2 measurements would be critical in testing the apparent scaling approach trend of the JLab E91-26 deuteron form factor data. Higher JLab energies in the range 6-12 GeV are essential for such measurements. To separate elastic from inelastic scattering and to suppress backgrounds, recoil deuterons should be detected in coincidence with scattered electrons. A possible scenario would be to use the proposed Super High Momentum and existing High Momentum Hall C spectrometers of JLab. Assuming a 15 cm long liquid deuterium target and a 10 GeV beam in excess of 100 μA, $A(Q^2)$ can be measured up to ~ 9 (GeV/c)^2, in one month of beam time, as shown in Figure 5.

The existing data on the ^{3}He $A(Q^2)$, [15] shown in Figure 6, are in good agreement with the meson-nucleon (IA+MEC) calculations [16] at low Q^2 but are fairly inconclusive at the largest momentum transfers. They are consistent

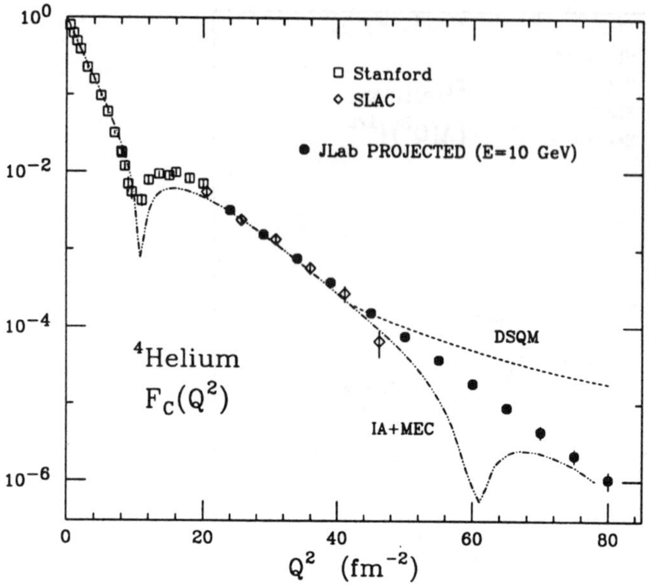

Figure 7: Existing and JLab projected data, using a 10 GeV upgraded beam, for the ^{4}He charge form factor $F_c(Q^2)$. Also shown are large momentum transfer predictions of IA+MEC and the dimensional scaling quark model.

with a change in slope, indicative of an onset of scaling, but, at the same time, cannot exclude the presence of a second diffraction minimum as predicted by the conventional meson-nucleon theory. As in the case of deuterium, more measurements at higher Q^2 would be crucial in testing the quark scaling prediction and any possible breakdown of the meson-nucleon framework. A future energy upgrade of JLab up to 8 GeV will also allow for new ^{3}He measurements that could double the Q^2 range of the existing data with a ten-fold better sensitivity. The sensitivity limit is defined as the lowest $[A(Q^2)]^{1/2}$ value that can be measured with $\pm25\%$ statistical error in a week of beam time and is shown in Figure 6 (this limit would be even lower with a 10-12 GeV beam).

Higher JLab beam energies will be also instrumental in extending the existing ^{4}He charge form factor $F_c(Q^2) \equiv [A(Q^2)]^{1/2}$ measurements [15] up to the highest momentum transfer possible. The availability of a 10 GeV beam would be sufficient to also double the Q^2 range of the available data, as shown in Figure 7, in approximately one month of beam time. The new data could reveal the existence of a second diffraction minimum as predicted by the IA+MEC theory [16] or show signatures of a scaling behavior as predicted by quark dimensional scaling. [11,17]

In summary, JLab is the ideal place to perform large momentum transfer measurements of the form factors of the deuteron and the helium isotopes. These measurements can provide unique tests of the few-body "standard model" based on the impulse approximation and meson-exchange currents, and valuable input for an expected transition from the conventional meson-nucleon to the quark-gluon description of the few-body nuclear systems.

References

1. J. Carlson and R. Schiavilla, *Rev. Mod. Phys.* **70**, 743 (1998); and references therein.
2. C.E. Carlson, J.R. Hiller, R.J. Holt, *Annu. Rev. Nucl. Part. Sci.* **47**, 395 (1997); and references therein.
3. E. Hummel and J.A. Tjon, *Phys. Rev. Lett.* **63**, 1788 (1989); *Phys. Rev.* C **42**, 423 (1990).
4. J.W. Van Orden, N. Devine and F. Gross, *Phys. Rev. Lett.* **75**, 4369 (1995); and references therein.
5. D.R. Phillips, S.J. Wallace and N.K. Devine, *Phys. Rev.* C **58**, 2261 (1998).
6. H. Dijk and B.L.G. Bakker, *Nucl. Phys.* A **494**, 438 (1989); Z-J. Cao and L. Kisslinger, *Phys. Rev.* C **41**, 647 (1990); and references therein.
7. S.J. Brodsky and G.R. Farrar, *Phys. Rev. Lett.* **31**, 1153 (1973); V.A. Matveev, R.M. Muradyan and A.N. Tavkhelidze, *Lett. Nuovo Cimento* **7**, 719 (1973).
8. S.J. Brodsky, C-R. Ji and G.P. Lepage, *Phys. Rev. Lett.* **51**, 83 (1983).
9. L.C. Alexa *et al.*, *Phys. Rev. Lett.* **82**, 1374 (1999).
10. R.G. Arnold *et al.*, *Phys. Rev. Lett.* **35**, 776 (1975).
11. S.J. Brodsky and B.T. Chertok, *Phys. Rev. Lett.* **37**, 269 (1976); *Phys. Rev.* D **14**, 3003 (1976).
12. N. Isgur and C.H. Llewellyn Smith, *Phys. Rev. Lett.* **52**, 1080 (1984); *Phys. Lett.* B **217**, 535 (1989); G.R. Farrar, K. Huleihel and H. Zhang, *Phys. Rev. Lett.* **74**, 650 (1995).
13. C. Bochna *et al.*, *Phys. Rev. Lett.* **81**, 4576 (1998).
14. P.E. Bosted *et al.*, *Phys. Rev.* C **42**, 38 (1990); and references therein.
15. R.G. Arnold *et al.*, *Phys. Rev. Lett.* **40**, 1429 (1978); and references therein.
16. R. Schiavilla and D.O. Risca, *Phys. Rev.* C **43**, 437 (1991); R. Schiavilla, V.R. Pandharipande and D.O. Riska, *Phys. Rev.* C **41**, 309 (1990).
17. B.T. Chertok, *Phys. Rev. Lett.* **40**, 1429 (1978).

TRANSITION FORM FACTORS OF THE PROTON AT HIGHER MOMENTUM TRANSFER

P. STOLER
Rensselaer Polytechnic Institute, Troy, NY, 12180, USA

1 Introduction

The anticipated energy upgrade of JLAB, as well as the possible construction of ELFE will enable us to conduct exclusive reactions at momentum transfers never before possible. Descriptions in terms of valence pQCD have typically been invoked as the primary tool to describe the physics at increasing momentum transfer. However, due to decreasing cross sections and the requirements for very high resolution make it questionable whether high enough momentum transfers would ever be accessed for the application of valence pQCD. Except in special cases non-leading non-perturbative contributions may play an important role at all accessible momentum transfers (see ref. [1] for a review). This would be especially true for reactions involving compared to mesons because of their intrinsically more complex structure. It appears then that one of the most important issues for the future studies of exclusive reactions is to determine which degrees of freedom are appropriate at experimentally accessible momentum transfers. The first task is to identify the experimental signatures for putative models, and how are models at low, medium and high Q^2 related? In this contribution I will discuss the progress we have made in this respect for baryon transition form factors.

Constituent scaling (see eg. ref. [2]) is the most widely known predictors of the onset of pQCD dominance. The apparent agreement of leading form factors and other reactions with pQCD calculations gave widespread hope that pQCD would be appropriate at experimentally accessible momentum transfers. However, this notion was also widely disputed [3]. More recent JLAB experiments involving non-leading order transitions, such as G_{EP} (ref. [4]), and the $N \to \Delta$ transition amplitudes, which is discussed below, have shown that pQCD does not suffice for those reacton in the several $\mathrm{GeV}^2/\mathrm{c}^2$ range. If pQCD is not dominant, we must search for descriptions which are appropriate at these momentum transfers. One obvious question is to determine just how high in Q^2 (small distance) are constituent quarks meaningful as fundamental entities. We will see that the upper limit in Q^2 of the validity of constituent quark model predictions have not yet so far been observed, even at Q^2 of a few $\mathrm{GeV}^2/\mathrm{c}^2$, ie. at distances as small as a ~ 0.1 fm.

In the 1980's the ***handbag*** mechanism was introduced [6] bas an alternative way to describe the exclusive reactions in which soft *Feynman* processes are assumed to dominate at accessible momentum transfers. The distribution of the complex non-perturbative structure of the hadron is characterized by an empirical, non-calculable distribution function in which a quark which carries most of the momentum of the hadron absorbs the virtual photon and reabsorbed in such a way that the original hadron remains intact. This distribution function is parameterised through the use of QCD sum rules. The three alternative mechanisms discussed above are schematically illustrated in fig. 1.

How far the validity of these models extend in Q^2 is still an open question. During the past few years a significant body of theoretical literature has stressed the importance of transverse/soft mechanisms in spite of the apparent constituent scaling. This region in Q^2 is perhaps the most difficult to model since it does not have the simplicities of either CQM or pQCD. However, this is just the region in momentum transfer which is, and will be, accessible

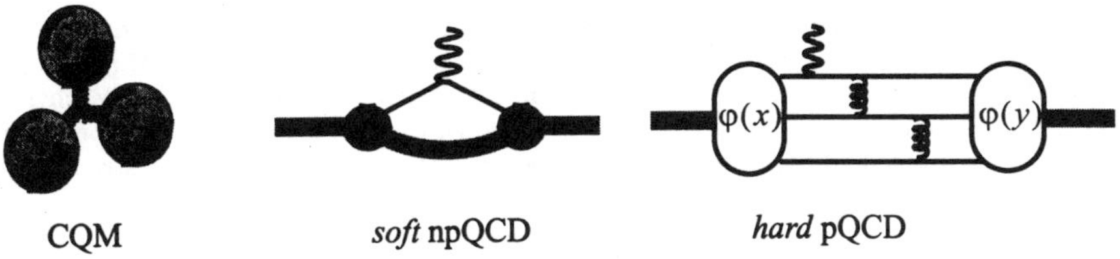

Figure 1: Models for describing exclusive reactions at low, medium and high momentum transfer, respectively.

to new facilities such the upgrded JLAB, and thus it becomes important for theorists to develop techniques for dealing with this kinematic region. Recently [7] there have been promising developments in bridging the high and low Q^2 extremes with a QCD quark-parton description of exclusive reactions which formalizes the properties of the handbag in terms of ***skewed parton distributions*** (SPD). The properties of these SPD's and their relationship to inclusive deep inelastic structure functions, as well as exclusive hadronic form factors have been discussed extensively at this workshop. In this approach the perturbative hard part of the reaction, which is calculable, is isolated from the non-perturbative soft, physics which is parameterized in terms of the SPD's, generically illustrated in fig. 2a. An attractive aspect of this is that the same SPD's are common to different exclusive reactions which involve the same set of hadrons. In the limit of forward scattering it is shown that the SPD's become the usual inclusive parton distribution functions, as illustrated for the case of virtual Compton scattering in fig. 2b,c. Fig. 3 illustrates how they are related specifically to baryon elastic and transition form factors, which is the subject of this article.

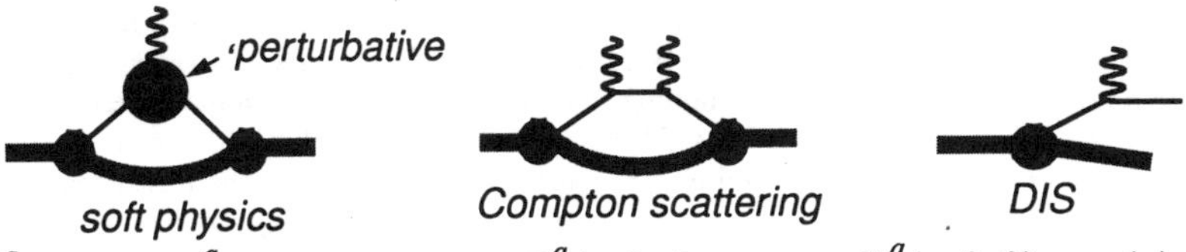

$H^q(x,\xi,t)$, $E^q(x,\xi,t)$, ... $\quad$ $H^q(x,\xi,t)$ .... $\quad$ $H^q(x,0,0) = g(x)$, ...

Figure 2: a. Schematic illustration of the use of off-forward parton distributions in exclusive reactions. The relationship between virtual forward Compton scattering and deep inelastic scattering (DIS) is illustrated in b. and c..

$$\int H^q(x,\xi,t) \ldots \rightarrow G_E, G_M, A_{3/2}, A_{1/2}, C_{1/2}$$

Figure 3: Schematic illustration of the application of off-forward parton distributions for describing elastic and transition baryon form factors.

I believe an important advance of the new formalism will be to explicitly show the connection between these SPD's and valance pQCD at the high Q^2 limit, and the CQM at low Q^2. It should be emphasized that this approach, while expressing the non-perturbative physics of exclusive reaction in terms of the traditional variables as other high energy reactions, does not solve the problem of how the SPD's are obtained from the non-perturbative structure of the hadrons.

2 Baryon transition form factors

Baryon resonance transition amplitudes are appropriate for testing theoretical approaches to exclusive reactions at different Q^2 regimes. The various versions of the CQM's, valence pQCD and SPD approach make very different predictions about the magnitudes, the Q^2 evolution and the relationships between the contributing amplitudes. For example, valance pQCD predicts $1/Q^4$ scaling for the leading ***helicity conserving*** form factors, and $1/Q^5$ and $1/Q^6$ dependence for next to leading ***helicity non-conserving*** form factors, which asymptotically vanish relative to the helicity conserving form factors.

The measurements of the properties of baryon elastic and inelastic amplitudes over a wide range of kinematic and physical conditions, using unpolarized as well as polarized beams and targets, form a major part of the JLAB experimental program. New results have appeared $G_{EP}P$ (ref. [4] above), the the $p \to \Delta$ (ref. [5] above) and $p \to S_{11} transition$form factors [8], which will be discussed here. The very extensive program using the CLAS in Hall B has begun to take copious data, and in all three halls experiments are on the books to extend the measurements to higher Q^2.

Figure 2 shows the virtual photon cross section in the resonance region obtained in inclusive electron scattering at $Q^2 = 1$ and 3 GeV^2/c^2. There are approximately 20 known non-strange resonances in this energy region. However, a few are particularly strong, the $\Delta(1232)$, $D_{13}(1520)$, $S_{11}(1535)$ and $F_{15}(1680)$, which appear as strong peaks in fig. 2. These peaks continue to dominate this region of W for all values of increasing Q^2, the ratio of resonance to non-resonant cross section remaining roughly constant. This scaling of resonance to non-resonance production, known as the ***Bloom-Gilman duality.*** This phenomena been the subject of extensive theoretical [9] and experimental [10] study during the past decade and has been shown to be a fundamental and significant manifestation of QCD.

2.1 Measurement of the $\Delta(1232)$ and $S_{11}(1535)$ at JLAB

I would like to discuss the first completed JLAB experiments on resonance production at high Q^2, the measurement of the the reactions $e + p \to p + \pi^0$ and $e + p \to p + \eta$ for a range in W from pion threshold to about 1.7 GeV^2/c^2, which were carried out in Hall C, and which were the bases of PhD theses of V. Frolov [11] and C. Armstrong[12], respectively.

In this range of W the π^0 final state is dominated by the $\Delta(1232)$ below the 2-pion threshold and the η final state is dominated by the $S_{11}(1535)$.

The experiments, carried out in Hall C utilized a high luminosity ($\sim 1\times 10^{38}$ cm^{-1}s^{-1}) cw beam. Scattered electrons were detected in the Short Orbit Spectrometer (SOS), and recoil protons were detected in the High Momentum Spectrometer (HMS). The incident beam energies were 3.2 and 4.0 GeV which resulted in $Q^2 = 2.8$ and 4.0 GeV^2/c^2 for the $\Delta(1232)$ measurement and $Q^2 = 2.4$ and 3.6 GeV^2/c^2 for the $S_{11}(1535)$. Since the experiment is kinematically complete, the undetected single π^0 and η final states were selected by missing mass. At higher Q^2 the protons emerge in a narrow cone around the $\vec{q}$ vector so that a significant fraction of 4π c.m. solid angle was obtained in the decay of the $\Delta \to p\pi^0$, and somewhat less in the decay of the $S_{11}(1535) \to p\eta$.

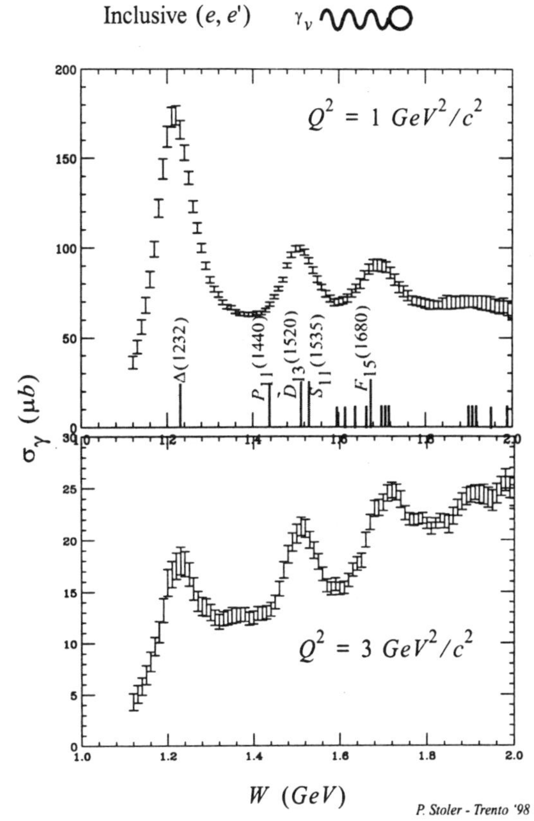

Figure 4: Inclusive electron scattering cross section in the resonance W region. Top; Q^2=1GeV^2/c^2. Bottom; Q^2=3GeV^2/c^2. The verticle lines are at the position of the kown resonances, and those highlighed in grey are for the dominant resonances most commonly studied.

Details of the experiments data analysis for the Δ and S_{11} experiments can be found in refs [11, 12] respectively.

Assuming the one photon exchange approximation, the differential cross section of single meson electroproduction is related to the center of mass differential cross section for meson production by virtual photons, $d\sigma/d\Omega^{cm}_{\pi,\eta}$, as follows :

$$\frac{d\sigma}{dW dQ^2 d\Omega_{\pi,\eta}} = \Gamma \frac{d\sigma}{d\Omega_{\pi,\eta}}$$

where Γ is the virtual photon flux. $d\sigma/d\Omega^{cm}_{\pi,\eta}$ was obtained at more than 2,000 intervals in $\cos\theta^{cm}_{\pi,\eta}$, $\phi^{cm}_{\pi,\eta}$, $Q^2$and W.

2.2 The $\Delta(1232)$

Examples of the pion angular distributions at $W = 1.235$ are shown in 2.2.

Theoretical models make predictions about the Q^2 dependence of the transition form factor, G^*_M, as well as the relative contributions of M_{1+}, E_{1+}, and S_{1+} multipole amplitudes. At low Q^2, according to the CQM the $N - \Delta$ transition is predominantly M_{1+}, and $R_{EM} \equiv E_{1+}/M_{1+}$ and $R_{SM} \equiv S_{1+}/M_{1+}$ at the resonance position W_R are small, which is

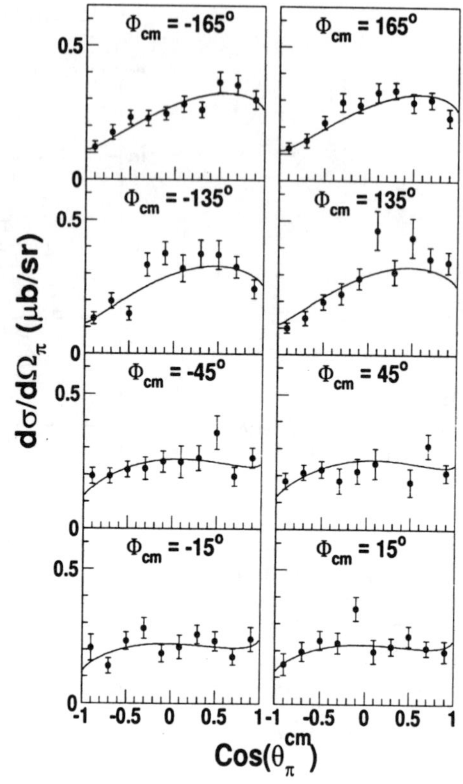

Figure 5: Examples of the angular distributions for π^0 production at W = 1.235 GeV Q^2=2.8 GeV^2/c^2. The curves are due to the empirical multipole analysis discussed in the text.

experimentally confirmed at the photon point[13], and at low Q^2 [14]. In the high Q^2 limit, according to valence pQCD only helicity conserving amplitudes should contribute, leading to the prediction $R_{EM} = +1$ and $R_{SM} = 0$. The earlier data from DESY at $Q^2 = 3.2$ GeV^2 have been evaluated [15], suggesting that R_{EM} is small, but with large errors.

The differential cross sections are not sufficient to extract the amplitudes without further assumptions. In order to obtain an empirical idea of the magnitude of the form factor G^*_M, and the relative contributions of the amplitudes R_{EM}, and R_{SM} , for each W bin the angular distributions were fit assuming S and P wave contributions and M_{1+} dominance as in ref. [5, 11]. The overall fit to the data was quite good (eg. at W = 1.235 GeV $\chi^2_{pdf} \sim 1.36$ and 1.22 for $Q^2 = 2.8$ and 4.0 GeV^2/c^2 respectively) and verifies the M_{1+} dominance at the resonance pole.

To extract more reliable quantitative values of the amplitudes an effective Lagrangian approach [16], which extracts the *dressed* amplitudes, was employed and is in good agreement with the above multipole expansion approach. The resulting resonance part of G^*_M, R_{EM} and R_{SM}, shown in fig. 2.2, are nearly identical with the combined resonant plus non-resonant values, and agree with the empirical analysis very well considering the approximations of that procedure. Also shown in fig. 2.2 are the results of earlier experiments and selected theoretical calculations.

The following features are noteworthy. The Δ form factor G^*_M exhibits a greater decrease with Q^2 than the dipole form as suggested by previous analysis of inclusive data [20]. The ratio R_{EM} remains very small, and consistent within errors with its value at $Q^2 = 0$. The ratio R_{SM} is non-zero and negative. All these results show that the hadron helicity is not conserved in this exclusive reaction and the valence pQCD description is not applicable for this reaction at these momentum transfers. As a caveat, I note that ref. carson has suggested that the faster than usual falloff of the G^* may suggest that pQCD dominance may set in at a higher Q^2 than for other reactions.

2.3 The $S_{11}(1535)$

This is the only strong resonance with a large η branching ratio b_η. The fact that the resonance pole is very close to the η threshold, and the $\eta - N$ scattering amplitude is strong has even led some to question whether it is really a resonance in the traditional sense [24]. However, the majority point of view that it is a well defined resonance will be adopted here. As in the case of the $\Delta(1232)$ an empirical partial wave fit was made to the data, assuming S and P wave contributions, and A_{0+} dominance. In addition a smooth non-resonant background with realistic threshold dependence was included in the fit. To extract the helicity conserving electromagnetic amplitude $A_{1/2}$ from the differential cross section requires the input of the η branching ratio b_η, as well as the resonance width Γ, neither of which, according to the PDG [25] have previously been unambiguously measured. We found that the best fit value of Γ (154$\pm$20 MeV) is very close to the PDG preferred value of 150 MeV. For the branching ratio, combining the present data with analysis of inclusive data [20, 26], we find a best fit $b_\eta = 55^{+5}_{-10}\%$. The obtained $A_{1/2}$, along with previously published data, and the results of a number of different theoretical quark model based curves, cited in [8], are shown in fig. 2.2. All the data, including the previous data, have been normalized assuming the above Γ and b_η, so if either is wrong all the data will scale together. Note the good agreement with the present results and the inclusive fit. If b_η were lower this would shift the present result up relative to the inclusive fit. None of the theoretical curves adequately describe the data [27].

3 Future plans at JLAB

Experiments to $Q^2\sim$ 7.5 GeV^2/c^2 with an electron beam energy of 6 GeV^2/c^2 are on the books for halls B and C. At these large Q^2 and short wavelengths it would be remarkable if the CQM results ($R_{EM} \sim 0$) continued to hold. We also would like to know whether the $\Delta(1232)$ form factor continues to anomalously fall relative to the dipole formula. For the $S_{11}(1535)$ it will be interesting to see if constituent scaling becomes apparent, as suggested by the inclusive data. In Hall B data is currently being obtained for the π^+ as well as the π^0 and η channels, for both unpolarized cross sections and polarized asymmetries, which should help resolve issues relating the extraction of the resonance multipoles. In the recent JLAB at Higher Energy workshop, it was pointed out [29] that with the appropriate detector upgrades, such measurements may be extended to $Q^2>$ 10 GeV^2/c^2.

ACKNOWLEDGMENT: The author thanks the organizers of the Exclusive Reactions at High Momentum Transfer conference for their excellent hospitality, and the opportunity to present this work.

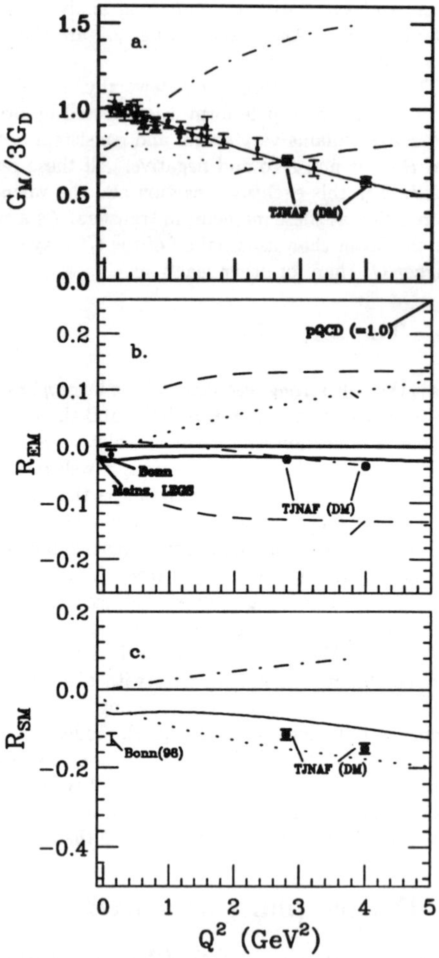

Figure 6: The $N \to \Delta(1232)$ transition form factor G_M^* in the notation of ref. [18], and the ratios R_{EM} and R_{SM} compared to the predictions. **a.)** G_M^*: The data for $Q^2 < 2$ GeV^2/c^2 are evaluations [19] of previous data. **b.)** R_{EM}: Only recent data are shown. The data at $Q^2 = 0$ are from ref [13]. **c.)** R_{SM}: Only recent data are shown. The datum at $Q^2 = 0.13$ GeV^2 is from ref. [13]. The data points at $Q^2 = 2.8$ and 4.0 GeV^2, denoted TJNAF are the effective Lagrangian [17] extracted values of R_{EM} and R_{SM} from JLAB experimental data. The dash-dot curve is due to a light-front calculation [22] in the framework of the relativistic CQM. The dotted curves are due to a light-front CQM calculation [21] including constituent quark form-factors. The dashed curve is the result of a pQCD sum rule calculation [23].

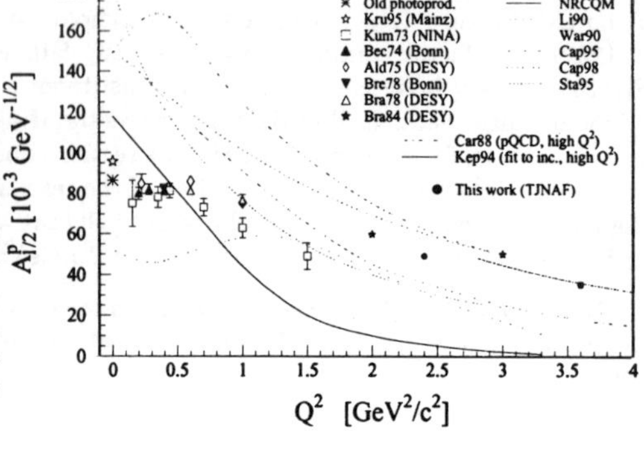

Figure 7: The amplitude $A_{1/2}$ for $N \to S_{11}(1535)$ assuming a resonance width $\Gamma = 150$ MeV, and η branching ratio 0.55. The error bars on the JLAB results at $Q^2 = 2.4$ and 3.6 GeV^2/c^2 are statistical only, and do not include systematic errors in the η branching ratio, or in our resonance width measurement. The other data points and theoretical curves are cited in ref. 11.

References

[1] G. Sterman and P. Stoler, Annual Reviews of Nuclear and Particle Science, 47, 193 (1997).

[2] S. J. Brodsky and G. P. Lepage, *Phys. Rev.* D 23, 1152 (1981);

[3] N. Isgur and C. H. Llewellyn-Smith, *Phys. Lett.* B 52, 1080 (1984); A. Radyushkin, *Nucl. Phys.* A 527, 53 (1991) and private communications.

[4] R. Gilman, these proceedings

[5] V. V. Frolov, *et al. Phys. Rev. Lett.* 82,45 (1999).

[6] A. V. Radyushkin, *Nucl. Phys.* A 527, 53 (1991); V.M.Belyaev, B.L. Ioffe, Sov.Phys.JETP 56 493 (1982)493.

[7] A. Radyushkin, *Phys. Lett.* B 385, 333 (1996); X. Ji, *Phys. Rev. Lett.* 78, 610 (1997). J. Collins, L. Frankfort, and M. Strickman, *Phys. Rev.* D56 2982 (1997).

[8] C. S. Armstrong, *et al. Phys. Rev.* D, to be published.

[9] E.D Bloom and F.J Gilman, *Phys. Rev. Lett*, **25**, 1140 (1970); *Phys. Rev.*,**D4**, 2901 (1971); C.E. Carlson and N.C. Mukhopadyay, *Phys. Rev.*,**D41**, 2343 (1990); *Phys. ReV. Lett.***74**,1288 (1995).

[10] C. Keppel, these proceedings, I. Niculescu et al, to be published.

[11] V. V. Frolov, PhD thesis, Rensselaer Polytechnic Institute; V. V. Frolov, *et al. Phys. Rev. Lett.*, to be published.

[12] C. S. Armstrong, PhD thesis, College of William and Mary; C. S. Armstrong, PhD *et al.* (submitted to *Phys. Rev. Lett.*).

[13] R.Beck *et.al.*, *Phys. Rev. Lett.* 78, 609 (1997); G.Blanpied et.al., *Phys. Rev. Lett.* 79, 4337 (1997).

[14] F. Kalleicher *et al.*, *Z. Phys.* A 359, 201 (1997); R. Gothe, these proceedings;

[15] V. Burkert and L. Elouadrhiri, *Phys. Rev. Lett.*75, 3614 (1995); R. Haidan, PhD Thesis, DESY report F21-79/03 (1979).

[16] R. Davidson, N.Mukhopadhyay, and R.Wittman, *Phys. Rev. Lett.* 56,804(1986);

[17] R. Davidson and N.C. Mukhopadhyay, private communication.

[18] H.F. Jones and M.D. Scadron, Annals of Physics 81, 1 (1979).

[19] S.S. Kamalov and S.N. Yang, nucl-th/9904072, private communication.

[20] P. Stoler, *Physics Reports* 226, 103 (1993);

[21] F. Cardielli et al., *Phys. Lett.* B 371, 7 (1996).

[22] S. Capstick and B.D. Keister, *Phys. Rev.* D 51,3598(1995);

[23] V.M. Belyaev and A.V. Radyushkin, *Phys. Rev.* D 53, 6509 (1997).

[24] G. Hoö hler and A. Schulte, $\pi N - Newsletter$, **7**,94 (1992).

[25] Review of Particle Physics, *Phys. Rev.* D 54, 1 (1996).

[26] C. Keppel, private communication.

[27] M. Aiello, M.M. Giannini and E. Santopinto, nucl-th/9801013, M.M. Giannini and E. Santopinto, Proceedings of the Joint ECT/Jefferson Lab Workshop on N* Physics. ECT, Trento, (1998).

[28] C. E. Carlson and J. L. Poor, *Phys. Rev.* D 38, 2758 (1988).

[29] P.Stoler, Proceedings of the 2nd workshop on JLAB at higher energies, S.Dytman and P. Roos, eds (1998) to be published.

PARTICIPANTS LIST

Andrei Afanasev
NCCU
1801 Fayetteville St.
Durham, MC 27707
Phone: 919 560 5559
Fax: 919 560 5104
afanas@jlab.org

Chris Armstrong
Jefferson Lab
12000 Jefferson Ave.
Newport News, VA 23606
Phone: 757 269 7168
Fax: 757 269 5800
csa@jlab.org

O. Keith Baker
Hampton Univ.
12000 Jefferson Ave.
Newport News, VA 23606
Phone: 757 269 7343
Fax: 757 269 7363
baker@jlab.org

Ian Balitsky
Jefferson Lab
12000 Jefferson Ave.
Newport News, VA 23606
Phone: 757 269 7383
Fax: 757 269 7002
balitsky@jlab.org

Zoltan Batiz
College of William & Mary
1330 S. Mt. Vernon Ave. #F
Williamsburg, VA 23185
Phone: 757 221 3544
zsbati@physics.wm.edu

Stanley Brodsky
SLAC
P.O. Box 4349
Stanford, CA 94305
Phone: 650 926 2644
Fax: 650 926 2525
sjbth@slac.standford.edu

Matthias Burkardt
New Mexico State Univ.
Gardiner Hall, Horsehoe Dr.
Dept. of Physics
Las Cruces, MN 88003-0001
Phone: 505 646 1928
Fax: 505 646 1934
burkardt@nmsu.edu

Volker Burkert
Jefferson Lab
12000 Jefferson Ave.
Newport News, VA 23606
Phone: 757 269 7540
Fax: 757 269 5800
burkert@jlab.org

Carl Carlson
College of William & Mary
Physics Dept.
P.O. Box 8795
Williamsburg, VA 23187-8795
Phone: 757 221 3509
Fax: 757 221 3540
carlson@physics.wm.edu

Paul Carter
Caltech
Kellogg Radiation Lab 106-38
Pasadena, CA 91125
Phone: 626 395 4370
Fax: 626 564 8708
pwcarter@krl.caltech.edu

J.P. Chen
Jefferson Lab
12000 Jefferson Ave.
Newport News, VA 23606
Phone: 757 269 7413
Fax: 757 269 5800
jpchen@jlab.org

John Collins
Penn State Univ.
104 Davey Lab
University Park, PA 16802
Phone: 814 863 0783
Fax: 814 865 3604
collins@phys.psu.edu

Marius Coman
Florida International Univ.
CP208, Physics Dept.
Miami, Fl 33199
Phone: 305 348 1710
Fax: 305 348 6700
comanm@fiu.edu

Dieter Cords
Jefferson Lab
12000 Jefferson Ave.
Newport News, VA 23606
Phone: 757 269 7526
Fax: 757 269 5800
cords@jlaab.org

Claudio Coriano
Istituto di Fisica
Universita' di Lecce
Via Per Arnesano
73100 Lecce, Italy
claudio.coriano@le.infn.it

Donald Crabb
Dept. of Physics
Univ. of Virginia
Charlottesville, VA 22901
Phone: 804 924 6790
Fax: 804 924 4576
dgc3q@virginia.edu

James Crittenden
Physikalisches Inst.
University of Bonn
Nussallee 12
Bonn, Germany 53424
Phone: 49 228 73 32 36
Fax: 49 228 73 78 69
crittenden@physik.uni-bonn.de

Kees de Jager
Jefferson Lab
12000 Jefferson Ave.
Newport News, VA
Phone: 757 269 5254
Fax: 757 269 5800
kees@jlab.org

Markus Diehl
DESY
Notkestr. 85
Hamburg, Germany 22607
Phone: 49 40 8998 2086
Fax: 49 40 8998 2777
markus.diehl@desy.de

Sonja Dieterich
Rutgers University
136 Frelingyuysen Rd.
Piscatway, NJ 08854
sonja@jlab.org

Michael Eides
Penn State University
104 Davey Lab. Physics Dept.
University Park, PA 16802
Phone: 814 863 0007
Fax: 814 865 3604
eides@phys.psu.edu

Kim Egiyan
Yerevan Physics Inst.
Alikhanian Brothers St 2
Yerevan 375 036 Armenia
Fax: 011 3742 352 261
egiyan@jlab.org

Latifa Elouadrhiri
CNU
12000 Jefferson Ave.
Newport News, VA 23606
Phone: 757 269 7303
Fax: 757 269 5800
latifa@jlab.org

Anton Empl
Jefferson Lab
12000 Jefferson Ave.
Newport News, VA 23606
Phone: 757 269 7100
Fax: 757 269 5800
empl@jlab.org

Rolf Ent
Jefferson Lab
12000 Jefferson Ave.
Newport News, VA 23606
Phone: 757 269 7373
Fax: 757 269 5235
harris@jlab.org

Jun Forest
Jefferson Lab
12000 Jefferson Ave.
Newport News, VA 23606
Phone: 757 269 7839
Fax: 757 269 7002
jforest@jlab.org

Salvatore Frullani
UROME
Instituto Superiore die Sanita
INFN/Sezione Sanita
Vaile Regina Elena 299
Italy
Phone: 011 39 6 49902234
Fax: 011 39 6 49387075

K. Garrow
Jefferson Lab
12000 Jefferson Ave.
Newport News, VA 23606
Phone: 757 269 6929
Fax: 757 269 5235
garrow@jlab.org

Ronald Gilman
Rutgers University
116 Frelinghuysen Rd.
Piscataway, NJ 08855
Phone: 757 269 7011
Fax: 757 269 5703
gilman@jlab.org

Jose Goity
Hampton Univ/Jefferson Lab
12000 Jefferson Ave.
Newport News, VA 23606
Phone: 757 269 7345
Fax: 757 269 7002
goity@jlab.org

Lionel Gordon
Jefferson Lab
12000 Jefferson Ave.
Newport News, VA 23606
Phone: 757 269 5854
Fax: 757 269 7002
lgordon@jlab.org

Ignati Grigentch
Old Dominion Univ.
1021 W. 49th Street
Norfolk, VA 23529
Phone: 757 683 4613
Fax:
grigentc@jlab.org

Paul Gueye
Hampton University
Jefferson Lab
12000 Jefferson Ave.
Newport News, VA 23606
Phone: 757 269 7167
Fax: 757 269 6273
gueye@jlab.org

Vadim Guzey
Penn State
104 Davey Lab, Box 191
University Park, PA 16802
Phone: 814 863 0133
Fax:
guzey@phys.psu.edu

Ross Hicks
Dept of Physics & Astronomy
University of Massachusetts
Amherst, MA 01003-4525
Phone: 413 545 2019
Fax: 413 545 4884
rhicks@phast.umass.edu

Roy Holt
Univ. of Illinois at Urbana-Champaign
1110 W. Green St.
Urbana, IL 61801
Phone: 217 244 6039
Fax: 217 333 1215
r-holt@uiuc.edu

Theodore Horton
North Carolina State Univ
704 East Whitaker Mill Rd.
Raleigh, NC 27608
Phone: 919 755 1217
Fax: 919 515 2471
tuhorton@unity.ncsu.edu

Paul Hoyer
Nordita
Blegdamsvej 17
Dk-2100 Copenhagen, Denmark
Phone: 45 3532 5222
Fax: 45 3538 9157
hoyer@nordita.dk

Charles Hyde-Wright
Old Dominion Univ.
4600 Elkhorn Ave.
Norfolk, VA 23529-0116
Phone: 757 683 5853
Fax: 757 683 3038
hyde@jlab.org

Nathan Isgur
Jefferson Lab, MS12H2
12000 Jefferson Ave.
Newport News, VA 23606
Phone: 757 269 7451
Fax: 757 269 7002
isgur@jlab.org

Sabine Jeschonneck
Jefferson Lab, MS 12H2
12000 Jefferson Ave.
Newport News, VA 23606
Phone: 757 269 7380
Fax: 757 269 7002
sabine@jlab.org

Chueng Ji
North Carolina State Univ.
Dept. of Physics
P.O. Box 8202
Raleigh, NC 27695-8202
Phone: 919 515 3478
Fax: 919 515 2471
ji@ncsu.edu

Xiangdong Ji
Univ. of Maryland
TQHN group, Dept of Physics
College Park, MD 20742
Phone: 301 405 7277
Fax: 301 405 6114
xji@physics.umd.edu

Mark Jones
College of William & Mary
7 Frond Ct.
Williamsburg, VA 23188
Phone: 757 269 5304
jones@jlab.org

Christopher Jutier
Old Dominion Univ.
12000 Jefferson Ave, MS16B
Newport News, VA 23606
Phone: 757 269 6386
jutier@jlab.org

Cynthia Keppel
Hampton Univ.
Physics Dept./Olin 102G
Hampton, VA 23668
Phone: 757 727 5823
Fax: 757 728 6910
keppel@jlab.org

Alexander Khodjamirian
Univ. of Wuerzburg
Inst of Theoretical Physics
Am. Hubland
Wuerzburg, Germany D-97074
Phone: 0049 931 888 5734
Fax: 0049 931 888 4604
ak@physik.uni-wuerzburg.de

Dubravko Klabucar
Zagreb University
Physics Dept.
Bijenicka c. 32
Zagreb, Croatia HR-10000
klabucar@phy.hr

Peter Kroll
Fachbereich Physik
Universitaet Wuppertal
Gaussstrasse 20
Wuppertal, Germany D42097
Phone: 49 202 439 2620
Fax: 49 202 439 2811
kroll@theorie.physik.uni-wuppertal.de

Elena Kuchina
Old Dominion University
201 Pointer Circle, Apt. 2
Newport News, VA 23602
Phone: 757 249 8371
Fax: 757 269 7002
kuchina@jlab.org

David Lawrence
Univ. of Massachusetts
12000 Jefferson Ave.
#16/136A
Newport News, VA 23606
Phone: 757 269 6356
davidl@jlab.org

Jean-Marc Laget
CEA/SACLAY
DAPNIA/SPhN
GIF-SUR-YVETTE Cedex,
France F91191
Phone: 33 1 69 08 75 54
Fax: 33 1 69 08 75 84
jlaget@cea.ft

Richard Lebed
Jefferson Lab
12000 Jefferson Ave.
Newport News, VA 23606
Phone: 757 269 7062
Fax: 757 269 7002
lebed@jlab.org

Garry Levman
Univ of Toronto
60 St. George St.
Toronto, Ontario,
Canada M5S 1A7
Phone: 416 978 1564
Fax: 416 978 8221
levman@physics.utoronto.ca

M. Liang
Jefferson Lab
12000 Jefferson Ave.
Newport News, VA 23606
Phone: 757 269 7176
Fax: 757 269 5800
mliang@jlab.org

Simonetta Liuti
Univ of VA
McCormick Road
Charlottesville, VA 22901
Phone: 804 082 2087
Fax: 804 924 4576
sl4y@virginia.edu

K. Loukachine
Jefferson Lab
12000 Jefferson Ave.
Newport News, VA 23606
Phone: 757 269 5007
Fax: 757 269 5800
costy@jlab.org

Lech Mankiewicz
Physik Dept
Technische Universitaet Muenchen
James-Franck-Strasse
Garching, Germany D-85747
Phone: 49 89 2891 4403
Fax: 49 89 2891 2325
lech@physik.tu-muenchen.de

Pete Markowitcz
Florida International Univ.
CP208, Physics Dept, FIU
Miami, Florida 33199
Phone: 305 348 1710
Fax: 305 348 6700
markowit@jlab.org

Maurizio Melis
INFN Sezione di Cagliari
Univers Monserrato
Casella Postale 170
Cagliari, Italy
Phone: 011 39 070 6754982
Fax: 011 39 070 573512
melis@ca.infn.it

Andreas Metz
Institut fuer Theoretische Physik
Philosophenweg 19
Heidelberg, Germany 69120
Phone; 49 6221 549442
Fax: 49 6221 549331
metz@frodo.tphys.uni-heidelber

Gerald Miller
Physics Dept.
University of Washington
Box 35-1560
Seattle, WA 98195-1560
Phone: 206 543 2995
Fax: 206 685 0635
miller@phys.washington.edu

Wally Melnitchouk
FZ-Juelich, IKP
Juelich, Germany C-52425
Phone: 49 2461 614560
49 2461 613930
w.melnitchouk@fz-juelich.de

Richard Milner
Bates Linear Accelerator Center
21 Manning Avenue
Middleton, MA 01949
Phone: 617 253 9200
Fax: 617 253 9599
milner@mitlns.mit.edu

Igor Musatov
Old Dominion Univ.
Dept of Physics
Norfolk, VA 23529
Phone: 757 269 7377
Fax: 757 269 7002
musatov@jlab.org

Alan Nathan
Univ. of Ilinois at Urbana-Champaign
Dept of Physics
1110 W. Green St
Urbana, IL 61801
Phone: 217 333 0965
Fax: 217 333 1215
a-nathan@uiuc.edu

Bogdan Niczyporuk
Jefferson Lab
12000 Jefferson Ave.
Newport News, VA 23606
Phone: 757 269 7251
Fax: 757 269 5800
bogdan@jlab.org

Nikolai Nikolaev
Inst f. Kernphysik
Forschungszentrum Juelich
Juelich, Germany D-52425
Phone: 49 2461 616472
Fax: 49 2461 613930
n.nikolaev@fz-juelich.de

Jonathan Osborne
Univ of Maryland
Nuclear Theory Group
College Park, MD 20742-4111
Phone: 301 405 6125
Fax: 301 405 6114
jao@physics.umd.edu

Kornelija Passek
Rudjer Boskovic Institute
Bijenicka c. 54
Zagreb, Croatia 10000
Phone: 385 1 4561 133
Fax; 385 1 4680 223
passek@thphys.irb.hr

Josef Pochodzalla
MPI fuer Kernphysik Heidelberg
Saupfercheckweg 1
Heidelberg, Germany 69117
Phone: 49 6221 516 331
Fax: 49 6221 516 604
jgp@bxnhd1.mpi-hd.mpg.de

Andrey Pronyaev
Virginia Tech
Dept. of Physics, Robeson Hall
Blacksburg, VA 24061
Phone: 540 231 5272
Fax: 540 231 7511
andrey@vt.edu

Anatoly Radyushkin
Old Dominion Univ/Jefferson Lab
12000 Jefferson Ave.
Newport News, VA 23606
Phone: 757 269 7377
Fax: 757 269 7002
radyush@jlab.org

John Ralston
Univ. of Kansas
Dept. of Physics
Lawrence, KS 66044
Phone: 785 864 4626
Fax: 785 864 5262
ralston@ukans.edu

Franck Jean Sabatie
Old Dominion University
Jefferson Lab
12000 Jefferson Ave.
Newport News, VA 23606
Phone: 757 269 6386
Fax: 757 269 6273
sabatie@jlab.org

Arun Saha
Jefferson Lab
12000 Jefferson Ave.
Newport News, VA 23606
Phone: 757 269 7605
Fax: 757 269 7363
saha@jlab.org

Cetin Savkli
College of William & Mary
1342-C South Mt. Vernon Ave.
Williamsburg, VA 23185
Phone: 757 221 5522
csavkli@physics.wm.edu

Wolfgang Schaefer
Institut fuer Kernphysik
FZ Juelich
Juelich, Germany D-52425
Phone: 49 2461 614167
Fax: 49 2461 613930
wo.schaefer@fz-juelich.de

Stefan Schlenstedt
DESY Zeuthen
Platannallee 6
Zeuthen, Germany 15738
Phone: 49 40 8998 2459
Fax: 49 40 8998 3092
stefan.schlenstedt@ifh.de

Wolfgang Schweiger
Institute of Theoretical Physics
University Graz
Graz
Styria, Austria A-8010
Phone: 43 316 380 5244
Fax: 43 316 380 9820
wschweig@balu.kfunigraz.ac.at

Sudip Seal
New Mexico State University
Dept. of Physics
P.O. Box 30001
Las Cruces, NM 88001
Phone: 505 646 3217
Fax: 505 646 1934
skseal@weizen.nmsu.edu

Silvano Simula
INFN – Rome
Via dell'a Vasca Navale 84
00146 Roma, Italy
Phone: 011 39 06 5517 7053
Fax: 011 39 06 5517 7059
simula@roma3.infn.it

Elton Smith
Jefferson Lab, MS 12H
12000 Jefferson Ave.
Newport News, VA 23606
Phone: 757 269 7625
Fax: 757 269 5800
elton@jlab.org

Elio Soldi
North Carolina Central Univ.
Physics Dept.
1801 Fayetteville Street
Durham, NC 27707
Phone: 919 560 5106
elio@sci.nccu.edu

Stepan Stepanyan
Jefferson Lab
12000 Jeffeson Ave.
Newport News, VA 23606
Phone: 757 269 7196
Fax: 757 269 5800
stepanyan@jlab.org

Paul Stoler
Rensselaer Polytechnic Inst.
Physics Dept.
Troy, NY 12180
Phone: 518 276 8388
stoler@rpi.edu

Steffen Strauch
Rutgers University
136 Frelinglugsen Rd.
Piscadaway, NJ 08854
strauch@jlab.org

Riad Suleiman
Kent State University
Jefferson Lab, MS 16B
12000 Jefferson Ave.
Newport News, VA 23606
Phone: 757 269 7035
Fax: 757 269 6273
suleiman@jlab.org

Adam Szczepaniak
Physics Dept.
Indiana University
Bloomington, IN 47405
Phone: 812 855 3977
Fax: 812 855 5533
aszczepa@indiana.edu

Mauro Taiuti
INFN
Via Dodecaneso 33
Genova
Italy I-16142
Phone: 39 10 353 6336
Fax: 39 10 31 3358
Taiuti@ge.infn.it

Bryan Tipton
MIT
77 Massachusetts Ave.
Cambridge, MA 02139
Phone: 617 253 7065
tipton@mit.edu

Luminita Todor
Old Dominion Univ.
Dept of Physics
Norfolk, VA 23529
Phone: 757 683 5807
Fax: 757 683 5809
luminita@jlab.org

Maurizio Ungaro
Departimento di Fisica
Via Dodecaneso 33ALY
I-16142 Genova, Italy
Phone: 011 3910 353 6336
Fax: 011 3910 313358

Nikolai Uraltsev
Physics Dept.
Technion – Israel Institute of technology
Haifa 32000, Israel
uraltsev@physics.technion.ac.il

Marc Vanderhaeghen
University Mainz
J.J. Becher Weg 45
Mainz, Germany D-55099
Phone: 49 6131 39 42 77
Fax: 49 6131 39 54 74
marcvdh@kph.uni-mainz.de

Branko Vlahovic
NCCU
1801 Fayetteville St.
Durham, MC 27707
Phone: 919 560 5559
Fax: 919 560 5104
branko@jlab.org

Christian Wahlquist
College of William & Mary
3187-A Lake Powell Rd.
Williamsburg, VA 23185
Phone: 757 220 9368
wahlquis@physics.wm.edu

Christian Weiss
Inst. F. Theoret. Physik II
Ruhr-Universitat Bochum
Bochum, Germany D-44780
Phone: 49 234 700 3722
Fax: 49 234 7094 248
weiss@tp2.ruhr-uni-boachum.de

Bogdan Wojtsekhowski
Jefferson Lab
12000 Jefferson Ave.
Newport News, Va 23606
Phone: 757 269 7191
Fax: 757 269 5800
bogdan@jlab.org